INTRODUCTORY ECOLOGY

Introductory Ecology

Peter Cotgreave

The Save British Science Society
London
UK

Irwin Forseth

Department of Biology
University of Maryland
USA

Editorial Offices:
Osney Mead, Oxford OX2 0EL
25 John Street, London WC1N 2BS
23 Ainslie Place, Edinburgh EH3 6AJ
350 Main Street, Malden
MA 02148-5018, USA
54 University Street, Carlton
Victoria 3053, Australia
10, rue Casimir Delavigne
75006 Paris, France

Other Editorial Offices:
Blackwell Wissenschafts-Verlag GmbH
Kurfürstendamm 57
10707 Berlin, Germany

Blackwell Science KK
MG Kodenmacho Building
7–10 Kodenmacho Nihombashi
Chuo-ku, Tokyo 104, Japan

Iowa State University Press
A Blackwell Science Company
2121 S. State Avenue
Ames, Iowa 50014-8300, USA

First published 2002

Set by Graphicraft Typesetters, Limited
Hong Kong
Printed and bound in Italy
by Rotolito Lombardo

A catalogue record for this title
is available from the British Library

ISBN 0-632-04227-3

Library of Congress
Cataloging-in-Publication Data

Cotgreave, Peter.
Introductory ecology/
Peter Cotgreave, Irwin Forseth.
p. cm.
Includes bibliographical references
(p.).
ISBN 0-632-04227-3 (alk. paper)
1. Ecology. I. Forseth, Irwin.
II. Title.
QH541 .C66 2001
577—dc21 2001043128

Distributors

Marston Book Services Ltd
PO Box 269
Abingdon, Oxon OX14 4YN
(*Orders*: Tel: 01235 465500
Fax: 01235 465555)

The Americas
Blackwell Publishing
c/o AIDC
PO Box 20
50 Winter Sport Lane
Williston, VT 05495-0020
(*Orders*: Tel: 800 216 2522
Fax: 802 864 7626)

Australia
Blackwell Science Pty Ltd
54 University Street
Carlton, Victoria 3053
(*Orders*: Tel: 3 9347 0300
Fax: 3 9347 5001)

For further information on
Blackwell Science, visit our website:
www.blackwell-science.com

Contents

Preface

The greatest task for humankind over the coming century will be to reconcile the needs of a growing population with the requirement to behave in ways that are environmentally sustainable. As the human species appropriates more and more natural resources for our own purposes, it will be increasingly important for us to understand the processes that drive the natural environment.

If we fail, the consequences will be literally unimaginable. To succeed, people in all walks of life will need an ever-deeper understanding of how the human species fits into the wider ecology of the planet; this will involve a detailed comprehension of the effects that our own species continues to have on the physical environment and on other life forms, and an equally detailed comprehension of their effects on us. This understanding will only come through a full appreciation of the fundamental principles of ecology.

Despite the importance of ecology, there is no requirement for a course in ecology in many college syllabuses, even those that involve some degree of specialization in science and biology. The result is that many students leave college or university with only a rudimentary idea of the science of ecology, and of its relevance to the political and societal problems facing countries throughout the world. We have attempted to address this by providing a straightforward text that can be used both by students taking elective courses and by those science students for whom this book may be their only exposure to ecology.

The book aims to give students the kind of basic ecological knowledge and comprehension on which they can build an understanding of how the human species fits into the wider ecology of the planet. We have also attempted to connect basic ecological principles and research with the everyday problems that are becoming increasingly pronounced in today's world. In doing so, we have always emphasized that applied ecology can only be understood in a wider context, in which we interpret ecological experiments, observations and theory to build up a picture of how ecosystems function.

Although only two authors are listed for this book, no text of this type can be developed in a vacuum. Therefore, we acknowledge the contributions that so many others have made to the development of the book, starting with our own teachers, lecturers and mentors, who were so important to our development as naturalists, as scientists and as biologists. Prominent among these are James Ehleringer, Roger Cotgreave and John Lawton.

The other people who have influenced the project are too numerous for us to name, but a small number deserve particular mention. Help, advice and useful ecological examples have come from Andrew Bourke, Brenda Casper, Tim Coulson, Sara Lourie, Mike Peek, Graham Stone and Rosie Woodroffe.

Various people have been kind enough to read sections of the manuscript, and we want to thank Alan Cotgreave, Claire Knapton, Jan Bakker, Martin Kent, John Spicer, Tom Crist, Alan Hastings and anonymous reviewers. Philippa Bayley undertook the unenviable task of giving detailed comments on the entire book.

At Blackwell Science, Ian Sherman was a particularly helpful friend and editor throughout much of the project, and when he moved on, his replacement, Sarah Shannon, demonstrated more than enough energy and enthusiasm to keep us on track. Without their constant advice and encouragement, we would have been even further behind schedule. Simon Rallison, Julie Wilson, Katrina McCallum, Cee Brandson and Jane Andrew all played important roles in seeing the book published.

Writing a book is not as simple as we imagined it would be, and we must thank those who have tolerated and supported us when we have been frustrated and tired. Sometimes we have simply been in need of a break from the plants, animals, and other organisms that so intrigue us, and more often than not, we have been so entranced by them that we have failed to realize it. For coping with us when this happened, we thank Cathy Cooper, Laura Erik and Kirsten Forseth.

Peter Cotgreave
Irwin Forseth
July 2001

Chapter 1

The diversity of life

1.1 A vast array of life forms

Every day, each of us encounters an incredible diversity of life forms (Fig. 1.1). Our stomachs contain bacteria, we catch colds caused by viruses and we wear leather shoes made from the skin of a mammal. We eat mushrooms, fish and vegetables sitting at a table made out of a tree, which we cover with a cloth made from the seeds of a cotton plant. We keep dogs, lizards and stick insects as pets, grow flowers in our gardens and use drugs that were first isolated from plants or fungi. Our bread is made from the seeds of grass plants, and we put yeast cells in it to make it rise. We use the juices of fruits so that our soaps and shampoos smell pleasant. The list of

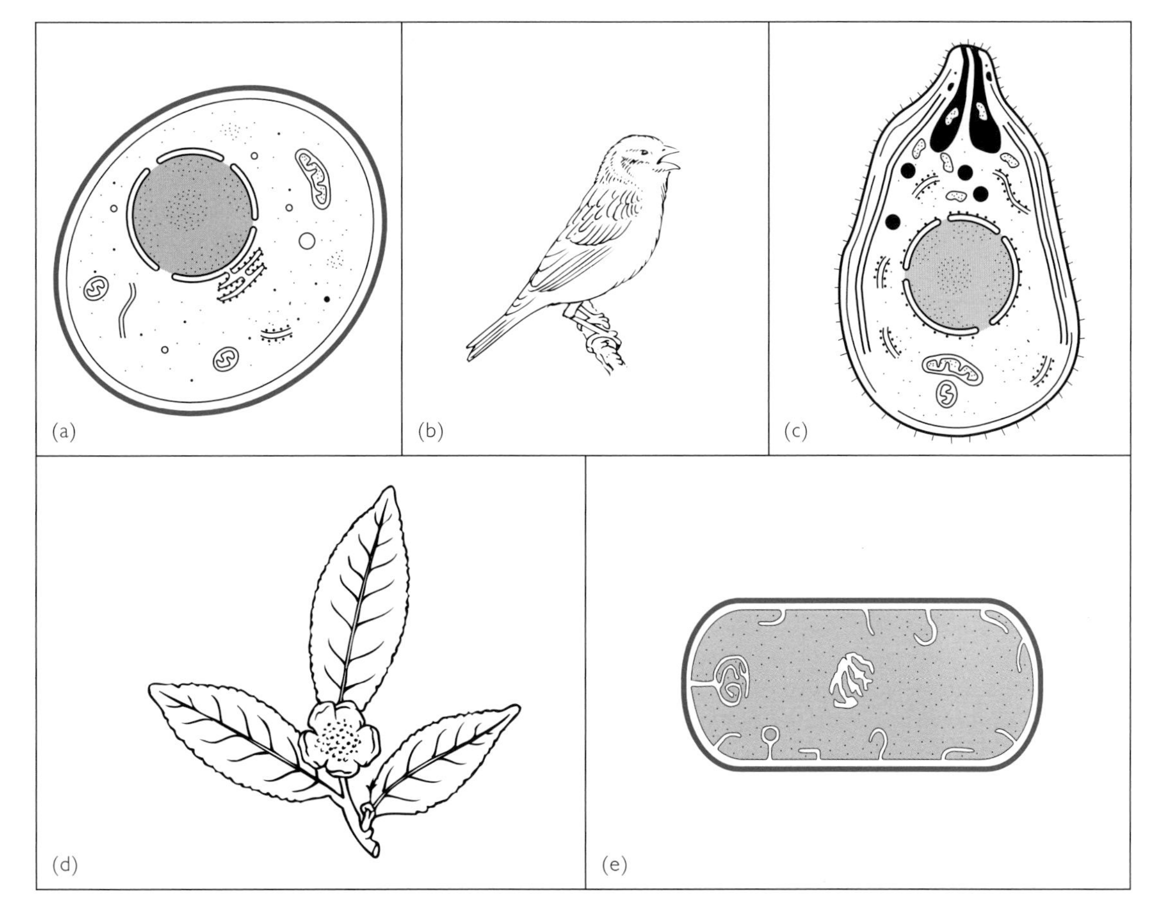

Fig. 1.1 People encounter and use organisms from all the major groups. (a) The fungus *Saccharomyces*, which is used as yeast in brewing and baking. (b) The canary (*Serinus canaria*), which is kept as a pet. (c) The parasite *Plasmodium falciparum*, which causes malaria. (d) The leaves of the tea plant (*Camellia sinensis*), which are infused to make a drink. (e) The bacterium *Lactobacterium*, which is used in the culture of yoghurts.

ways in which we meet organisms, or materials made from them, is seemingly endless.

One reason that people come across organisms in such a variety of ways is that the world has such a vast array of different life forms. Currently, the best estimates of the number of different kinds of organisms in the world vary from about 2 million upwards to 50 or 100 million, and these estimates do not even include bacteria or viruses. Ecology is concerned with every one of these types of organisms, and it is also concerned with the physical environment in which they live, so that an ecologist needs to know not only about biology but also about chemistry, physics and geography. The task of the ecologist is made more difficult by the fact that we do not even know for certain how many different types of living organism have so far been discovered and named, because there is no central list. The only thing that can be said with certainty is that the number of known kinds of organism is well in excess of a million.

It may seem remarkable that humans know so little about the ecology of their planet, especially since it must be one of the oldest subjects of human investigation—men and women must have been trying to understand the life around them ever since they first evolved conscious thought. Moreover, they have always practised applied ecology, in the form of hunting, agriculture and other ways of obtaining food.

An apparent lack of knowledge can be frustrating for the ecologist, who sees scientific colleagues in other fields (such as medicine, physics or chemistry) developing intricate theories and experiments that lead to cures for diseases, exploration into space, or useful inventions such as versatile plastics or labour-saving machines. Nevertheless, the potential ecologist should not despair, because inquisitive ecologists have one major advantage over these other scientists. The fact that ecological scientists have so far discovered so little of what there is to know means that every interested ecologist can add to the sum of human understanding and knowledge and, as often as not, he or she can do so without spending vast sums of money.

More significantly, ecologists know that their science is ultimately just as important as anything else any human has ever done. As the number of people in the world continues to rise, and as increasing numbers of people come to expect the privileged lifestyle enjoyed in places like Western Europe, Japan and North America, the pressures on our planet threaten to become intolerable. This is not simply a matter of the threat of extinction for tigers and giant pandas (which humans happen to find attractive), it is the possibility of serious human and environmental disasters occurring worldwide (Fig. 1.2).

If the world's human population does not properly understand the ecological system in which it lives, it will never really understand how to solve any of its problems. This does not imply that the average ecologist is trying to feed millions of starving mouths, nor does it mean that this book is in any way intended to be political, because it is not. This is a book that will introduce the reader to the fascinating array of different questions that ecologists study, and which make the life of the scientific ecologist exciting, fun, frustrating and fulfilling.

1.2 What is ecology?

If ecology is about every kind of living organism, in every place on the planet and at every time, then it is clearly an extremely large topic. Ecologists could not hope to make any progress in understanding their subject without taking the time to define some sensible limits to what ecology is. Broadly speaking, scientific ecologists tend to have two definitions of their subject, each of which captures something different about what we mean by ecology. The first definition is that ecology is concerned with the interaction between organisms and their environment. The second stresses that ecologists are trying to understand the distribution and abundance of organisms. Each of these definitions has strengths and weaknesses, and it is necessary to understand the two definitions in more detail before progressing any further in trying to understand the subject.

1.2.1 Interacting with the environment

One of the commonest descriptions of ecology is that it is the study of the interactions between organisms and their environments. The beauty of this definition is that it starts with the organism. Since all ecology is about organisms, and since evolution acts through the survival and death of particular individuals, ecologists should never forget that their theories and experiments must be explained with reference to **individual** plants, animals, fungi or micro-organisms. The components of

(a)

(b)

(c)

Fig. 1.2 Some ecological problems. (a) The giant panda, an endangered species. (Copyright John Cancalosi/Still Pictures.) (b) Erosion in the Peak District. (Courtesy of M.R. Ashman.) (c) The Rangitata River, New Zealand, in flood, January 1994. (Courtesy of G. Browne, Institute of Geological & Nuclear Sciences.)

an organism's environment fall into two categories—the physical and the biological environments. The physical environment includes rocks, soils, rainwater, sunshine, minerals and pollution, while the biological environment includes an organism's food, its parasites, its mate, its offspring and its competitors—all of the other organisms it ever encounters, whether they are of its own species or not.

Ecologists call the living, biological element of the environment the **biotic** environment and the physical element the **abiotic** environment.

1.2.1.1 A problem

The drawback of this first definition of ecology is that it is very broad. In effect, every aspect of every organism involves an interaction with something. Walking, for example, is an interaction with the physical environment, since it involves an animal creating friction with the ground. In other words, if ecologists were to take this definition too literally, they would end up studying every aspect of biology. That would be fascinating, and indeed ecologists should be careful never to ignore any aspect of biology—we can never know when something apparently irrelevant will turn out to shed light on an ecological question. However, ecologists generally find it more useful to restrict their study to interactions that affect the distribution and abundance of organisms.

1.2.2 Distribution and abundance

The second popular definition of ecology is much more limited. By this definition, ecology is the study of the distribution and abundance of organisms. The kind of question that an ecologist might ask about distribution is: Why do we see penguins in the Antarctic but not in the Arctic? Why are bromeliad plants found almost exclusively in South America, while plants in the buttercup family can be found almost throughout the world? Questions of abundance might be something like: Why are there twice as many doves in my garden as there are robins? Why are there fewer pandas in China than there used to be?

The advantage of this second definition of ecology is that it is focused—it allows ecologists to ask specific questions, which is what science is all about. The disadvantage with this definition is that it deals with whole groups of organisms (e.g. all the pandas in China, all the buttercups in the world), not with individual organisms. This is important because of the way the biological world is shaped by natural selection, which is the process by which evolution has created the current ecology of the world, and by which that ecology continues to change as organisms experience selection pressure in each generation. To gain a full understanding of any aspect of ecology, investigators must be certain they understand this process.

1.2.3 Linking the two definitions

In order to tie together the two different definitions of ecology, is it necessary to investigate different **levels of biological diversity**. This allows ecologists to perceive the ways in which individual organisms affect the groups of which they are part, and helps to draw links between the definition of ecology that is based on individuals and the definition that is concerned with whole groups. This concept will be studied in Section 1.3.

The final link joining the two different definitions of ecology will come from an understanding of **evolution by natural selection**, which is the process by which the births, deaths and reproduction of individual organisms combine to govern the composition of a population. This process will be discussed in Section 1.4.

1.3 Levels of diversity

Evolution has created an incredible diversity of form and function in the natural world. There are enormous organisms such as blue whales (*Balaenoptera musculus*) and giant redwood trees (*Sequoia sempervirens*) and also tiny life forms such as viruses. Some organisms, like green plants, make their own food by using the sun's energy to break down gases in the air, while others, such as fungi, digest parts of other organisms. There is life at the bottom of the ocean and at the top of the highest mountains. In fact, the Earth's organisms are so variable that a human lifetime is far too short to appreciate them all fully. A word has been coined that aims to describe this amazing variation—the word is **biodiversity**. But it says much more than a simple statement that there are millions of different kinds of organisms, because biological diversity exists at many different levels.

Perhaps the easiest level to understand is the diversity of **species** on the planet. Most people have some idea of what is meant by the word 'species'. It is normally defined as a set of organisms that are genetically very similiar, and can thus interbreed with one another to produce fertile offspring. This definition works well for most animals and plants. There is a species of badger in Europe and Asia (*Meles meles*) and a related species, also

Fig. 1.3 Hybrids like the liger, a cross between a lion and a tiger, are sterile. (Copyright the Zoological Society London.)

known as the badger (*Taxidea taxus*), in North America. Any female Eurasian badger can interbreed with any male Eurasian badger but not with a male American badger. Likewise, any American badger could in theory breed with any other American badger of the opposite sex but not with a Eurasian badger.

Sometimes, in unusual circumstances, two different species will interbreed, but they cannot normally produce fertile young. Horses (*Equus caballus*), for example, will mate with donkeys (*Equus africanus*) to produce infertile mules. Lions (*Panthera leo*) will breed with tigers (*Panthera tigris*) if they are caged together in zoos or circuses; the offspring, known as tigons or ligers (depending on which species is the mother), are infertile (Fig. 1.3).

In using such a definition, it is essential to recognize that two kinds of organisms may never interbreed, simply because, living in different places, they never have the opportunity. If they did so, however, they might be able to produce fertile young. For example, polar bears (*Ursus maritimus*) live only in the Arctic, and grizzly or brown bears (*Ursus arctos*) live further south in Europe, North America and Asia, so that the two species are separated geographically and rarely have the opportunity to come into contact in the wild. However, when they are brought together in captivity, they can interbreed to produce offspring that are fertile and can themselves go on to produce young of their own. Thus, it appears that by the strict definition of a species, the polar bear and the brown bear may be the same species, but, in fact, they are still classified separately because they live very obviously different lives, and because they never interbreed in the wild.

Human activity may change the degree to which populations have the opportunity to interbreed. For example, the introduction of the ruddy duck (*Oxyura jamaicensis*) into Europe has allowed it to interbreed with the white-headed duck (*O. leucocephala*), a native of Spain and other parts of the Mediterranean. Before human intervention, the ruddy duck was confined to the Americas, and there was no possibility of hybridization occurring. In other areas, as habitats are destroyed and fragmented, organisms may become separated where they would formerly have formed part of the same population.

In reality, as with most definitions in the biological sciences, there are many exceptions to the idealized definition of a species; for example, it is more difficult to define some plant species. In some kinds of plant, for example, each individual can fertilize only itself or a genetically identical individual, so that each genetic type could technically be thought of as a separate species. But for most animals and many plants, the normal definition of a species is a good one, and works well in practice for most ecologists.

The definition works less well for some other kinds of organisms. Bacteria, for example, reproduce in very different ways from animals and plants, with the result that the species concept is less clearly applicable. Nevertheless, such organisms can be roughly classified and, as a framework, the idea of a species tends to be suitable for most things that most ecologists want to think about most of the time.

Each species may be divided into populations. A **population** is a group of individual organisms of the same species living together in the same place and usually at the same time. Different populations of the same species may show variation—the African elephants (*Loxodonta africana*) that live on the plains of East Africa are larger than the forest-dwelling elephants of West Africa, although they belong to the same species and can interbreed. Populations are different because they have a different genetic make-up, so variation at the level of the **gene** is very important to the ecologist.

This brings home an important point about evolution. Although natural selection acts through the life, death and reproduction of **individual organisms**, it is **populations** of organisms that evolve. It is a general feature of all West African forest elephants that they are small—it is a **population** characteristic. However, they are like that because natural selection favoured smaller

individuals in the past, and allowed them to produce more offspring, while larger individuals fared less well. All kinds of biologist, including ecologists, must always remember that natural selection is the major reason why an organism has its particular anatomy, physiology and behaviour. So ecologists must always be careful not to postulate theories about populations, or whole species, that do not take account of individual organisms.

Populations of different species in the same place form **communities**. Thus, all the organisms living together in the Serengeti National Park in Tanzania might form a community. Another community may be all the organisms living in a pond in a garden in Tokyo. An **ecosystem** consists of this biological community and the physical, non-living, or abiotic, environment—the rocks, soils, water and climate.

In a sense, communities and ecosystems are human concepts that we have invented to make our scientific lives easier. In general, we define them at a scale that we happen to find convenient—the scale of a garden pond or a national park, for example. Organisms, of course, live their lives at different scales—to a lion, the Serengeti National Park may seem like a single habitat, but to a grass plant it is a mass of slightly different kinds of soils, some of which are suitable to grow in, while others are not.

In fact, ecologists frequently also define populations at a scale that suits their own purposes. A population may simply be defined as 'all the yeast cells in an uncooked loaf of bread', 'all the squirrels in a single forest', or 'all the redwood trees in California'. Because of this, ecologists tend to use the word population rather loosely, so when they are talking about the distribution and abundance of populations, they might sometimes find it convenient to define an entire species as a population. For example, if people are worried that some kind of organism is in danger of becoming extinct, they may study the distribution and abundance of the whole species.

1.4 Evolution by natural selection

Because there are so many different kinds of organisms and because they do so many different things, it would be easy to be daunted by the complexity of ecology. Indeed, as professional ecologists progress through their careers, they discover that there are many complex aspects of the biological world that they cannot yet even begin to explain. However, ecologists have a single, beautifully simple reference point to which they can always return. Ever since life first evolved, more than 3000 million years ago, the living world has been shaped by the process of natural selection. Charles Darwin (1859) described the process in the verbose language of the nineteenth century but his ideas were very simple in essence.

All organisms need resources—animals need food and shelter, green plants need water and sunlight, and so on. Sometimes, there are not sufficient resources for all the organisms in a locality to obtain enough to survive, so some of the organisms die without ever reproducing. Alternatively, they may not die but may be sufficiently impoverished that they produce a smaller number of offspring than others. **Thus, individuals do not all make the same contribution to the next generation**.

The first important step in Darwin's argument is the observation that the organisms that survive and leave most offspring will be the ones that happen, by chance, to be best suited to the particular environment in which they find themselves. For example, if someone were to take some tawny owls (*Strix aluco*) and put them in the snow-covered habitats of the far north of Europe, they would be unlikely to produce as many offspring as the native snowy owls (*Nyctea scandiaca*), for many reasons. One of these reasons is that snowy owls are better camouflaged in the ice and snow and are thus better able to catch prey. The ill-suited tawny owls, which would be easily seen by the rodents they were chasing, would either die of starvation, or at the very least would fail to provide adequately for their chicks.

The next crucial step in Darwin's argument relies on offspring being similar to their parents—red-flowered pea plants (*Pisum sativum*) often (but not always) produce seeds that grow into red-flowered plants, while plants with white flowers are more likely to produce white-flowered offspring. Darwin had to guess at the mechanism for this inheritance, but it is now known that offspring are like their parents because of the genetic code stored in DNA, and that natural selection acts on the genes that make up this code. Some pea plants contain genes for red flowers and others contain genes for white flowers. These genes are passed into the seeds, so offspring inherit some of their parents' genes and, in consequence, share some of their parents' characteristics (Fig. 1.4).

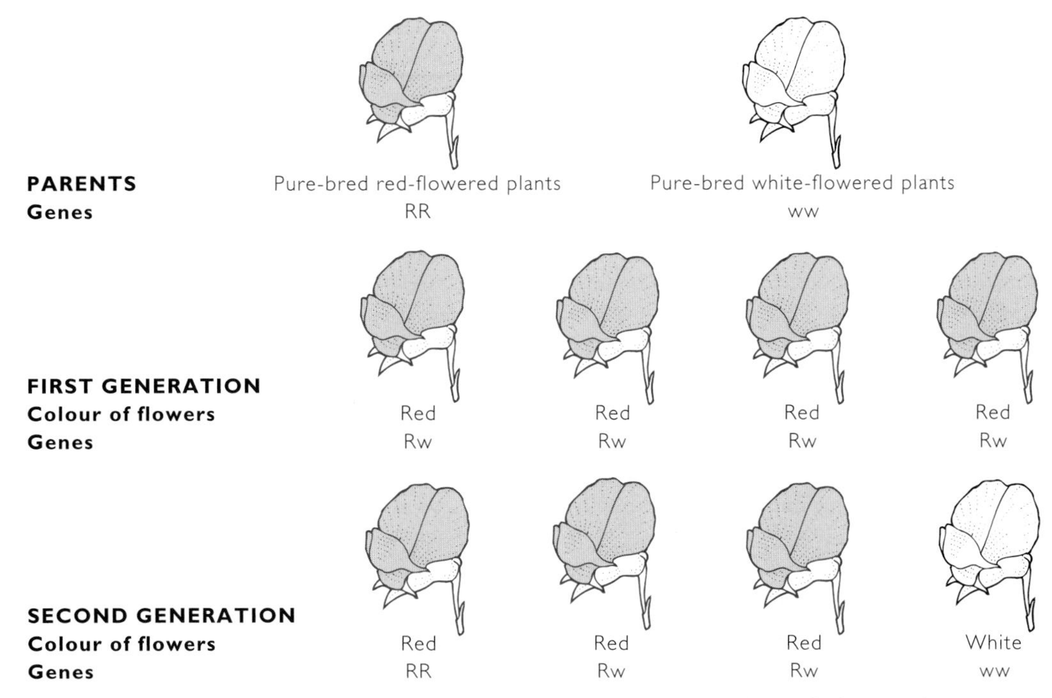

Fig. 1.4 Gregor Mendel (1822–1884) discovered the basis of modern genetics by experimenting with cross-fertilizing pea plants.

Thus, evolution by natural selection can be understood in three points:

1 Some organisms leave more offspring than others.

2 The organisms that leave most offspring are those that are best suited to their environment—they have the highest 'fitness'.

3 The offspring inherit genes from their parents, which means that they also inherit at least some of the characteristics that made their parents well-fitted to the environment, so they too tend to be well-suited.

It is important to note that the word 'fitness' in this context is concerned with 'fitting the environment', not with being 'fit' in the sense of being able to exercise for a long time without getting tired.

Obviously, the environment is not constant—it is always changing in some way. The biological environment could change when a new disease spreads into an area, or if all the local predators became extinct, and the physical environment might change because of, say, global warming. When this happens, evolution will tend to favour those individuals best suited to the new environment, which will probably not be the same ones that would have fared well in the old environment. Thus, natural selection can 'change its mind' about which individuals to favour—there is no single, idealized form that each species is evolving towards. For example, until 65 million years ago, natural selection favoured the set of characters enjoyed by the dinosaurs. But then the environment changed and other animals were favoured at the expense of the dinosaurs. Perhaps a large meteorite struck the Earth and caused a huge cloud of dust and debris that blocked out much of the sun's energy. The colder conditions that would have followed would not have been suitable for the huge lizards, but allowed other, quite different, animals to dominate.

1.4.1 Cadmium tolerance in plants

As an example of how natural selection works, we can look at the many kinds of plants that have evolved tolerance to high levels of cadmium, which is normally extremely toxic both to plants and to animals, including humans. Cadmium is a metal used in a variety of industrial processes and is found at much higher concentrations in the soils of areas that have suffered industrial pollution than in soils of unpolluted areas. Some individuals of some kinds of plants happen to be more tolerant of cadmium poisoning than others, because they have genes that give them a slightly different physiological make-up. In areas of high cadmium pollution, these tolerant plants survive while others simply perish, so that in the next generation, many of the offspring inherit the genes for tolerance and can live in the polluted environment.

However, not all kinds of plants have the same genetic variation and they do not all deal with cadmium in the same way. Some plants, like the soybean (*Glycine max*) and tomato (*Lycopersicum esculentum*), manage to move all the cadmium into a small number of cells so that it does not interfere with the working of the majority of the other cells, while other plants, like rice (*Oryza sativum*) or the water hyacinth (*Eichhornia crassipes*), produce proteins that bind to the cadmium and neutralize its effects (Prasad 1995).

It is important to understand that the particular mechanism that operates in a particular kind of plant depends entirely on which genes the plant has. It so happens that some of the rice plants in polluted areas happened to have genes that produced the binding proteins, but they could equally well have had genes for another kind of mechanism. Equally, it could have been the case that they had no genetic variants capable of dealing with high cadmium levels, in which case that particular kind of plant would not have been able to adapt to the conditions and would have become extinct in areas of cadmium pollution. Natural selection can only operate on the random genetic variation that happens to exist, which is created by mutations in the genetic material of individual organisms.

1.4.2 Evolution is concerned with individuals

An important feature of evolution that must be kept in mind is that it affects the survival and reproduction of individual organisms. Section 1.3 described some of the ways in which those effects are manifest in the composition of populations.

However, it is crucial to avoid the perception that features of organisms can be interpreted in the context of the 'good of the population' or the 'good of the species'. If an organism appears to be generous towards others, it is not concerned with the good of the species. It can almost certainly be shown to be acting in the interests of its own genes.

Many animals, for example, appear to be generous towards others by foregoing their own opportunities to reproduce, and instead helping others to raise their young. This behaviour is not unselfish—it has evolved as the best way, in the circumstances, of increasing the number of copies of the helpers' own genes in the next generation.

Scrub jays (*Aphelocoma coerulescens*), for example, will help their parents to raise more young rather than reproduce themselves. This is because, in some places, their habitat does not provide enough territories for them to have a high chance of raising their own offspring. In these circumstances, while waiting for a suitable territory to become available, the young jays secure more copies of their own genes in the next generation by increasing the survival of their siblings, because two siblings share, on average, one half of their genes. The young jays are not concerned with the good of the species, or the good of the population, and they move away and secure their own territory when they can. In some areas, where the habitat is not fully occupied, the young birds behave in a way that appears much more selfish—they occupy their own territories straight away and do not spend time helping their parents.

1.4.3 Similar solutions to similar problems

Because they have evolved by the same process, different types of organism that live in similar conditions often share many characteristics. For example, there are many places in the world where the temperature falls below the freezing point of water—the Arctic, the Antarctic, and the tops of mountains on all continents. The organisms that live in each of these locations have evolved in isolation from one another but they share many characteristics. In most organisms, the cells are broken if their contents are frozen, so many kinds of organisms, particularly plants and invertebrates, are known to produce chemicals that act to prevent freezing. Many insects (in cold places all over the world) produce glycerol, which lowers the freezing point of the fluid in their cells, exactly as the similar chemical ethylene glycol does when used as antifreeze in the engines of motor vehicles. Plants have evolved an almost identical strategy, using a variety of related chemicals. Some green plants, such as the apple (*Malus*

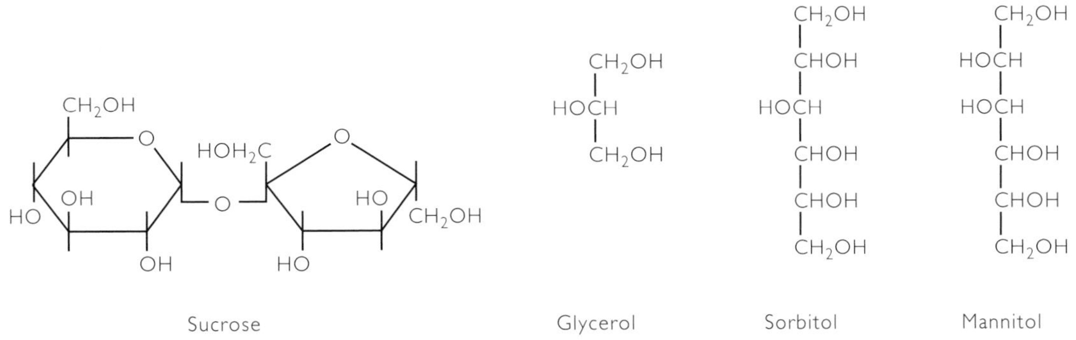

Fig. 1.5 The chemical structure of some molecules used by organisms to prevent their cells from freezing. Many insects produce glycerol and some plants produce chemically similar compounds such as sorbitol and mannitol, but other plants use very different compounds, such as sucrose.

pumila), use the same chemical as most insects or very similar chemicals, such as sorbitol or mannitol. Others, such as ivy (*Hedera helix*) use different sugars, such as sucrose or raffinose (Fig. 1.5).

Chapters 4 and 5 will examine the physical environment of the world in greater depth, and will describe in more detail how evolution has found the same solutions to similar problems in different parts of the world.

1.5 A working definition of ecology

With an understanding of the process of evolution by natural selection, and with a clear idea of what is meant by populations and communities, it is possible to revisit the two different definitions of ecology and unite them into one working definition for the rest of the book.

Recall that the first definition was about organisms interacting with their biological and physical environment, and that the second was about distribution and abundance. The first definition benefits from being centred on the individual organism, but is too unwieldy because it could include any aspect of biology. The second is less cumbersome but has the disadvantage of not concentrating on the individual organisms whose lives we can actually study. Instead it focuses on groups of organisms, such as populations.

These two aspects of ecology are interlinked. Populations evolve because of the action of natural selection on individuals. Thus, in order to preserve the advantages of both trains of thought, it is possible to create a new definition of ecology.

Ecology is the study of how the distributions and abundances of populations (and species) are determined by the interactions of individual organisms with their physical and biological environments.

1.6 Ecological niches

People who live in the tropics may be familiar with day-flying bats, but inhabitants of the temperate zones see bats less frequently, although sometimes on a summer evening, they may notice bats flying around their houses. Because the light is fading, they often have to look twice before they are sure whether they have seen a bat or a bird, or even a large moth. But they do not stop to wonder whether what they have seen was a mouse or a toadstool, because mice and toadstools cannot fly. Likewise, when someone sees something swimming underwater in a pond, they look more closely to see whether it is a frog, a fish or a dragonfly larva but it never crosses their minds that it might be a sparrow or a grass plant. If the water is not a pond but a fast-running stream, they can probably rule out the possibility that what they have seen is a frog. All of this is obvious to the point of being almost trivial.

What is less obvious is the reason whereby people can narrow down what they might have seen. In essence, it is because everyone knows something about ecological niches. Niches are descriptions of what organisms do and where they do it. Usually niches describe the overall attributes of a whole species, although they could refer to populations or even individual animals. Theoretically, the niche occupied by a species defines everything about its needs. Whatever resources are required by organisms—food, shelter, water, space and so on—form part of the niche of a species.

1.6.1 Fundamental and realized niches

The **fundamental niche** of a species defines the places where its members are physiologically capable of living. Most fish cannot live out of water, so dry land is not part of their fundamental niche. In other words, the fundamental niche depends on the physical environment.

In practice, of course, members of a species do not necessarily occur in all the places where they are physiologically capable of doing so. There may be a number of reasons why organisms do not live everywhere that they could theoretically exist. One reason is geography, which is part of the explanation for the lack of wild marsupial mammals, such as kangaroos, in Europe. Bromeliad plants evolved in the Americas and would have had to cross huge oceans to colonize Asia. This interface between geography and ecology is known as biogeography and its effects on biodiversity will be examined in more detail in Chapter 14.

Another reason why organisms of a particular species do not occur in all the places that they might do is that they are excluded by some form of biological interaction. For example, another similar species may already be established and may happen to be a superior competitor. In the prairies of the upper midwest of the United States, the grasses known as little bluestem (*Schizachryium scoparium*) and big bluestem (*Andropogon gerardii*) outcompete grasses such as Kentucky bluegrass (*Poa pratensis*) in obtaining nitrogen from poor, sandy soils. Kentucky bluegrass cannot establish itself in areas where either of the bluestem grasses is already present. However, it can grow in these habitats after a fire creates open space. Alternatively, a piece of habitat may contain a very high density of predators that would very soon eat any member of the species that ventured into the area.

Thus, the **realized niche** of a population is the part of its fundamental niche that it actually occupies, where it is not excluded by predators, competitors, geographic history or anything else.

Both fundamental and realized niches are dynamic, not static—they can change as the biological and physical environment changes. A good example comes from the past ecology of humans and their close relatives. Until about 130 000 years ago, Europe was populated by the Neanderthals, who were either a race of humans or a different but similar species, named after the Neander Valley in Germany, where Neanderthal fossils

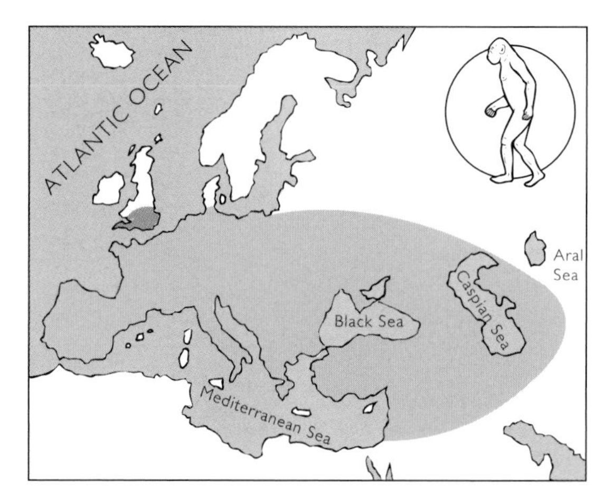

Fig. 1.6 The distribution of the Neanderthal people (*Homo sapiens neanderthalensis*), who were displaced by modern humans (*Homo sapiens sapiens*) about 40 000 years ago. The extent to which the two races interbred is not known, but the realized niche of the Neanderthals certainly contracted as a result of competition from the previously unknown modern humans.

were first discovered in 1856. Similar fossils are known from a variety of places in Europe, so we know that the fundamental niche of the Neanderthals was wide. But when modern humans evolved and emerged from Africa, they replaced the Neanderthals rather suddenly. The exact degree to which competition played a part is not clear, but there can be little doubt that it was an important factor. As modern humans moved northwards from the Middle East and southern Europe, the realized niche of the Neanderthals receded until they finally became extinct (Fig. 1.6).

1.6.2 Pitfalls with the niche concept

One way of looking at ecological niches is to say that organisms live in environments to which they are suited. Organisms are adapted to their environment because evolution has selected individuals with characteristics that enable them to survive in the particular conditions that exist. The niche of a species, therefore, reflects the set of conditions to which its members are adapted. However, there are two pitfalls that ecologists must be careful to avoid.

First, it should never be assumed that every aspect of every organism is perfectly adapted for some function. Take, for example, the bactrian camel (*Camelus bactrianus*) and its relative the dromedary (*Camelus

Fig. 1.7 The bactrian camel (*Camelus bactrianus*) has two humps and the dromedary (*Camelus dromedarius*) has only one, but this may just be an accident of history, and the difference may have no adaptive value to individual camels and dromedaries. Ecologists do not need to assume that every piece of variation in the natural world has necessarily been caused by natural selection.

dromedarius). The dromedary, which comes from Arabia, has one hump and the bactrian camel, a native of central Asia, has two (Fig. 1.7). These humps, which are full of fat, are adaptations to life in the desert. They act as a store of energy when food is scarce, and breaking down the fat may also be used as a source of water, although this is doubtful. This is equally true in both species.

An inquisitive person may ask why the dromedary has just one hump, while the bactrian camel has two. There is no harm in asking such questions, so long as we are content if there turns out to be no adaptive explanation. It is possible that a one-humped version of the bactrian camel would outcompete the existing two-humped form, but no such animal has ever evolved, so it is impossible to say. The number of humps is just as likely to be an accident of history. Millions of years ago, when the dromedary evolved, the individuals that happened to have the best suite of characteristics for life in the African desert also happened, by chance, to have genes for one hump rather than two. It is possible that there was no selective advantage in having one hump and it is conceivable that they could equally well have had genes for two humps or even three or four, but that is not the way things happened.

The second piece of thinking that ecologists must be careful to avoid is to imagine that the niche of a species represents its 'role' in the system, in the way that a taxi driver, a farmer or a schoolteacher has a role in a human community. This train of thought suggests that the system would be incomplete, or could not function, without the species, just as a human community could not function properly without teachers or farmers. In fact, what happens when a species is removed from a system is that the remaining organisms find themselves in an altered biological environment. This means that some populations are subjected to new pressures by natural selection.

If these pressures are considerable, then other species might become extinct and the area might become less rich in terms of its biological diversity. However, the system would still exist, and new populations might even invade, or existing populations may evolve to create new species. It is unhelpful to think of ecological systems as fixed entities; they are always changing, as the component populations undergo evolution.

Of course, this logic is not an excuse for humankind to bring about extinctions without concern. There is little doubt that human activity has the capacity to cause extinctions so rapidly that the remaining species could be subjected to such fierce selection pressures in such short spaces of time that they could not evolve quickly enough to avoid extinction themselves.

1.7 Four concepts that form a basic framework for the ecologist

Sections 1.3–1.6 have described a powerful set of ideas with which to study the ecology of our planet. The

definition of ecology in Section 1.5 takes account of different strands of thinking and allows ecologists to take a wide overall view without ignoring the importance of individual organisms. Equally important, the study of ecology cannot afford to forget the concept of other **levels of diversity**, such as genes, populations, species and communities, and these were described in Section 1.3. Ecologists must also have an understanding of the process of **evolution by natural selection** (Section 1.4) as a sound theoretical base from which it is dangerous to stray. They also rely on the concept of the **ecological niche**, which places organisms in the context of their physical and biological environments (Section 1.6); in other words, it describes (in ecological terms) each of the different life forms that natural selection has produced.

1.7.1 A final concept: making comparisons

The final idea that is needed before an ecologist can really begin the study of the ecology of planet Earth is one that is common to almost all fields of study—ecologists must develop a habit of making comparisons. It is difficult to learn anything of value, and difficult to be fascinated by anything, without making comparisons. This may sound like an odd statement until one realizes that we compare things all the time without even noticing that we are doing it.

For example, we may study a large oak tree and observe that it produced a total of five acorns this year. This fact would tell us nothing until we compare our tree with oaks in other places, or with the same oak tree in previous years, or even simply make a comparison with a general knowledge of what large trees normally do—drawn from experience of trees in a variety of places and at different times. Knowing about any of these situations would tell us that oak trees usually produce thousands of acorns each year. Armed with this knowledge we can say that our oak tree produced an unusually small crop of acorns this year. If we were being more precise, we would say that the crop is small **compared with our expectation based on previous knowledge**.

1.7.2 Patterns

When making comparisons, ecologists discover patterns in the behaviour, anatomy, distribution and physiology of the organisms around them. Deer and antelope have eyes on the sides of their heads but most monkeys have eyes that face forward. Many plants that live in Australia have woody stems but fewer species from the high Arctic do so. Toadstools are common in the countryside of Wyoming but absent from the Great Barrier Reef. Tuberculosis, sore throats and other diseases caused by bacteria can be cured using antibiotics (like penicillin and tetracycline) but viral illnesses, such as AIDS (acquired immune deficiency syndrome) and influenza, cannot.

If one person sat for a thousand years writing down such patterns about animals, plants, fungi and microorganisms, he or she would not exhaust the possibilities. Ecologists can restrict the list—they are likely to be more interested in why toadstools are rare at the bottom of the sea than they are in why penicillin does not cure common colds. But even if they did their best always to remember a limited definition of ecology and concentrated only on how interactions with the environment affect the distribution and abundance of biodiversity, they would barely have begun their list at the end of a millennium of cataloguing patterns in nature.

Maybe at some far distant point in the future, a group of people will feel the need to spend a lifetime listing ecological patterns and the questions that must be answered if those patterns are ever to be explained. They will perhaps do this when they think that all the big questions have been answered, when all the obvious, universal patterns have been explained. But for the time being, there are plenty of unexplained patterns that are obvious to even the most casual observer.

1.8 Examples of ecological patterns

For the remainder of this chapter, it will be helpful to examine a few of the kinds of patterns in which ecologists are interested. The patterns that will be investigated will represent a small selection, in an attempt to try to illustrate as much variation as possible in the types of questions that ecologists ask. They do not represent the whole spectrum of questions that could be asked, or even of those that researchers are currently studying, but aim to give a flavour of the variety of different spatial and temporal scales with which ecologists are concerned. Some of the examples are of patterns at the scale of the whole globe, while others involve patterns over a few square metres. Some of the patterns describe ecological changes over the course of a single year, others deal with several generations of a population, and yet more are concerned with the effects of changes over millions of years.

1.9 Spatial patterns in ecology

The definition of ecology in Section 1.5 stresses the importance of the distribution and abundance of organisms. These words are simply another way of talking about **where** organisms are situated. The distribution of a population is a description of where in the world its individuals live, while abundance is about where those organisms are placed relative to one another—are they generally placed near to other members of their own species or at a greater distance from one another?

To understand ecological processes, it is necessary to understand spatial patterns in the distribution of organisms.

1.9.1 Large-scale geographical patterns

There are birds almost everywhere there is life on Earth. Emperor penguins (*Aptenodytes forsteri*) tolerate harsh blizzards as they nest on the ice of Antarctica, while white-eyed vireos (*Vireo griseus*) live in the equable woodlands of North America, and malleefowl (*Leipoa ocellata*) incubate their eggs in piles of rotting vegetation in Australia. Birds have been seen flying above the peak of Mount Everest and some species can dive underwater to depths of 50 m. However, despite their universal presence, birds are not distributed evenly over the available habitats. It is possible to make useful insights into ecology by comparing at a broad geographical scale the variety of different birds that live in different places.

About a quarter of all living species of birds are restricted to South America, and it is well-known that an area of tropical rainforest will contain more different types of birds (and other organisms) than an equal area of temperate forest (Fig. 1.8). Likewise, although coral reefs are not especially valuable habitats for birdlife, they harbour a greater diversity of other organisms than do coastal areas at higher latitudes.

Another kind of geographical pattern might relate not to the variety of different species of organisms in each place, but to the characteristics of those organisms. For example, most mammals in Australia have pouches in which to carry their young, while no Japanese mammal has any such pouch. There is plenty of tropical rainforest in West Africa but none in the east of the continent.

The explanations for such large-scale patterns are likely to be complex, and it may often prove most practical to study some smaller aspect of the larger system. But it is always useful to remember that if seemingly simple spatial patterns cannot be explained, then an understanding of ecology will lack any basic generalizations. Under these circumstances, ecologists would end up with a patchwork of interesting results from the four corners of the world, but these results would not make an intellectually satisfying whole.

1.9.2 Small-scale geographical patterns

Imagine that a group of ecologists is looking at the barnacles living on boulders on a rocky shore anywhere in the world. The ecologists have learned to identify the different species and are only interested in one of the species; they draw a map that plots the position of each individual of their chosen species. The distribution could be described as one of three broad types. The barnacles could be spread out regularly, with each one separated from its neighbours by a standard distance, or alternatively they might be scattered randomly across the surface of the rock, with no apparent pattern. The third possibility is that the barnacles might show a clumped distribution, with large aggregations of barnacles living close together, with spaces in between where no barnacles live.

Broadly, all populations can be described in such a manner. Rainforest trees often have either clumped or over-regular distributions. Many rainforest tree species in Costa Rica, Australia, Malaysia and West Africa have been found to be dependent on canopy openings to grow to be larger than saplings. For example, *Trema micrantha* on Barro Colorado Island in Panama requires high light levels for its growth, and is found only in forest gaps larger than 376 m^2. Thus, it has a clumped distribution, as large gaps fill up with trees of this species. In Borneo and Malaysia, trees of the dipterocarp family tend to have clumped distributions because their large seeds fall only short distances from the parent tree.

By contrast, many other rainforest tree species have distributions that are overdispersed; individuals are located further away from one another than might be

Fig. 1.8 The approximate number of bird species on each of the world's continents.

predicted if seeds were distributed and germinated randomly. In the case of *Platypodium elegans* in Central America, fungal diseases kill so many seeds near to the parents that no saplings are found within 20 m of the parent tree.

1.10 Temporal patterns in ecology

In defining ecology in Section 1.5, it was necessary to understand the difference between the interactions involving an individual organism and the processes that determine the distribution and abundance of a population. One of the most obvious differences is to do with the timescales over which these processes occur.

Individual organisms can only be affected by events that take place within their own lifetimes, whereas populations can be affected by processes and events that operate over many generations. The process of evolution by natural selection, for example, has been occurring for about 3.5 billion years.

One of the most important skills for an ecologist is integrating the effects of short-term and long-term processes.

1.10.1 A timescale of millions of years

Evolution has produced tens of thousands of species of fungi, but it does not appear to have been entirely fair to the different kinds of fungus. Why, for example, are there 16 000 species of Basidiomycotina fungi (such as mushrooms and toadstools) but fewer than 1000 species of Zygomycotina, a different group of fungi that tend to be parasitic and cause infections in other organisms (Fig. 1.9)?

This kind of question can be asked again and again. Why do rodents account for one-half of all mammal species? Why are there millions of animal species but only a few hundred thousand plants? More striking still, why is it that more than one-third of all the known species of animals are beetles? The processes

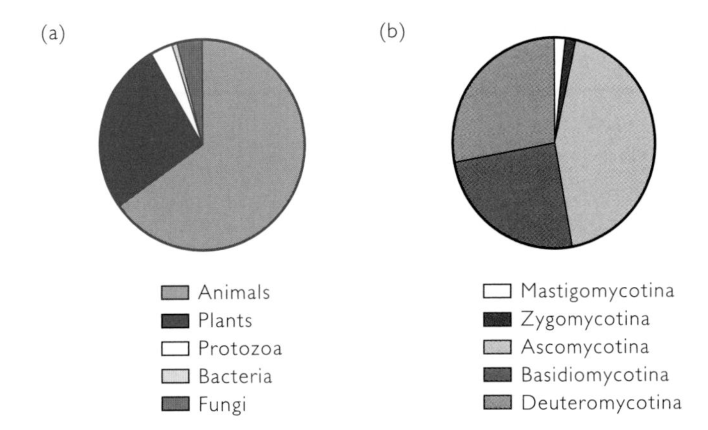

Fig. 1.9 Piecharts showing the relative number of known species in (a) each of the five kingdoms of organisms, and (b) each of the five orders of fungi.

that created these patterns operated over many millions of years.

There are many possible answers to these sorts of questions and, on the whole, they will not be considered in this book. Explaining such patterns is the work of evolutionary biologists, but ecologists cannot ignore the patterns or pretend that they are unimportant.

1.10.2 Timescales that relate to the lifespan of the organisms being studied

Imagine that someone wishes to carry out an ecological study of a population of protozoa in a garden pond. They wish to know how the population changes over time—when the number of protozoa increases and when it decreases. They believe that there may be some patterns that describe what happens to the protozoan population. They consider that temperature might be important—when the weather is hot, the population rises. It is also possible that pollution plays a part in determining the number of protozoa—when pesticides or fertilizers are accidentally spilled into the water, the population declines.

To know whether these hypotheses are true, it would be essential to study the protozoa continuously for some time. To have a real feel for whether the temperature is important, it would be as well to study the population for several years. That way, the investigation would be likely to include some warm summers and some cool ones, as well as harsh and mild winters. Similarly, to understand the importance of fertilizer accidents, it is not going to be enough to show that one April, some chemicals were spilled and the number of protozoa in the pond went down. However convinced the investigator may be that the population crash was caused by carelessness with garden chemicals, validation of the theory will only come if the population is studied during the April of other years, to test that a sudden and dramatic crash is not part of some natural, regular cycle. And the investigator would need to spill some chemicals at other times of the year, over several years.

In short, studying populations of organisms takes time, and the amount of time needed for any particular study depends to some extent on the organisms in question. Organisms with short generation times like bacteria could be studied in months, weeks or even days, but those with longer generations will require years or even decades of study.

Even with the simple example of protozoa in a garden pond, the time needed to describe adequately the many factors affecting population change is impractical. Therefore, ecologists tend to manipulate populations, or use simpler versions of the real world, to try to find answers to their hypotheses. For example, they might create a model system of a garden pond by setting up several glass tanks using samples of pond water. Then, they could alter the temperature, or concentration of fertilizers in different tanks, and observe the changes made to the populations of protozoa.

It is not always easy to define the best timescale over which any population should be studied, and in practice, the length of time over which something is studied may depend on the funding available, or the degree to which new ideas or constraints take the investigators away from the study. But there are many ecological patterns to be seen over timescales that people can appreciate. Nine studies of plant ecology reported in a

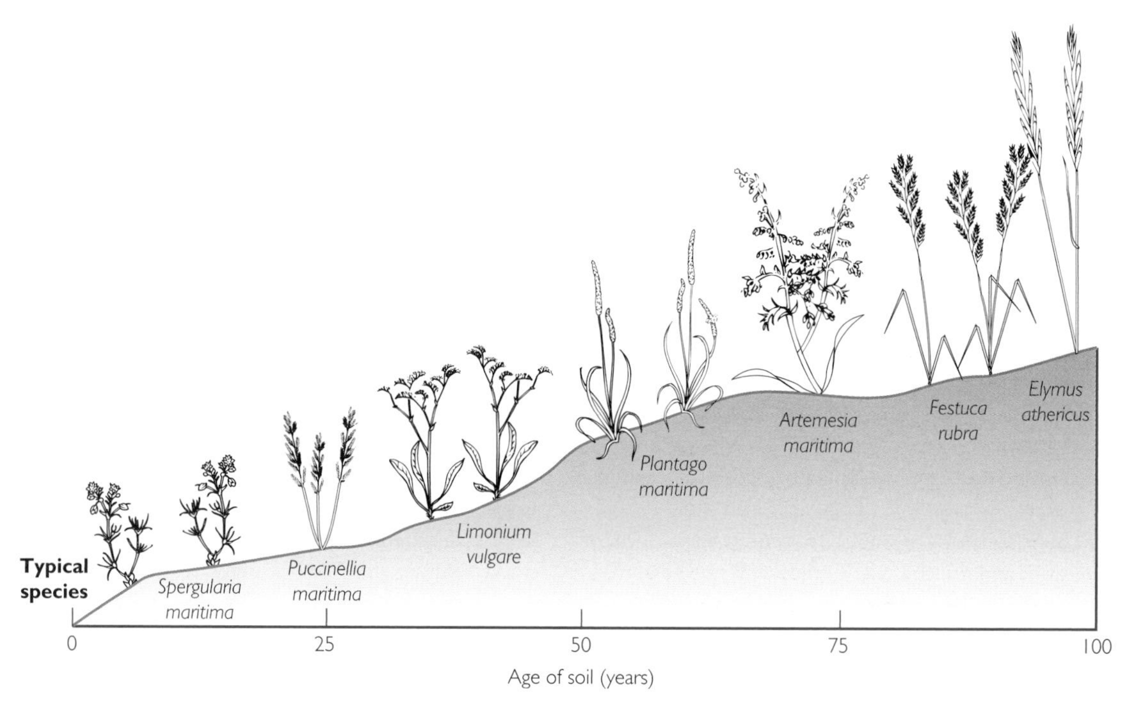

Fig. 1.10 Ecological succession at Schiermonnikoog in the Dutch Frisian islands. When the soils are first deposited, the plant community is characterized by small short-lived plants such as the sea spurrey (*Spergularia maritima*), but as the soil builds up over a century, the community is typified by larger, longer lived species such as red fescue grass (*Festuca rubra*).

single issue of the *Journal of Ecology* vary in timescale from a 'snapshot' study, attempting to make inferences from the structure of an ecological community in Brazil at a single point in time (Fragoso 1997), to an analysis of layers of pollen in a peat bed that give information about the European climate over 3000 years (Tallis 1997). There is one other study of very long-term processes, covering several centuries (Kelly & Larson 1997), but the remaining investigations all cover periods of a few years at most.

One example of a common pattern that occurs on the scale of human lifetime is ecological succession. Suppose that a new habitat opens up, because a fire kills off all the existing vegetation, or because humans abandon a plot of land that they have previously used as farmland. As time progresses, the structure of the ecological community that lives on the land will change, and will often do so according to a predictable pattern.

On the island of Schiermonnikoog in the Dutch Frisian islands, new land is created as the sea deposits silt on the sandy subsoil (Olff *et al.* 1997). Over a period of 100 years, the plant community on the resulting saltmarshes changes in a way that is common to each new saltmarsh. After about 10 years the sea spurrey (*Spergularia maritima*) is one of the dominant plants, but 25 years into the ecological succession, it has almost disappeared, to be replaced by the sea plantain (*Plantago maritima*) and sea lavender (*Limonium vulgare*). After the succession has proceeded for half a century or more, the sea spurrey and sea plantain are completely absent from the saltmarshes and the sea lavender is rare. The dominant plants now form a grassland, with couch grass (*Elymus athericus*) and red fescue (*Festuca rubra*) among the commonest species (Fig. 1.10).

If ecologists want to document the patterns of ecological succession that might occur over a span of 25–1000 years, it is obviously not possible by direct observation. However, many studies of succession have benefited from records and observations taken by different people over historical timescales. The retreat of glaciers in Glacier Bay, Alaska has been documented over the past 200 years and has allowed ecologists to reconstruct the pattern of ecological succession in this area. Often, spatial patterns are used to reconstruct the temporal pattern of succession. Thus, the development of vegetation with increasing distance from the shore of Lake Michigan was used to reconstruct the ecological succession of sand dunes formed by the side of the lake. Likewise, forest succession on the coastal plain of southeastern North America was described by studying a series of abandoned agricultural fields that had been out of cultivation for different periods, ranging from 1 to 200 years.

To understand the processes that act at the scale of a few years in the imaginary population of protozoa described above, or over a human lifetime in the ecological succession on the island of Schiermonnikoog, ecologists need to know both about the ways in which organisms interact with the physical environment and the ways in which they interact with the biological environment. For example, to understand succession on the Dutch islands, it is necessary to know how sea lavender thrives in different kinds of soils and with different concentrations of salt, and also to know how it fares in competition with sea plantain and couch grass, and how all the plant species are affected by grazing by geese or rodents.

When ecologists begin to understand these interactions, they begin to understand the patterns that emerge when they study a single site but make comparisons between different points in time. Once someone has really begun to understand ecology, these explanations for temporal patterns will dovetail neatly with the explanations that they discover for spatial patterns.

1.11 Chapter summary

Ecology is all about the vast diversity of life forms that share our planet. Scientific ecologists classify this biodiversity into different levels and then study it by attempting to understand the ways in which organisms interact with their biological and physical environments. Their explanations never forget the importance of evolution by natural selection.

But before someone can begin to explain the diversity of life, he or she must be able to find those patterns that are common and repeatable. Ecologists uncover these patterns by making comparisons in space and in time. The spatial and temporal variation that they discover is the basis of all ecological study, and it will be investigated in more depth throughout the rest of this book.

Recommended reading

Ghilarov, A. (1996) What does 'biodiversity' mean—scientific problem or convenient myth? *Trends in Ecology and Evolution* **11**: 304–306.

Orr, M.R. & Smith, T.B. (1998) Ecology and speciation. *Trends in Ecology and Evolution* **13**: 502–506.

Rosenzweig, M.L. (1995) *Species Diversity in Space and Time.* Cambridge University Press, Cambridge, UK.

Westoby, M. (1997) What does 'ecology' mean? *Trends in Ecology and Evolution* **11**: 166.

Wilson, E.O. (1992) *The Diversity of Life.* W.W. Norton & Co. Inc., New York.

Chapter 2

Global patterns of biodiversity and productivity: biomes

2.1 Major terrestrial biomes

Chapter 1 presented a definition of ecology, examined how the diversity of organisms on Earth today has been generated by evolution through natural selection, and addressed concepts of the niche and the spatial and temporal patterns of the distribution of organisms. That chapter also explored how the diverse array of organisms in the world can be organized into populations, species, communities and ecosystems. The ecology of each of these levels of organization will be examined in more detail throughout the rest of this book, but this chapter will focus on a broad level of organization that has been used historically by ecologists, geographers and evolutionary biologists, all of whom try to place their research into a global context.

This approach, aimed at describing the major types of ecological system that we can observe, has as its basis the explorations by plant geographers in the eighteenth and nineteenth centuries. These scientists had an advantage that no one had before them—the ability to travel regularly and relatively easily to widely separated parts of the world in wooden sailing ships. As these European geographers explored North and South America, Africa, India, Asia and Australia, they were struck by the broad relationships between the types of vegetation and animals that they encountered and the regional climate. They were also amazed by the overwhelming biodiversity of the tropical regions of the world, and the strange plants and animals they found in Australia. As part of their explorations, they collected thousands of plant and animal specimens from these regions to send back to Europe for classification in the newly developed binomial taxonomic system of Carolus Linnaeus (Box 2.1).

Early plant geographers used the outward, external appearance of the dominant plant types to describe the communities of organisms that they found. Thus, they may have used terms such as 'temperate deciduous forest' to describe a forest that dropped its leaves in the winter and was located at latitudes that were between the hot tropics and the cold polar regions. A 'tropical evergreen rainforest' would describe a community found near the equator where the trees kept their leaves all year round.

Notice that these classifications included some information on climate, and, in fact, these early plant geographers were among the first scientists to correlate vegetation with climate, especially temperature and rainfall. The relationship between climate and vegetation will be examined in detail in Chapter 4. Early explorers also recognized that the presence of mountains, oceans and different soils was very important to the biological communities that would develop in an area. These attempts to classify regions of similar organisms led to the system of classifying **biomes**. This system is still in use today, and a map of the major biomes of the world is given in Fig. 2.1.

Within each of these major biomes are many smaller, more restricted, community types. However, we shall focus on only the largest biomes, including tropical forest, savanna, desert, temperate grassland, temperate deciduous forest, chaparral, coniferous forest and tundra.

2.2 Tropical forest biomes

2.2.1 Distribution

Tropical forest biomes are found in three main areas of the world, centred on the equator (Fig. 2.2). In South and Central America, the huge forests comprise about 50% of the world's total tropical forest. Africa has about 20% of the tropical forest areas, with the remaining 30% found primarily in South East Asia and Malaysia. Each of these areas has several different types of subbiome, including tropical evergreen rainforest, tropical and subtropical seasonal forests, subtropical evergreen forests and mangrove forests. These subtypes develop

Box 2.1 The binomial classification system

Taxonomy is the theory and practice of naming organisms, and a **taxon** is any recognizable group of organisms. The classification of organisms into groups has different goals in different societies. One of these goals is to express relationships among organisms, and with this in mind, the evolutionary biologist Ernst Mayr identified 138 species of birds in the rainforest of New Guinea. The native New Guineans also had a naming system for the local birds, recognizing 137 different birds that corresponded almost exactly with those of Mayr. However, their classification system was designed to provide information on the suitability of birds for food and for feathers used in ceremonies or clothing.

There are at least three roles for biological classification systems. They are an aid to memory, they increase predictive power (for example, even without ever seeing a particular animal, knowing that it is a mammal provides information about its reproductive features and the presence of certain structural features such as fur or mammary glands), and they improve the ability to explain relationships among organisms.

To fulfil these roles, a classification system must provide stable, unique names for organisms. This was not always the case in biology. Originally, scientific names were given to organisms in Latin, or in Latin versions of Greek words. The starting point was the **genus**, a group of morphologically similar organisms. Added to the genus name were descriptive terms to describe a particular species. This system was a **polynomial** system, one with many descriptors of each species' appearance. These early scientific names were long and sometimes cumbersome, as biologists kept adding descriptive terms to a species' name. There was also a confounding lack of uniformity in the names given to particular organisms.

Carolus Linnaeus (1707–1778) developed a shorthand version of the polynomial system. His was a **binomial** classification system consisting of the genus name followed by a single species name. The species name was usually descriptive, for example, *Canis familiaris* is the name for domesticated dogs. *Canis lupus*, the wolf, is in the same genus, but is a different species. The names were established by a rigid set of rules that reduced confusion and provided a uniform means of communicating about organisms.

The current system of taxonomy used by biologists is a **hierarchical** system. This means that there are progressively larger taxons that include all lower taxons within them (e.g. families contain a number of genera). Currently, the system proceeds from species to genus to family to order to class to phylum to kingdom.

The goal of **systematics** is to delineate the evolutionary relationships among organisms and to express them within a taxonomic system. Applying this to the taxonomic hierarchy, each succeeding group would represent a greater time since the taxa last shared a common ancestor. Thus, members of the same species would be most closely related to each other, a genus would include one or more related species, and a family would include one or more related genera. An example of the classification of a red oak, a house fly, a lion and a wolf are given in Table I.

Table I Taxonomic classification of four different species.

Category	Red oak	House fly	Lion	Wolf
Kingdom	Plantae	Animalia	Animalia	Animalia
Phylum	Tracheophyta	Arthropoda	Chordata	Chordata
Class	Angiospermae	Insecta	Mammalia	Mammalia
Order	Fagales	Diptera	Carnivora	Carnivora
Family	Fagaceae	Muscidae	Felidae	Canidae
Genus	*Quercus*	*Musca*	*Panthera*	*Canis*
Species	*rubra*	*domestica*	*leo*	*lupus*

due to changes in seasonal patterns of rainfall. For example, an area that experiences dry periods of the year may develop forests with trees that drop their leaves in the dry season. On mountains, the community type changes as temperatures decrease with increasing elevation.

The tropical rainforest is characterized by several layers of vegetation. There is an emergent layer of very tall trees, higher than 40 m, under which an overstorey of slightly smaller trees develops (about 30 m tall). Beneath the overstorey is an understorey of tall shrubs and subcanopy trees, and finally a ground layer of

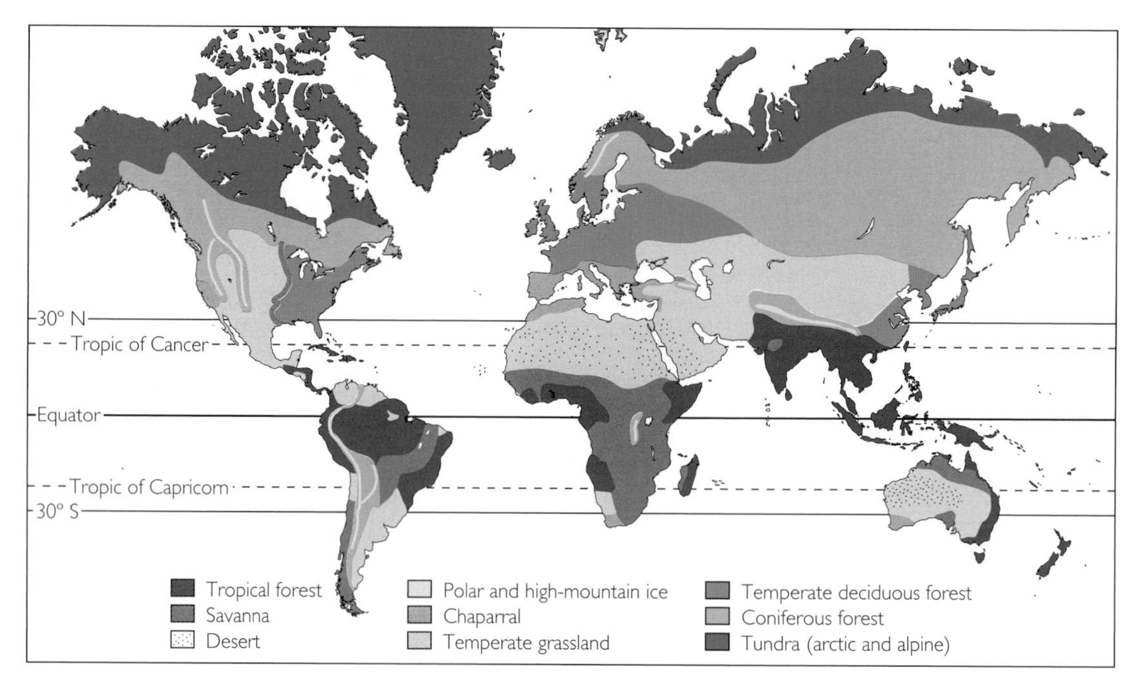

Fig. 2.1 The major biomes of the world.

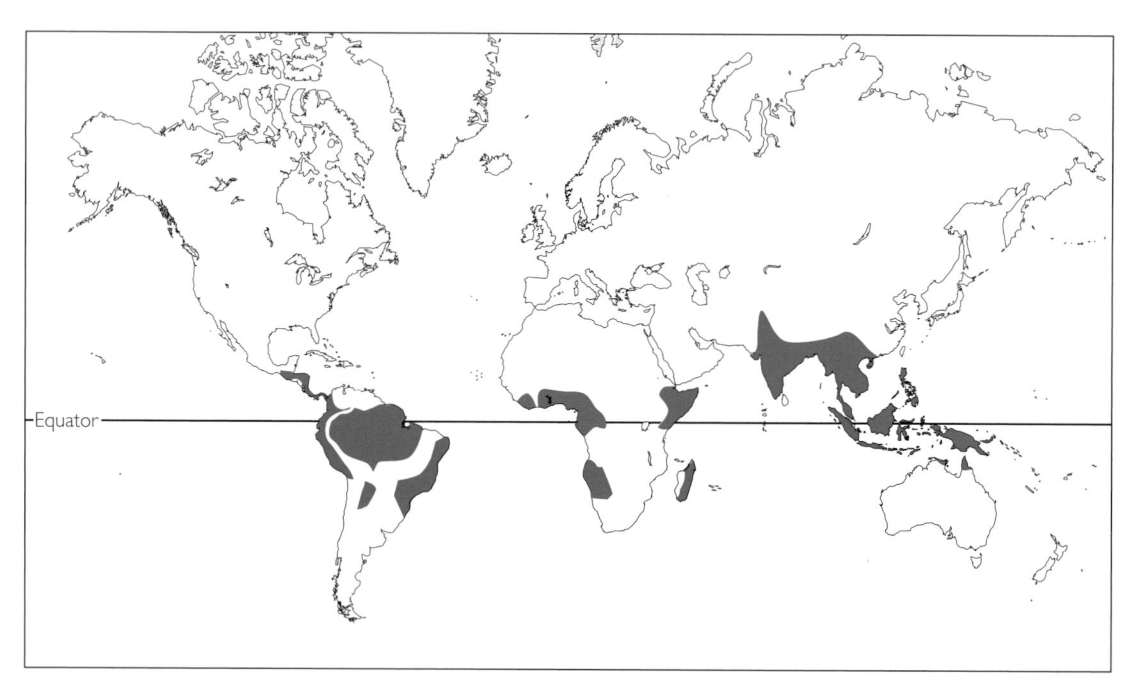

Fig. 2.2 Map of the global distribution of tropical forest biomes.

herbaceous vegetation. This layering results in dramatic differences in sunlight in different strata of the forest, and in patches where overstorey trees have fallen. As a result, this forest is characterized by a large array of different life forms, including epiphytes (plants that grow on top of other plants) and vines (plants that use tree trunks and branches for support to reach the upper layers of the canopy) (see Plate 2.1, facing p. 24). Because of the variability of the forest canopy, there are many chances for organisms to specialize in specific habitats (niches), such as small or large canopy gaps, different forest layers, on or under leaves, in trunks, in dead wood, in the soil, etc.

2.2.2 Productivity

The tropical forests of the world are among the most species-rich and productive of all biomes. Ecologists use the term **primary productivity** to describe how productive plant communities are. Gross primary productivity is the total amount of carbon gained through photosynthesis in the community. It is usually expressed in terms such as grams of carbon per square metre of ground per year (g carbon m^{-2} $year^{-1}$).

However, not all of this carbon can be used for plant growth; much of it is used in respiration by the plant. Net primary productivity is the amount of gross primary productivity remaining after the loss of some carbon to respiration. Net primary productivity is then used to build new plant structures and for reproduction. Some of it is also lost to herbivores and parasites that gain nourishment from plants (see Chapter 9). Although there is great variability, typical rates of net primary productivity for tropical rainforests can exceed 3500 g carbon m^{-2} $year^{-1}$. This high level of productivity is sustained despite nutrient-poor soils in many areas of the tropics. Because of high rainfall, soils are often heavily leached (as water passes down through the soil, it takes dissolved minerals and nutrients with it).

Productivity is maintained despite leached soils because plant **biomass** (the amount of living plant material present at any one time) represents a major store of nutrients. For example, evergreen leaves, tree trunks, roots and branches contain nutrients that remain in the plant through several years. When leaves or branches fall to the forest floor, they become **litter**. This litter is decomposed very rapidly through the action of small insects, protozoa, bacteria and fungi (**detritivores**). The decomposition of litter is a major pathway for recycling nutrients in this ecosystem (see Section 12.4). The warm temperatures and high moisture conditions are ideal for decomposers that rapidly break down litter and release nutrients. Rapid uptake from the litter is facilitated by **mycorrhizae**, which are mutualistic associations between plant roots and fungi. The plant benefits from this **mutualism** (see Section 9.6) through increased uptake of nutrients, while the fungus benefits through a supply of carbohydrates from the roots, i.e. the relationship is mutually beneficial.

2.2.3 Adaptations

Tropical rainforests are estimated to contain more than half the species on Earth, and the range of adaptations that they show stretches the human imagination. For example, of the 250 000 known vascular plants (flowering plants, ferns, club mosses, horsetails and quillworts), 170 000 are found in the tropics and subtropics. In Columbia, Ecuador and Peru alone, there are an estimated 40 000 vascular plant species. In a tropical rainforest near Iquitos, Peru, 300 tree species have been recorded in 1 ha of land (100 m × 100 m). This number is even more remarkable when we consider that there are only 700 native tree species found in all of Canada and the United States combined. This amazing biodiversity is present in almost all other life forms as well, including birds, insects (especially ants, beetles, butterflies and termites), amphibians and mammals; 1209 butterfly species have been identified in the 55 km^2 Tambopata Reserve in southeastern Peru. This compares to 380 species in Europe, the Mediterranean coast and North Africa combined. In this same Tambopata Reserve, 43 species of ant were identified from the canopy of a single tree, equal to the total number of ant species in the British Isles. There are estimates of 18 000 species of beetle in 1 ha of Panamanian rainforest, compared to 24 000 beetle species in all of the United States and Canada. Box 2.2 explores some of the threats faced by the biodiversity of the world's forests.

The tropical rainforests are part of a larger pattern of increasing biodiversity with decreasing latitude worldwide. The pattern holds for almost all habitats and groups of organisms. Even shallow water, open sea and bottom-dwelling marine organisms follow this trend. Biologists have developed a number of hypotheses to

Box 2.2 Threats to biodiversity in the world's forests

A major challenge for ecologists and conservation biologists is how to preserve the biodiversity found in the world's forests. By 1989, tropical rainforests occupied only half the total area that they covered in prehistoric times. Between 1% and 2% of the total area of rainforests in the world is cleared each year! This results in a great amount of habitat destruction, and the creation of smaller plots of forest, called **habitat fragmentation**. As habitat is lost, so are species. The current rate of habitat loss results in the forests losing about 0.5% of their species each year. However, the total rate of species loss is even higher, since factors other than habitat destruction also contribute to the loss of species. With increased human presence, more exotic, non-native species are introduced to new habitats. These new species compete with, eat or infect the native species, resulting in the disappearance of vulnerable species.

Recent work by R.K. Didham and co-workers has focused on how the reduction in forest area and the increasing level of habitat fragmentation are affecting beetle species in central Amazonia. Almost 50% of the species that were sampled showed changes in population density in response to forest fragmentation: 49.8% of common species were lost from forest fragments of 1 ha (100 m × 100 m) in size, 29.8% from 10 ha plots and 13.8% from 100 ha forest fragments. Species losses may be even higher than these estimates, because these data were taken from the most common beetle species. The rarer species, with lower population sizes, may be more subject to problems caused by random events such as disease, drought or increased predation.

Temperate forests are facing some of the same environmental hazards as tropical forests. Many of them are being cleared for human habitation and roads, there is a great deal of fragmentation occurring, and some tracts are being converted into stands of a single commercially valuable species. Biodiversity is decreasing, and exotic, alien species are becoming increasingly widespread. Additional challenges to these forests occur due to pollution—acid rain and nitrogen depositions from fossil fuel burning both have detrimental effects upon tree growth and survival. Often these effects can occur at great distances from the source of the pollution. For example, forests in Scandinavian countries have suffered large losses due to acid precipitation originating from the burning of fossil fuels in other European countries such as the United Kingdom.

explain this trend, including climate, available solar radiation, habitat variability, length of time the habitat has been available and the level of environmental disturbance. A number of factors probably affect this pattern and these themes are explored in more detail in Chapter 14.

The high biodiversity of tropical biomes is also reflected in the large number of biotic interactions between species (see Chapters 8–10). For example, there are over 900 species of fig tree (*Ficus* spp.) in tropical forests, and each different species has its own unique wasp species that pollinates its flowers. A similar level of specialization is seen in the thousands of orchid species and their bee pollinators in the tropics. There are over 500 species of tropical passion flower, and each species has a host of other organisms with which it interacts. Different animals pollinate its flowers, eat its fruits, disperse its seeds and eat its leaves. Because of the great number of potential herbivores passion flower species are exposed to, this group of plants has evolved a wide array of defensive chemicals. The stems and leaves of certain passion flower species are among the most toxic known to humans. Many passion flower species have developed mutualistic associations with ants. The plants have nectar-secreting organs on their stems, outside their flowers. The nectar these organs secrete is rich in sugars and amino acids. Ants use this nectar as a major food source. They patrol the surface of the plant, aggressively collecting and destroying butterfly eggs and caterpillars. Thus, the ants defend the plants from potential herbivores, while the plants provide a source of energy for the ants.

2.3 Savanna biomes

2.3.1 Distribution

Moving north and south away from the equator, rainfall decreases and becomes more seasonal (see Chapter 4). As distinct wet and dry seasons develop, tropical rainforests give way to tropical deciduous forests, where the trees drop their leaves in the dry season. As rainfall decreases even more, these tropical deciduous forests are replaced by thorn forests and finally by a mixture of widely spaced trees and grassland, or savanna (Fig. 2.3).

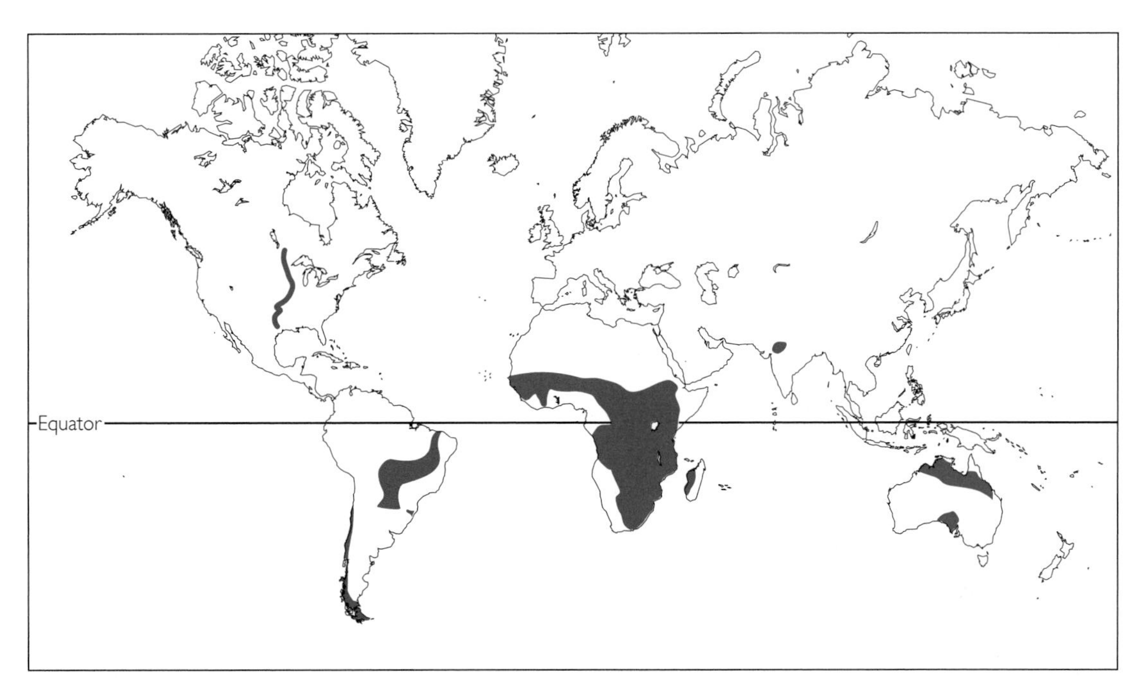

Fig. 2.3 Map of the global distribution of savanna biomes.

Savannas cover 60% of the land area in Africa, and are found in India, Australia and South America, especially Brazil. Savannas represent a transition between tropical forests and deserts. There are several subtypes of savanna, distinguished by the amount of rain they receive. Generally, lower rainfall results in fewer trees and more grass. The trees in savannas are usually deciduous in the drought, and are often legumes with thorny branches, such as *Acacia* (see Plate 2.2, facing p. 24).

Fire plays a major role in the formation and maintenance of savanna biomes. Repetitive fires have occurred in the African savanna over the last 50 000 years. The most common fires are relatively cool, fast-moving surface fires. The soil provides a good insulator, so most plant seeds and underground plant parts are protected from damage.

2.3.2 Productivity

The productivity of savannas is dependent upon the soil depth and rainfall. Net primary productivity can range from 400 to 600 g carbon m^{-2} $year^{-1}$.

Decomposition occurs rapidly and all year round in the savanna. The turnover rate of leaf litter is about 60–80% per year. Fire also breaks down lignin and cellulose in decay-resistant grass leaves, aiding the release of nutrients back into the soil. Termites and earthworms are important decomposers in this biome. Termites can break down cellulose and wood due to the presence of symbiotic micro-organisms in their gut. Earthworms help aerate and incorporate organic material into the soil, and are important factors in soil structure.

Perhaps the most striking aspect of the savanna is the rich diversity of large, grazing animals and their predators. Up to 60% of the plant productivity in the savanna is eaten by ungulate grazers, small mammals and arthropods. Of course, with this rich herbivore fauna comes a variety of predators and scavengers, including lions (*Panthera leo*), cheetahs (*Acinonyx jubatus*), African wild dogs (*Lycaon pictus*) and vultures like the Ruppell's griffon (*Gyps rueppellii*).

2.3.3 Adaptations

The enormous diversity and numbers of herbivores in savannas have created evolutionary pressures on the plant life. Many trees and shrubs that are typical of savanna habitats, such as those in the genera *Acacia* and

Commiphora, have enormous thorns to deter giraffes (*Giraffa camelopardalis*), impala (*Aepyceros melampus*) and other ungulates. Invertebrate diversity of the savannas is also very high, so many plants have also evolved mechanisms for reducing damage from insect herbivores. In South America, the bull's horn acacia (*Acacia cornigera*) produces nectar all year around to attract the ant *Pseudomyrmex ferruginea*, which protects the plant by killing other insects that land on it. In Africa, *Acacia zanzibarica* has large hollow structures called pseudogalls, which provide shelter for the ant *Crematogaster sjöstedt*, which kills insect pests.

Rainfall in savannas can be highly seasonal or unpredictable, and all organisms that inhabit this biome will at times experience long periods of severe drought. The baobab tree (*Adansonia digitata*) has evolved a thick barrel-shaped trunk, which stores large quantities of water. To reduce the wastage of water, the small mammals known as hyraxes (in the family Procaviidae) have very efficient kidneys, which can create very high concentrations of urea and other dissolved chemicals. They even excrete large amounts of undissolved calcium carbonate, which gives a white appearance to the rocks where they live. Termites, such as *Macrotermes natalensis*, build mounds that are architecturally very complex, with an elaborate system of air ducts that allows the termites to maintain levels of humidity that are extremely high. This high humidity may also be partially maintained by the respiration of fungi such as *Xylaria furcata*, which are tended by the termites.

Hot, open savannas are ideal habitats for cold-blooded reptiles, and some bird species, such as the secretary bird (*Sagittarius serpentarius*), have evolved to specialize in feeding on this abundant food source. Other birds, such as the kori bustard (*Ardeotis kori*) and von der Decken's hornbill (*Tockus deckeni*), take advantage of frequent fires by following the moving edge of the fire to catch some of the thousands of insects that are trying to escape the flames.

2.4. Desert biomes

2.4.1 Distribution

At latitudes further north and south of the savannas, precipitation decreases, and savannas are replaced by semi-arid grasslands and deserts. Deserts are located primarily between 15° and 30° north and south latitude (Fig. 2.4), and cover about 26–35% of the Earth's land surface. This is an area where high-pressure systems and

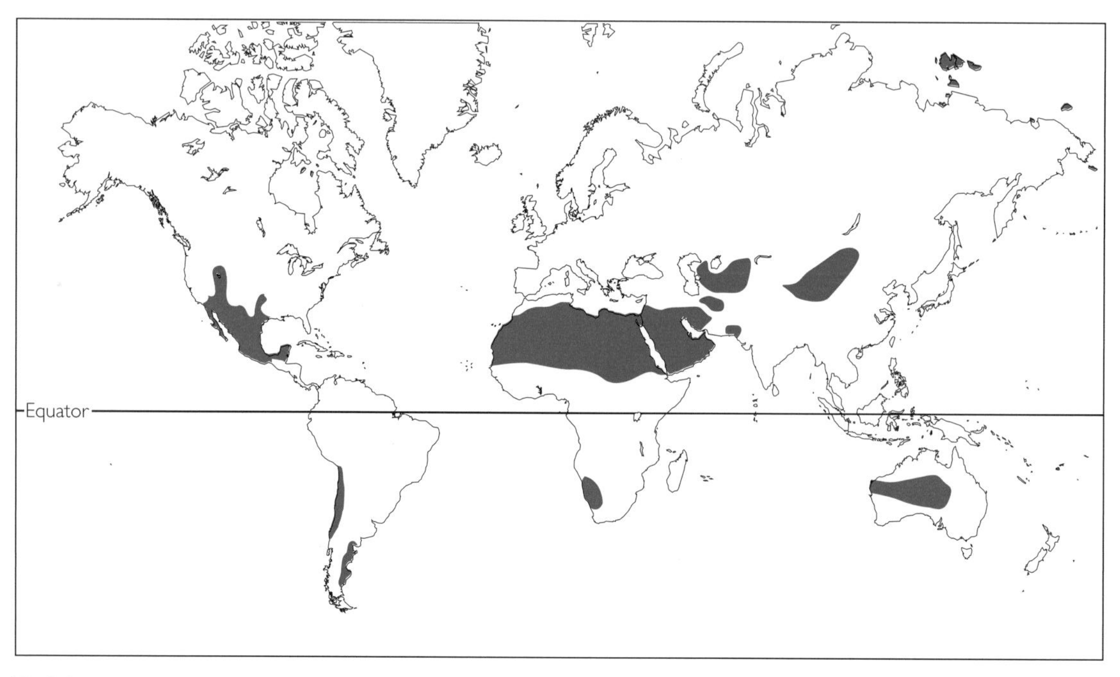

Fig. 2.4 Map of the global distribution of desert biomes.

(a) (b) (c)

Plate 2.1 (a) Tropical forest in Costa Rica. (Courtesy of Gerald S. Wilkinson.) (b) Tropical forest epiphyte—a plant that grows on top of other plants. (Courtesy of David W. Inouye.) (c) Tropical forest lianas, climbing vines that use trees as structural support. (Courtesy of David W. Inouye.)

Plate 2.2 Two views of savanna biomes with *Acacia* trees. (Courtesy of Gerald S. Wilkinson.)

Plate 2.3 The effect of local drainage patterns on plant productivity in deserts. (Courtesy of I. Forseth.)

Plate 2.4 Annuals can dominate the plant biomass and diversity in deserts after rainfall. Here, the yellow-flowered *Geraea canescens* and the pink-flowered *Abronia villosa* grow on sandy soils in Death Valley, California. (Courtesy of I. Forseth.)

Plate 2.5 (a) The overstorey and moss layer of a northern coniferous forest. (b) The shallow roots and heavily leached soil (light colour) characterizing much of the northern coniferous forest. (Courtesy of I. Forseth.)

Plate 2.6 A bog in northern coniferous forest caused by poor drainage. (Courtesy of I. Forseth.)

Plate 2.7 Tundra with reindeer, Siberia. (From Willmer *et al.* 2000.)

(a)

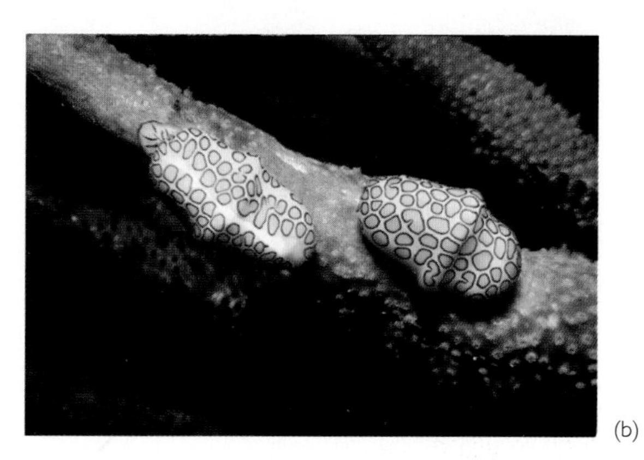

(b)

Plate 2.8 (a) A Caribbean coral reef. (b) The carnivorous flamingo-tongued snail, *Cyphoma gibbosum*, which feeds on corals. (Courtesy of Kenneth P. Sebens.)

Plate 2.9 Deep-sea vent fauna: giant, tubicolous, pogonophoran worms, *Riftia pachyptila*, found abundantly around deep-sea fumaroles. (Courtesy of R. R. Hessler.)

Plate 5.1 White leaf hairs on the leaf surfaces of this Sonoran Desert plant reduce leaf absorption of solar radiation. (Courtesy of I. Forseth.)

warm, dry air masses dominate the climate. However, deserts can also form on the side of mountains opposite that of prevailing winds. These are called rainshadow deserts and include the high elevation (3500–4500 m) Puna desert in Argentina.

The Sahara Desert is the world's largest desert, covering over 9 million km^2 of North Africa. It expanded by 7% between 1980 and 1995. In South America, the Atacama Desert along the coast of Chile is the most severe, having the lowest and most unpredictable rainfall regime. The largest desert region in the southern hemisphere occurs in Australia, where 40% of the land is classified as desert.

2.4.2 Productivity

The desert biome is characterized by low rainfall (less than 250 mm per year) and sparse plant cover. Once rainfall exceeds about 25–75 mm per year, there is a positive, direct relationship between productivity and precipitation in deserts (Fig. 2.5). Thus, while overall annual productivity is low in deserts, it is also extremely variable. The average productivity in North Africa can vary anywhere from 0 to 120 g carbon m^{-2} $year^{-1}$. The local drainage conditions greatly influence productivity, with the lower slopes of basins and depressions being most productive (see Plate 2.3, facing p. 24). Annual species are especially responsive to wet years, and their productivity varies a great deal from year to year.

The timing of precipitation and seasonal temperatures can have a large influence on the type of desert that

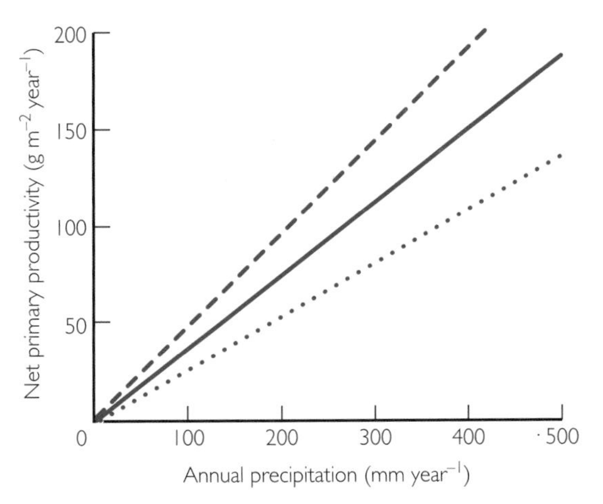

Fig. 2.5 Relationship between productivity and rainfall in the deserts of southwestern North America. (Data from Smith *et al.* 1997.)

develops. For example, four desert types can be distinguished in North America—the Great Basin Desert, the Mojave Desert, the Sonoran Desert and the Chihuahuan Desert.

The Great Basin Desert is a high elevation desert with snow and frost in the winter. It receives 70% of its annual precipitation in the winter. The combination of cold and low water availability results in a single-layered shrub community with low species diversity.

The Mojave Desert, west and south of the Great Basin Desert, occurs at lower elevations and receives 65% of its rain in winter. Temperatures in the Mojave are generally higher than the Great Basin, and it rarely snows. Here, a two-layered plant community develops with an overstorey of shrubs and an understorey of subshrubs. Because of the warmer temperatures, more annual species grow here than in the Great Basin, resulting in higher biodiversity.

The Sonoran Desert is a warm, low elevation desert that receives only 55% of its rain in winter. It rarely freezes in the Sonoran, and snow is hardly ever recorded. The Sonoran Desert has up to four vegetation layers, an overstorey of 3–5 m tall cacti and trees, a tall shrub–cacti layer of 2–3 m, subshrubs of about 1 m in height, and perennial grasses and annual plants forming a ground cover. There are two classes of annual plants, one that germinates after winter rains and one that germinates after summer rains. Biodiversity is high relative to the Mojave and Great Basin Deserts.

The Chihuahuan Desert is a warm desert, located at higher elevations than the Sonoran, that receives only 35% of its rain in winter. It is a three-layered community, lacking the tall, 3–5 m cacti and tree layer of the Sonoran Desert. Because of the summer rainfall, it has many summer annuals and grasses.

2.4.3 Adaptations

Deserts are dominated by shrubs, most with extensive roots and small leaves. Much of the biodiversity and productivity in the vegetation of deserts, especially warm deserts, exists in the form of short-lived, annual plants (see Plate 2.4, facing p. 24). These plants survive the dry season as seeds, and then germinate and grow after rainfall. They have adapted to germinate after a specific amount of rain, usually enough to provide a reasonable chance for them to grow to flowering stage and set seed. Because of the unpredictability of rainfall, many of these annual plants have seeds that remain viable in the soil for several years, until a triggering rainfall occurs. Some species also have mechanisms to prevent all of the seeds produced in 1 year from germinating at the same time. This adaptation allows reproduction to be spread over several years, reducing the probability that a harsh year will eliminate the entire seed crop of a plant. For example, *Aegilops ovata* in the Saharan desert has as many as six seeds in a fruit. One seed will germinate immediately after maturity if water and temperatures are favourable. Germination of the second seed will be delayed until the following year, and several years may elapse until all six seeds have germinated.

Animal life in the deserts is usually active only during specific times of the day or season. Many animals possess suitable adaptations, such as living in moist, deep burrows, and emerging only at night. Insects often have water-resistant cuticles and exoskeletons to retard water loss, and many of them **aestivate** to avoid hot, dry conditions. Aestivation is a period of dormancy or inactivity in the summer months often spent in burrows or below ground. The African lungfishes (*Protopterus* spp.) build up large quantities of nitrogen-based waste products during periods of aestivation, which they then excrete once the rains come and water is plentiful. More desert adaptations will be discussed in Chapter 5.

2.5 Grassland biomes

2.5.1 Distribution

Temperate grasslands are common in the interior of North and South America and Eurasia (Fig. 2.6). In these areas, the climate is characterized by hot summers and long, cold winters, with a large seasonal variation in temperature. The annual precipitation is 400–600 mm, and has a strong summer peak. These conditions have led to communities dominated by grasses, with a mixture of other herbaceous plants.

The type of grassland that develops in a particular area depends partly on precipitation. Higher precipitation results in taller grasses, up to 2 m tall, that often form continuous cover. Lower precipitation leads to what is sometimes called mid-grass prairie, about 1 m tall. The lowest precipitation results in short grasses that grow in bunches and do not form a continuous cover over the soil surface.

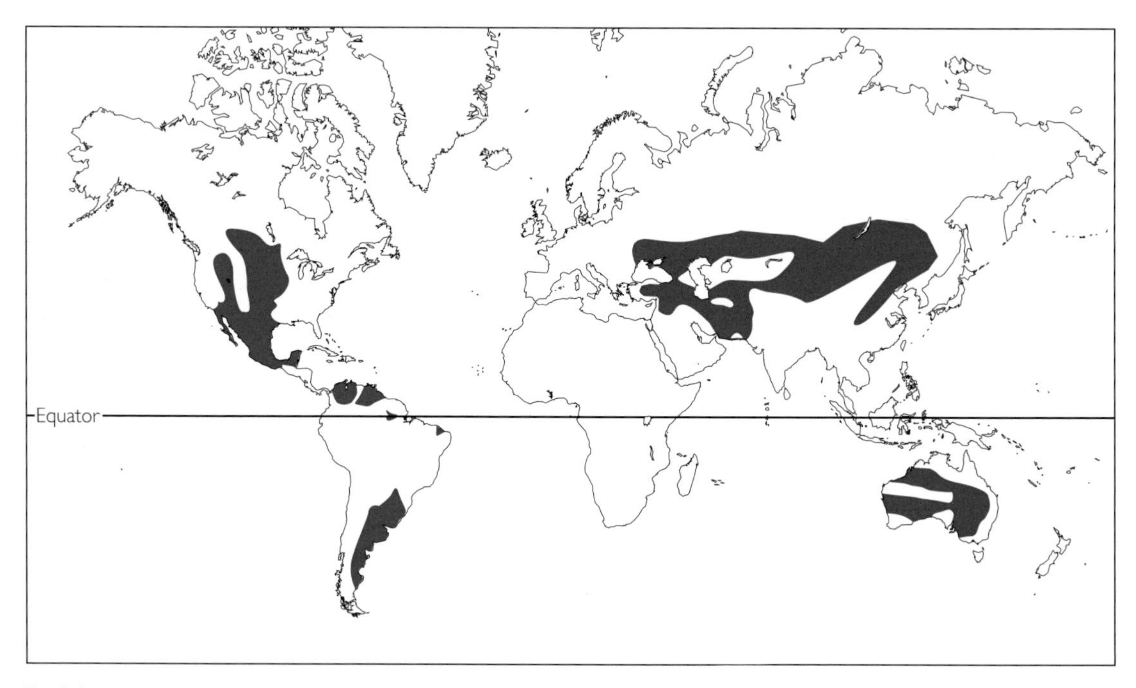

Fig. 2.6 Map of the global distribution of temperate grassland biomes.

2.5.2 Productivity

The net primary productivity of grasslands varies with precipitation. At the dry end of the spectrum, net primary productivity may be only 400 g carbon m^{-2} $year^{-1}$, while at the wet end, net primary productivity can exceed 1000 g carbon m^{-2} $year^{-1}$. On the wetter margins of grasslands, forests often develop, and there is a continuous shifting of the border between these two communities depending upon the occurrence of fire and drought, both of which favour grassland. On the drier borders of grasslands, deserts develop.

2.5.3 Adaptations

The organisms in grasslands have evolved under three major selective forces, the presence of regular fires, the occurrence of periodic droughts, and grazing by mammals and insects. The effects of these interacting factors result in a plant community where much of the plant biomass is below ground, in the roots and storage organs. By having their growing points (meristems) underground, plants protect these sensitive organs from herbivory, damage by fire and harsh winter temperatures.

2.5.3.1 Fire and drought

The interaction between grasslands and fire is especially interesting. Due to a combination of summer droughts that dry out the vegetation and summer lightning storms, grassland fires are common. Fire will destroy shrubs and trees, reducing competition from these life forms. Because most grasses have their buds and meristems below ground, they are generally protected from damage by fast-moving, cool, surface fires. Many of the non-grass (forb) plants in grasslands need to have their seeds exposed to the high temperatures of fire in order to germinate. If fires are suppressed for a long time (as might happen near human settlements where fires are quickly extinguished before they destroy crops or houses), litter builds up. This eventually results in hotter, more slow-moving fires that can be more destructive to grass meristems and forb seeds.

Fires can also aid in nutrient recycling in grasslands. Many grasses are decay-resistant and have tough, silica-containing tissues. These tissues evolved in response to grazing—the silica in the tissues wears down insect mandibles and the teeth of grazing mammals. Fires break down this decay-resistant tissue and release nutrients back into the soil. Fires of the appropriate frequency stimulate productivity and flowering of some grassland species, and increase biodiversity of the grassland by allowing species that specialize in openings and post-fire habitats to persist. Too long an interval between fires can result in a build-up of large litter masses that tie up nitrogen, resulting in a loss of productivity. Conversely, too short an interval can reduce nitrogen through increased leaching and volatilization (conversion of nitrogen to gaseous forms).

2.5.3.2 Grazing

Grasslands, along with savannas, have seen the evolution of some of the largest terrestrial animals. From grey kangaroos (*Macropus giganteus*) in Australia to bison (*Bison bonasus*) and horses (*Equus* spp.) in Eurasia and North America, grasslands are characterized by a rich assemblage of grazing animals, together with their predators, and scavengers. Many of these animals have disappeared as a result of hunting by humans and of the loss of grasslands to agricultural activities.

In present-day grasslands, only 5–10% of the biomass is lost to natural herbivores. It is hard to estimate what the situation was like prior to human settlement, but because of the long co-evolution of grazers and grasses, biomass losses were probably higher. There are journal reports from early pioneers in North America that food for horses was scarce in the presence of large herds of buffalo (*Bison bison*). However, these herds were migratory, moving to areas of plentiful grass as local areas were exhausted. Studies of remnant buffalo herds in North America show that the disturbances caused by buffalo can play an important role in increasing biodiversity; the buffalo cause openings in the grass community that rare species can colonize. Smaller mammals, such as black-tailed prairie dogs (*Cynomys ludovicianus*), were also important components of pre-European settlement grasslands in North America. Current studies have shown that prairie dogs, through their digging of burrows and mixing of soils from different depths, can have significant impacts on species composition and structure of grassland communities. Their colonies are characterized by decreases in grasses and increases in forbs and shrubs.

Grasslands often serve as major food-producing areas for humans, and there have been many studies of the effects of cattle grazing on grasslands. Many grass species that have evolved under selection by the grazing of large mammals have characteristics that make them resistant to the effects of cattle grazing at low densities. However, the effects of cattle on grasslands increase dramatically when the cattle are living at high densities. As overgrazing occurs, species composition of the grassland changes, with an increase in invading species that are non-native to the grassland. These weedy invaders are often less desirable to the cattle than native grasses. Soil properties also change, with increased compaction due to trampling, decreased root growth due to loss of above-ground tissues, and decreased water infiltration. The water-holding capacity of the soil decreases, runoff increases, and eventually increased erosion occurs. The spread of deserts into the dry margins of grasslands is often associated with overgrazing by domesticated animals.

2.6 Temperate deciduous forest biomes

2.6.1 Distribution

Temperate deciduous forests are found in North America, South America, Europe, Asia and Australia (Fig. 2.7). Their climate is characterized by mild winters, with warm wet summers. Precipitation is high all year round, averaging anywhere from 600 mm to 2000 mm per year. The winters, with freezing temperatures, create a distinct seasonal variation, and most of the trees in these forests drop their leaves in winter. Animals have also evolved responses to winter temperatures, with many of the insects, reptiles, amphibians and mammals going into **hibernation** or periods of inactivity during the winter.

2.6.2 Productivity

Temperate forests are moderately productive, with net primary productivity ranging from 600 to 1500 g carbon m^{-2} $year^{-1}$. Litter production is high in deciduous forests and is a major pathway for the recycling of nutrients back to the soil and growing plants (see Chapter 12). The turnover rate of litter in temperate forests is slower than in tropical forests, with a **half-life** of about 3 years. Half-life is a convenient measure of turnover time because it refers to the average amount of time it takes for half of the original amount of litter produced at any one time to decompose. Turnover rate varies with tree type—evergreen or deciduous—and with tree

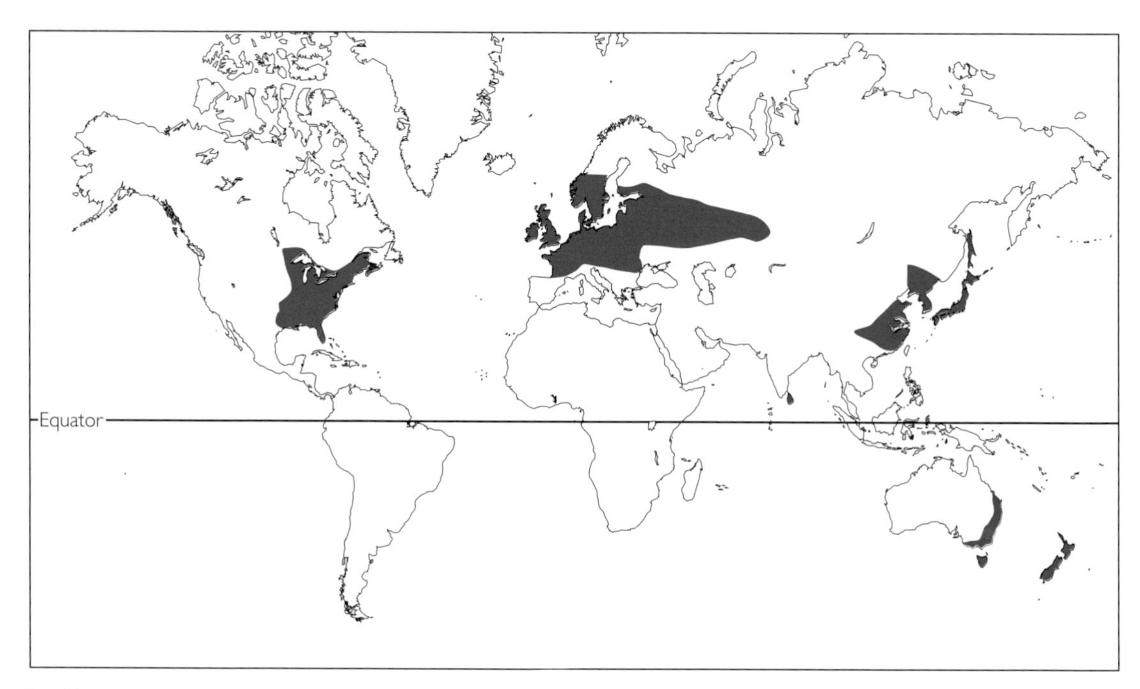

Fig. 2.7 Map of the global distribution of temperate deciduous forest biomes.

species, depending upon the resistance of tree leaves to decomposition.

2.6.3 Adaptations

The distinct seasons create a marked change in the ability of sunlight to penetrate throughout the forest, depending upon the presence or absence of leaves (Fig. 2.8a). This has led to a spring flora on the ground that exploits the brief period of high sunlight before leaves develop. Some of these plants are annual, but most are perennials (they live for several years). The perennials usually have storage organs, such as rhizomes and bulbs, to store carbon for reproduction and future growth (Fig. 2.8b). In this way, they compensate for the low light and low growth rates they experience during the summer months in the understorey. In many species, flowering and sexual reproduction is tightly tied to sunlight availability.

Other ground plants specialize in canopy gaps, where sunlight levels are much higher than in surrounding canopy-covered areas. Many of the ground layer plants have mechanisms to exploit brief periods of sun called sunflecks, which may last anywhere from a second to several minutes. The majority of the daily photosynthetic carbon gain of these plants may be due to these sunflecks.

As in tropical forests, there are several layers of vegetation in temperate deciduous forests, each varying in available sunlight. There are often four layers, a 20–30 m tall overstorey, a 5–10 m tall understorey, 1–2 m tall shrubs and a ground layer. The layers of vegetation allow niche partitioning among animals, and more complex forests support a greater number of animal species. For example, the number of bird species found in a forest is often positively correlated with the height and number of layers in the forest.

The warm summers and potentially cold winters of temperate forests could cause problems for invertebrates, but various solutions have evolved. One example comes from many butterflies and moths in temperate forests, which do not survive the winter as fragile adults, but as other life stages, which can remain in sheltered sites or underground to avoid the worst of the winter weather. The oak beauty moth (*Biston strataria*) overwinters as a pupa and the oak lutestring moth (*Cymatophorina diluta*) as an egg.

There is often great variation in the dominant tree species in temperate forests. This is a complex function

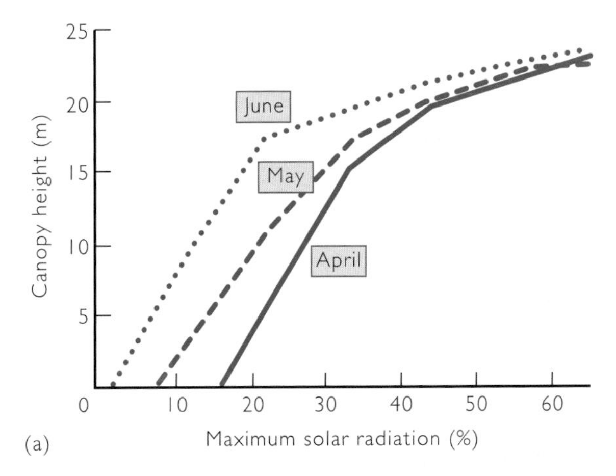

Fig. 2.8 (a) The change in the distribution of solar radiation at different heights in a forest canopy in the nothern hemisphere between April and June. In April, prior to the canopy trees expanding their leaves, there are higher amounts of solar radiation available at the forest floor. By June, with full overstorey development, light at the forest floor is reduced to very low levels. (Data from Hutchinson & Matt 1977.) (b) Corms of the understorey plant *Arisaema triphyllum*. Reproduction in this North American forest species is related to the size of the corm, which is an underground storage organ. Vegetative plants have small corms, males have intermediate-sized corms and females have large corms. A particular individual may change sex from year to year, depending upon carbon storage reserves in its corm. (Courtesy of I. Forseth.)

of past glaciation, soil types, water availability, local drainage patterns, fire history and disturbance. The present assemblages of forest trees in Europe, Asia and North America are probably not very old (less than 2000 years). These continents were characterized by repeated glaciations in the past, resulting in repeated range shifts by plants. Species move at different rates, so the reinvasion of previously glaciated areas over the last 10 000 years has occurred at different times for various species.

2.7 Mediterranean climate biomes (chaparral)

2.7.1 Distribution

The Mediterranean biome, or chaparral, covers about 1.8 million km^2 worldwide, and is dominated by evergreen shrubs and **sclerophyllous** trees (Fig. 2.9). Sclerophyllous refers to a small, tough, leathery type of leaf that is resistant to drought and decomposition.

The Mediterranean biome has a climate of hot, dry summers and cool, moist winters (similar to the Mediterranean coast of Europe, hence the name). It occurs in five widely separated regions at latitudes of between 30° and 40° north and south. Portugal, Spain, southern France, Italy, Greece, islands in the Mediterranean Sea and Turkey are the location for the original name of this biome. On the western coast of southern Africa, a rugged terrain of folded sandstone hosts an area of vegetation called fynbos, again with evergreen, sclerophyllous shrubs and trees. On the southwestern and southern coasts of Australia, called the karri, forests of eucalyptus occur along with a community called the mallee, a mixture of eucalyptus and mid-height shrubs. Along the coasts of central Chile, we find a community termed the matorral, and finally, along the west coast of California in North America, the chaparral.

2.7.2 Productivity

The productivity of Mediterranean climate biomes varies from 300 to 600 g carbon m^{-2} year^{-1}, and is affected by season, location, water availability and age of the vegetation. The productivity of an area of this biome starts to decrease after the initial 10–20 years of growth. Litter starts to accumulate, increasing the amount of fuel available to fire. Due to their pleasant

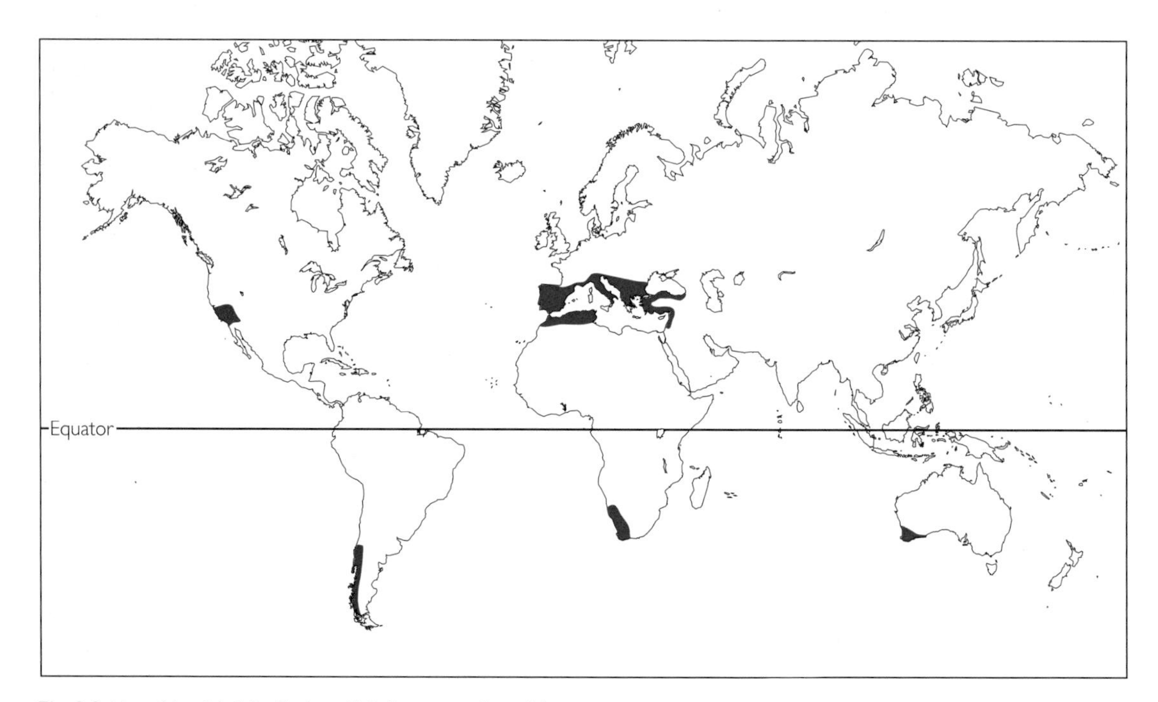

Fig. 2.9 Map of the global distribution of Mediterranean climate biomes.

climate and proximity to coastal areas, many Mediterranean climate biomes around the world have dense human populations. Because of this, fires are suppressed, and litter accumulation is heavy. This can lead to catastrophic, uncontrollable fires. Since many of these areas are on coasts, with steep hillsides, fires are often followed by landslides. Without plant cover, the bare soil cannot handle the rains that occur in the wet season. Once the soil has become saturated, slides can occur.

2.7.3 Adaptations

Ecologists have conducted many studies in this biome because of the striking similarity among life forms on different continents. Although life forms are similar, the species are not related. Thus, this is an example of **convergent evolution**, similar selective pressures leading to similar adaptations between widely separated, unrelated species. The small, evergreen leaves with thick waxy cuticles on their surface are viewed as adaptations to high temperatures, high light, low nitrogen levels and drought. Many shrubs display their leaves at steep angles to reduce midday solar radiation (see Chapter 5). Many also have stiff, upright branches with smooth bark that direct water falling on the leaves to the base of the plants (stemflow), increasing the amount of water available to the plant.

Mediterranean climate biomes have also been subjected to recurring fires. Plant features that have evolved in response to this selection include fire-induced or increased flowering, and bud protection and resprouting. Once fire has killed above-ground plant parts, underground buds sprout, and rapidly form new branches that are supported by the old roots of the plant. Since a well-developed root system is already present, plants can begin growth immediately. Other species have decay-resistant seeds that lie dormant in the soil until fire stimulates germination, while other species require fire to liberate seeds enclosed within cones or closed fruits.

The hot, dry conditions of the chaparral in summer are unsuitable for many larger animals like vertebrates. In winter, however, the Mediterranean area can be less cold and wet than most of Europe, and many birds migrate to this region. Small insectivorous birds like the meadow pipit (*Anthus pratensis*) are common in the area in winter, and most European duck species have winter ranges that centre on the Mediterranean Sea.

2.8 Coniferous forest biomes

2.8.1 Distribution

A damp, thick, evergreen forest occurs from about 45° to 70° north (Fig. 2.10). Coniferous forests can also be found in mountains, where they are termed subalpine forests. The northern coniferous forest climate is one of long, cold winters with cool, short, wet summers. The growing season is short, with a mean temperature just over 10°C. Generally, biodiversity is low, with only a two-layered forest community (see Plate 2.5a, facing p. 24). There is an overstorey of trees with a ground layer of herbs or mosses. In North America, the overstorey is often comprised primarily of only one or two species, white or black spruce.

2.8.2 Productivity

Due to the low temperatures and short growing season, productivity in the coniferous forest is relatively low, about 600 g carbon m^{-2} $year^{-1}$. There is a great deal of variation about this mean, due to differences in climate, precipitation, length of the frost-free period and local drainage. Much of the tree biomass is in the form of support tissues such as trunks. Because of the high proportion of woody material and the evergreen leaves on the trees, a large proportion of the nutrients in the system is stored above ground in plant biomass. Additionally, due to low temperatures, decomposition is slow and litter accumulates. The half-life of litter varies from 3 to 5 years and the turnover rate of nutrients is very slow. Up to 60% of the community biomass can be in litter and humus (dead plant material mixed with soil).

The soils in coniferous forests are often highly leached and are underlain with **permafrost** (frozen soil that does not melt in the summer). Tree roots are shallow, and generally do not extend down to the layer of maximum nutrient concentration in the soil (see Plate 2.5b, facing p. 24). The trees of the coniferous forest are heavily dependent upon mycorrhizal associations. These mycorrhizae extend to deeper soil zones and provide increased uptake of nutrients, especially nitrogen and phosphorus. The permafrost can also result in soil upheaval due to melting and freezing cycles. This leads to many areas of tree fall and poor drainage (see Plate 2.6, facing p. 24). This can affect the distribution

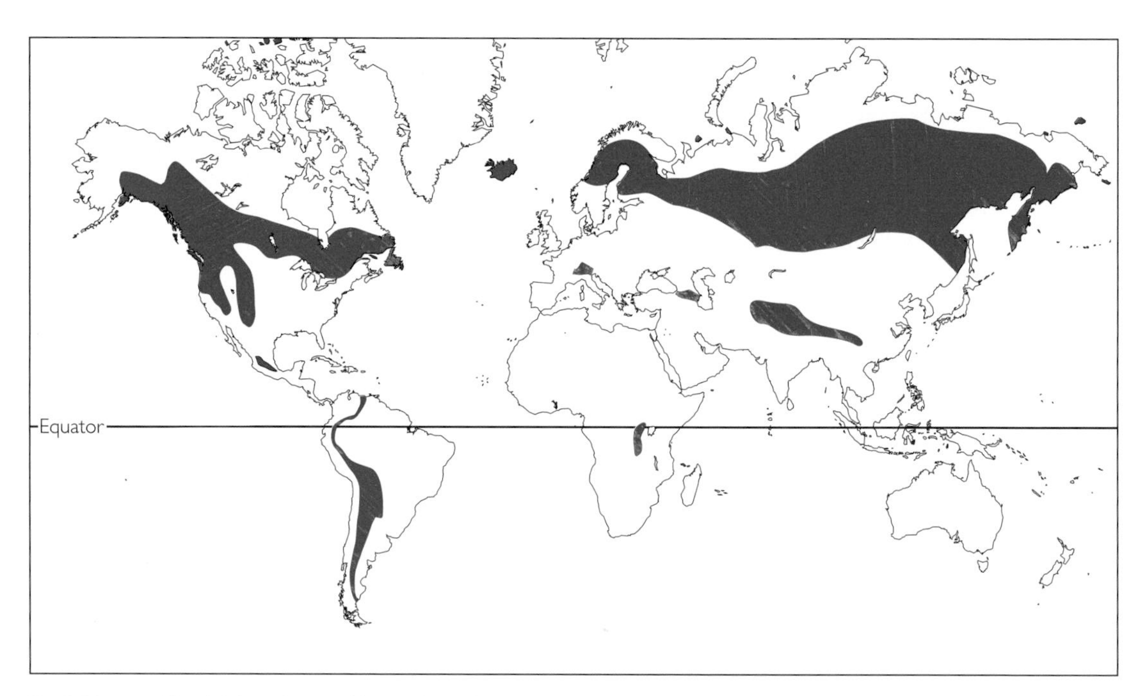

Fig. 2.10 Map of the global distribution of the coniferous forest biomes.

of many species, for example black spruce generally occurs in more poorly drained areas than white spruce. In addition, many areas of poor drainage can form bogs where sphagnum moss dominates. Acidic sphagnum tissue and flooded conditions make decomposition very slow in these areas, resulting in the formation of peat bogs.

Scientific research into high-latitude biomes has increased recently due to concerns about climate change. It is thought that cold biomes such as these will respond to global warming and that atmospheric carbon dioxide will increase more rapidly here than in other biomes. Thus, questions about the effects of these global changes on plant productivity, the release and uptake of carbon dioxide, melting of permafrost and species distribution are currently under investigation.

2.8.3 Adaptations

Many organisms in coniferous forests exhibit adaptations for surviving extremely cold temperatures and also for surviving periods when the weather is very dry. Conifer trees themselves have thin, needle-shaped leaves that have a small surface area, and so do not lose vital water during dry periods. In addition, the pores that allow carbon dioxide and oxygen to pass in and out of the leaves are often located in grooves or pits, to minimize the unnecessary loss of water. The leaves of most conifers have thick waxy cuticles, and contain a very low proportion of water, so that freezing causes relatively little harm.

Animals deal with the cold conditions of northern coniferous forests in a number of ways. Many insects, such as some aphids and wasps, produce glycerol in their cells, so that the fluid does not freeze (and hence damage the cells) at 0°C, but stays liquid to much lower temperatures (see Section 1.4.3). Some mammals in conifer forests, such as the yellow pine chipmunk (*Eutamias amoenus*), **hibernate** as a way of reducing the need for large reserves of energy during the cold winter. Hibernation involves a dramatic reduction in the rate of metabolism. Other animals, such as the black bear (*Ursus americanus*) do not hibernate in a strict sense, but retreat into their dens and conserve energy by sleeping for as much as 7 months of the year.

Snow is a significant feature of conifer forests to which organisms have adapted. Many conifers, such as the western yellow pine (*Pinus ponderosa*), have branches that slope downwards, creating a pyramidal shape, so that heavy snow slides off the tree rather than weighing down the branches, and potentially breaking them. Some mammals, like the mountain hare (*Lepus timidus*) and the stoat or ermine (*Mustela erminea*), which have brown fur in summer, moult in winter to produce white fur, as a camouflage against the snowy landscape.

2.9 Tundra biomes

2.9.1 Distribution

The tundra is a marshy, unforested area, beyond the **treeline** at high latitudes and altitudes (Fig. 2.11). Treeline is a term used to describe the boundary between areas where the tree life form can occur, and areas where trees are replaced by shrubs and shorter life forms. This limit is usually set by a complex of environmental factors such as length of the growing season, freezing temperatures and pruning caused by ice particles carried in the wind. Alpine tundra is a zone of low-growing shrubs that occurs above the treeline on mountains at lower latitudes.

2.9.2 Productivity

Net primary productivity in the tundra is low, 100–200 g carbon m^{-2} $year^{-1}$. Snowfall depths, drainage patterns and local relief can have large effects on species diversity and productivity. Rocky fields and dry meadows will be less productive than moist low-lying areas and wet meadows. Soils are so thin in many areas that sites without adequate snow cover will dry rapidly, and thus cannot support plant growth throughout the entire summer.

Soils are nutrient-poor in the tundra. Decomposition is slow because of low temperatures, and the evergreen tissues of plants are maintained for several years. Thus, nutrient cycling is slow. Often plant growth is stimulated around burrows of small mammals, due to the nitrogen supplied by animal droppings and urine. Much of the nitrogen that comes into the system enters from atmospheric nitrogen fixation by lichens and mosses hosting cyanobacteria. Studies on carbon dioxide enrichment have shown that tundra vegetation has a limited ability to respond to high carbon dioxide levels. This is because the plants are limited by the availability of phosphorus or nitrogen.

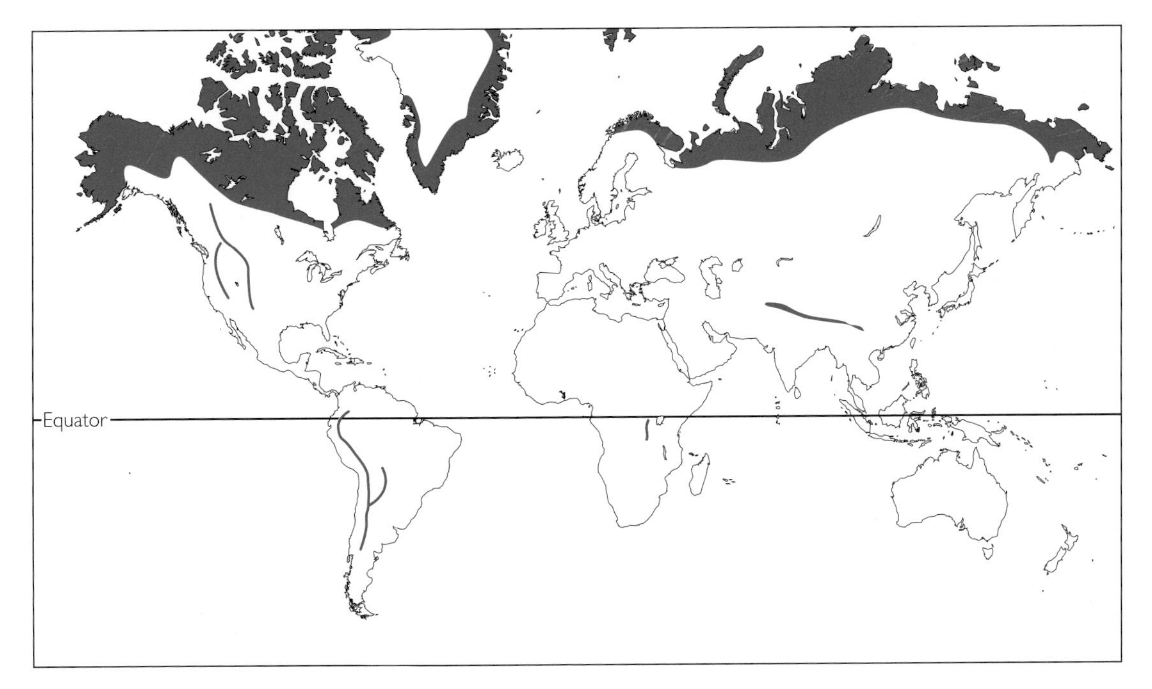

Fig. 2.11 Map of the global distribution of tundra biomes.

Despite the low productivity of tundra regions, large herds of reindeer (*Rangifer tarandus*) are found here (see Plate 2.7, facing p. 24). These herds are constantly migratory, moving to new supplies of food throughout the year. As humans populate tundra regions, roads and pipelines become increasingly problematic to the migratory path of these large herds of grazing mammals.

2.9.3 Adaptations

The tundra has a shorter growing season than coniferous forests, and the mean temperature of the warmest month is below 10°C. There is a low diversity of species, and the communities are dominated by mosses, lichens and low-growing perennial shrubs. A lichen is considered as a single organism, but is actually made up of a fungus and an alga in a symbiotic relationship. The fungi provide anchoring and protection while the algae provide carbohydrates from photosynthesis (see Chapter 9). There are very few annual plant species in the tundra due to the short growing season and low temperatures. The shrubs are primarily long-lived perennials that have their buds at or slightly below soil level. The reason why plants do not protect their buds deep below soil is that the soil is underlain throughout the tundra by permafrost.

Herbs in the tundra, like mountain avens (*Dryas octopetala*), almost all have a creeping, low-growing habit, which avoids burning by cold, dry winds. Shrubs in the tundra adopt a tightly packed, rounded canopy with closely spaced leaves and branches. This cushion or mound type of morphology quickly reduces wind speeds in the plant canopy. The canopy is then able to maintain temperatures well above the surrounding air temperature. On sunny days, plant canopies can be more than 10°C warmer than air temperature. Wind and ice damage may help form this shape by pruning off branches that develop outside the mound.

Most of the birds that nest in the tundra, such as the dunlin (*Calidris alpina*), avoid the harshest conditions by migrating south for the winter. Many vertebrates, such as the snowy owl (*Nyctea scandiaca*), are white as a camouflage against the snow. Arctic foxes (*Aloplex lagopus*) in different places are different colours (white, bluish grey or brown), depending on how snowy the local climate is.

2.10 Non-terrestrial biomes

Aquatic habitats can be broadly grouped into categories with shared properties. Because of the large differences between seawater and freshwater, ecologists usually discuss these two habitats separately.

2.10.1 Marine systems

2.10.1.1 Physical structure

Oceans cover about 70% of the Earth's surface. The distribution of organisms in the oceans is controlled by the interaction of water depth, latitude and distance from shore. Most species are found in shallow water near continents, although this represents only about 8% of the total ocean area. There are three major ocean habitats, each with subhabitats (Fig. 2.12).

The **neritic** zone is made up of the shallow ocean waters along coasts. It is inhabited by a comparatively large number of species. The neritic zone encompasses depths from 0 m to 200 m, and is characterized by the interaction of land and sea. This zone is influenced by wave action, and receives high amounts of nutrients from land surfaces. These nutrients help support the great fisheries of the world. Coral reefs also occur in neritic zones in tropical regions (Fig. 2.13). Coral reefs are very complex ecosystems that occur in nutrient-poor, clear waters. They are sometimes called the 'rainforests of the ocean', because of their high species diversity and complexity (see Plate 2.8, facing p. 24).

The **pelagic** zone is the open water region of the oceans. The **surface** (or **photic**) zone is the part of the pelagic that extends to roughly 100 m depth. This is the depth where enough light penetrates so that photosynthesis can occur. Organisms in the surface zone make up the microscopic **plankton** and macroscopic **nekton**. Plankton may adjust their depth, but generally can only float along with the ocean currents. Nekton are made up of organisms that can swim. Photosynthetic plankton account for about 40% of the Earth's photosynthesis.

The **abyssal** (or **aphotic**) zone is the area of the pelagic below 100 m depth. The **benthic** region is that part of the abyssal zone that is made up of the sea floor. The abyssal zone was initially thought to contain little life because of high pressures, low temperatures and low

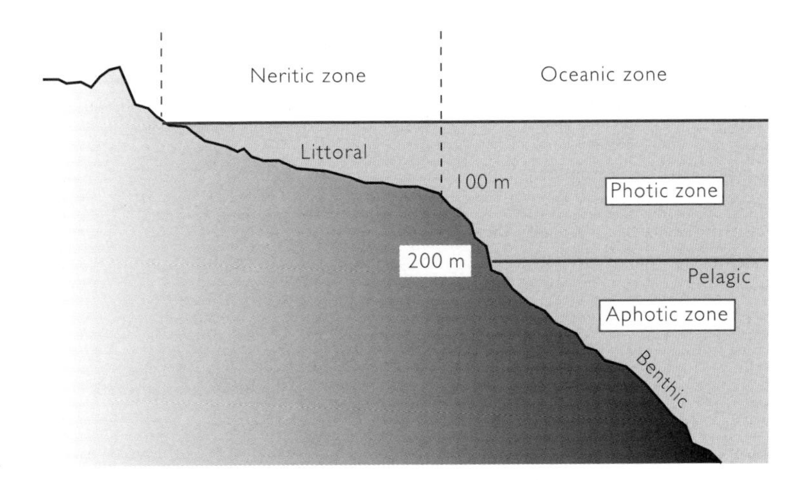

Fig. 2.12 The major habitats of the ocean.

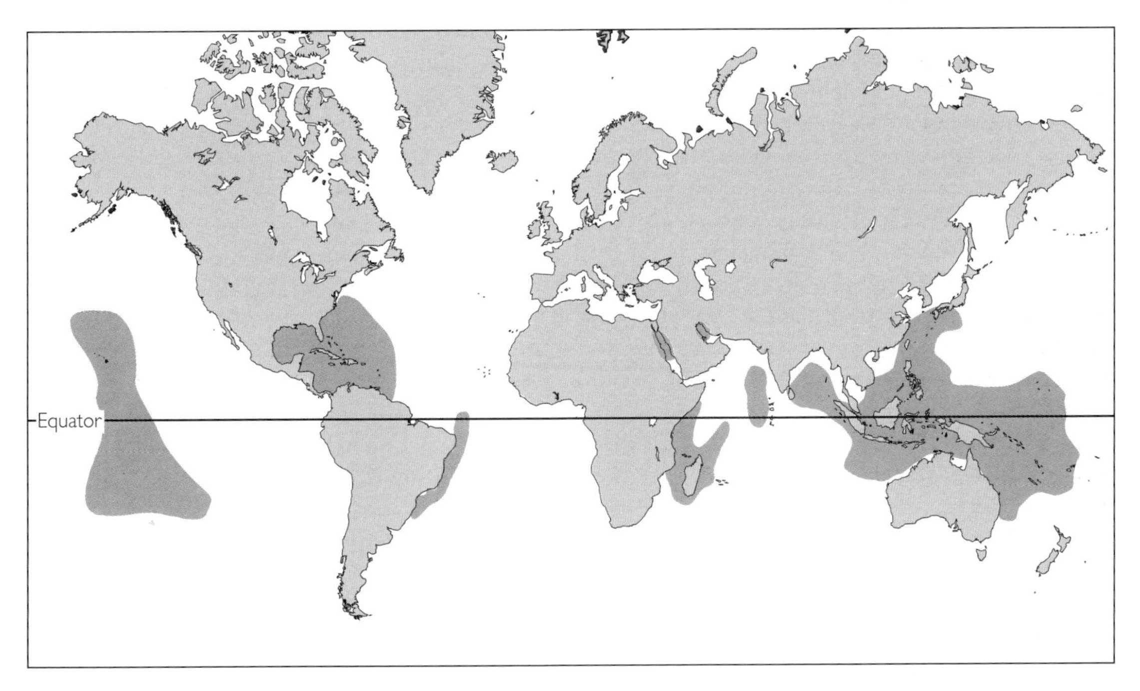

Fig. 2.13 Global distribution of coral reefs.

light conditions. In the last 25 years, however, scientists have discovered a wide array of unique life forms in this habitat. Life in the abyssal zone is dependent on organic matter and nutrients sinking from the surface zone. In addition, there are amazingly diverse communities assembled around deep-sea thermal vents (see Plate 2.9, facing p. 24). These are cracks in the ocean floor where magma superheats the water. Sulphur dioxide is emitted, which is used by chemosynthetic bacteria to form the base of the food chain. These bacteria essentially replace plants in this ecosystem. The bacteria are often incorporated into the tissue of animals around the sea vents to provide carbohydrates.

2.10.1.2 Productivity

Two primary factors limit productivity in the oceans, sunlight and nutrients. Sunlight in clear waters usually

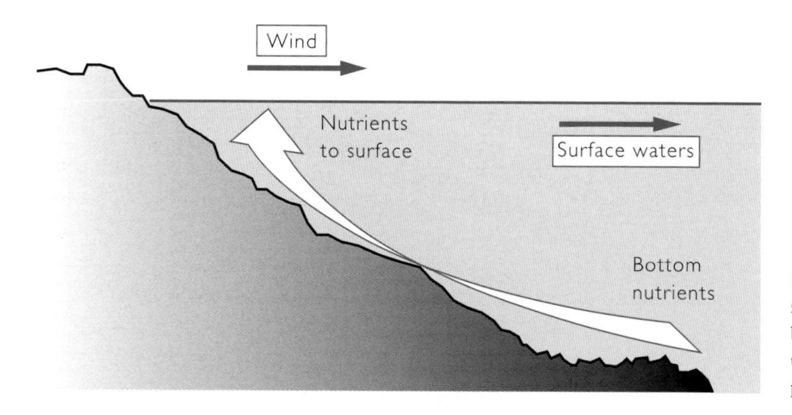

Fig. 2.14 Diagram of a coastal upwelling zone. Nutrients from bottom sediments are brought by currents up to the surface, where they support a productive community of plankton and fish.

supports photosynthesis down to depths of about 100 m. However, near the shore, with increased turbidity, this is reduced to a few metres. The concentration of nutrients in oceans is lower than on land, and productivity is often limited by this more than it is by sunlight.

Plankton productivity in the pelagic zone is dependent upon rapidly recycled nitrogen. Microbes in the surface waters rapidly decompose dead plankton that are not eaten by nekton, making nitrogen available for uptake by plankton. New nitrogen is added by cyanobacteria that fix nitrogen from the atmosphere; or by mixing of sediment from the bottom. Bottom sediment mixing is especially active in areas of upwelling, where deep, cold currents bring bottom waters to the surface. These areas are extremely productive, and can support a variety of sealife (Fig. 2.14). It is not surprising, therefore, that these areas coincide with the world's most important fishing grounds.

Because there is little mixing with deeper waters, tropical pelagic zones have low productivity, ranging from 50 g to 150 g carbon m^{-2} $year^{-1}$. However, coral reefs in the tropics have a net primary productivity ranging from 200 to 1000 g carbon m^{-2} $year^{-1}$. This is due to higher concentrations of phytoplankton around reefs, seagrasses with epiphytic cyanobacteria that fix nitrogen in soft sediments inside the reef, and dense mats of nitrogen-fixing filamentous cyanobacteria on reef rocks and coral sands.

Arctic pelagic waters have a net primary productivity of less than 25 g carbon m^{-2} $year^{-1}$ due to low light levels, and the loss of surface nutrients as cool waters sink. In contrast, Antarctic waters can have a net primary productivity up to 100 g carbon m^{-2} $year^{-1}$ due to continuous upwelling of nutrient-rich water.

In the neritic zone, the combination of upwelling and runoff of nutrients from nearby land surfaces results in higher net primary productivity from 200 to 1000 g carbon m^{-2} $year^{-1}$.

2.10.1.3 Adaptations

Light does not penetrate easily into deep water, so many organisms have evolved mechanisms for dealing with low levels of light. Some species of whales are able to communicate using complex systems of producing sounds, despite the fact that they cannot see one another over anything other than short distances. A large number of marine organisms have evolved mechanisms of producing their own light. **Bioluminescence** is the name given to this phenomenon, which is found in the latternfish (in the family Myctophidae), and appears to be present in invertebrates such as the blue-rayed limpet (*Patina pellucida*) even in relatively shallow water.

Maintaining buoyancy is a problem for larger animals, such as vertebrates and large molluscs, and a variety of solutions have evolved. The nautilus (*Nautilus* spp.) has a shell with a series of chambers, into which it can excrete gas; its buoyancy depends on how many of the chambers are full of gas and how many are full of water. Most fish have a swimbladder, into which they excrete gas to keep them buoyant. Fish with no swimbladder, such as sharks and tuna (*Thunnus* spp.) must keep swimming constantly or they sink.

2.10.2 Coastal habitats

2.10.2.1 Different kinds of coastal habitats

The **littoral** zone is a habitat that occurs within the neritic zone. It extends from the area of coast exposed during low tide down to the area that is influenced by wave action, approximately 100 m depth. The littoral region is characterized by large variations in temperature, salinity and turbidity (the clarity of the water column, which is determined by the amount of suspended material in the water). In this zone, organisms are usually anchored in some way to the substrate in order to resist the effects of wave action. This zone is transitional between ocean and land, and many organisms in it are able to withstand some drying and exposure to air.

On rocky shorelines, large red and brown algae (such as kelp) are found. These algae are exposed to periods of exposure, desiccation, turbulence and changes in temperature and salinity. Differences among species in the ability to handle this environmental variability can lead to a distinct zonation of different species on these rocky shores. These kelp forests can provide food and shelter for a variety of shellfish and fish, and in turn for the predators that feed on them (e.g. otters (*Lutra canadensis*), eagles and gulls).

Saltmarshes occur in mid to high latitudes where fine sediment accumulates along a sheltered coast. Saltmarshes are periodically inundated by seawater, as tides rise and fall. The plants and animals in this habitat have evolved mechanisms to regulate the salt content of their tissues.

In tropical and subtropical coastal areas with fine sediment deposition, mangrove forests form. Mangrove is a broad term that actually refers to about 50 different species of woody plant that can tolerate flooding by salt water. These different plant species belong to several different, unrelated plant families, and represent good examples of convergent evolution under the similar selective pressures caused by life in a saline, flooded environment.

2.10.2.2 *Productivity*

Net primary productivity in kelp beds is high, ranging from 500 to 1000 g carbon m^{-2} $year^{-1}$. On sandy shores, diatoms and other algae attached to the sand grains can produce 20–200 g carbon m^{-2} $year^{-1}$. This represents an important food source for bottom filter-feeders and bacteria. Bacteria and microscopic algae are eaten by protozoa, nematodes, copepods and oligochaete worms. These in turn are captured by larger, predatory worms. As molluscs siphon water through their digestive system, they also take up microscopic algae and animals. Shorebirds consume a considerable amount of the shellfish and small invertebrate animals of the intertidal zone .

Despite varying salinity, saltmarshes are extremely productive, ranging from 400 to 3000 g carbon m^{-2} $year^{-1}$. Only about 5% of the plant productivity is lost to herbivores, although large flocks of ducks and geese can take more. The vast majority of net primary productivity in marshes goes to decomposers (about 80%), except in places where humans use the productivity for their own advantage by grazing their livestock on the productive grasslands.

Net primary productivity in mangrove forests ranges from 500 to 1200 g carbon m^{-2} $year^{-1}$. These forests support a diverse community of marine life, including immature shrimps, many bird species, flying foxes, monkeys, arboreal snakes, turtles, sea crocodiles and many fish species.

2.10.2.3 *Adaptations*

Coastal organisms, particularly those in the littoral zone, can experience not only extreme physiological stress but also rapid and massive changes in conditions. Molluscs and crustacea on rocky shores, such as limpets and barnacles, are inundated with salty water twice a day, and in between may be completely dry, and possibly very hot, for extended periods. Limpets have very strong muscles that enable them to hold their shells down hard against rocks, to trap water inside.

Plants in coastal environments experience physiological stresses as a result of changes in the salinity of water as the tide comes in and goes out. Trees in mangrove swamps may stand in water that is virtually fresh at low tide and in seawater when the tide is high. They have evolved several solutions to the problem; some members of the family Rhizophoraceae have a system of transporting most of the salt to the oldest leaves, which will be dropped soon anyway, while others excrete the salt from glands on the surface of their leaves.

2.10.3 Freshwater ecosystems

2.10.3.1 *Distribution and structure*

Freshwater lakes are limited to about 2% of the Earth's surface. They are strongly influenced by the surrounding land, with marshes and swamps representing

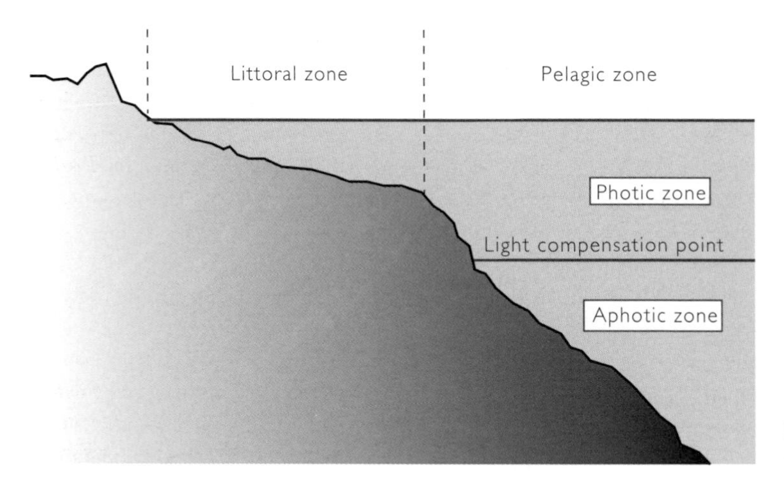

Fig. 2.15 Diagram of the different zones in a lake.

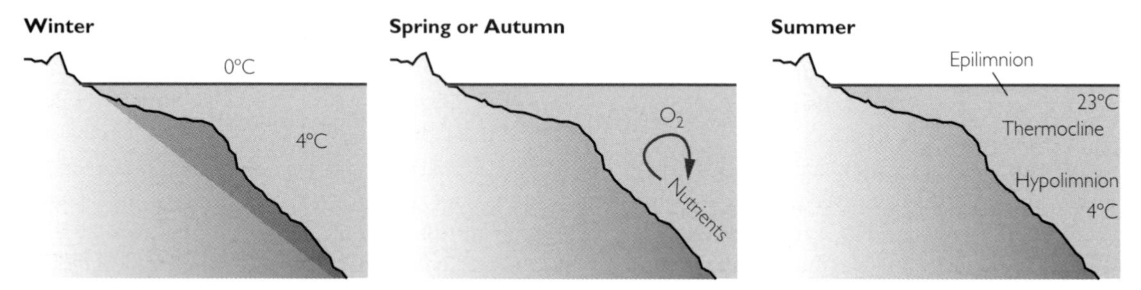

Fig. 2.16 Thermal stratification in a temperate lake. Changes in temperature and increased wind speeds during the autumn and spring result in mixing of the different layers. Temperature profiles during the summer result in a thermocline that prevents mixing of surface and deep waters.

intermediate zones. We can divide lakes into zones similar to those of the ocean. There is a **littoral** zone of shallow water close to shore; a **pelagic** zone of open water; a surface **photic** zone in the pelagic where light penetrates; and a deep **aphotic** zone in the pelagic with little light penetration (Fig. 2.15).

Ecologists also recognize two broad categories of freshwater lake, depending upon their nutrient status. **Oligotrophic** lakes are low in nutrients and organic matter, while **eutrophic** lakes are enriched in nutrients and organic matter. As in the ocean, light and nutrients can limit productivity in lakes. Additionally, the supply of carbon dioxide and oxygen can also limit organisms in lakes. The importance of these factors will vary with latitude and the type of lake, oligotrophic or eutrophic.

During the summer, surface waters in high and mid latitude lakes warm up. A temperature gradient (called a **thermocline**) then develops between surface and deep layers (Fig. 2.16). Density differences due to differences in temperature prevent surface waters from mixing with deeper waters. Deeper waters, where decomposition of dead plant and animal material is taking place, may therefore become depleted in oxygen, which can lead to fish mortality. Eutrophic lakes, with higher nutrients and greater phytoplankton production are especially sensitive to this oxygen depletion. In the spring and autumn, as temperatures cool, density differences are eliminated, and mixing occurs, accelerated by surface winds. Thus, most temperate lakes experience a mixing of deep and surface waters twice a year. In tropical lakes, the thermocline remains unless strong winds or cooler conditions occur that can cause mixing.

2.10.3.2 Productivity

Freshwater lake net primary productivity varies with plant type, latitude, season and nutrient supply. Submerged macrophytes (anchored, higher plants) can produce between 10 and 500 g carbon m^{-2} $year^{-1}$ in temperate lakes, and can exceed 1000 g carbon m^{-2}

year^{-1} in tropical areas. Phytoplankton production varies from 100 to 600 g carbon m^{-2} year^{-1}. The low end of these ranges occurs in oligotrophic lakes, while the high end is associated with eutrophic lakes.

Grazing of phytoplankton by zooplankton (microscopic animals such as rotifers, crustaceans, protozoa and larval insects) can account for as much as 82% of phytoplankton production. Because of the feeding efficiency of zooplankton and high rates of decomposition of dead organisms by bacteria, dead phytoplankton settle slowly to the bottom layers of the lake. Much of the organic material is recycled before it leaves the euphotic zone, maintaining productivity. A turnover time for nutrients of the order of a few days has been recorded in some lakes.

2.10.3.3 Adaptations

Animals that live underwater must have ways of obtaining oxygen that differ from those by which terrestrial animals obtain it from the air. Many freshwater animals use the same or similar mechanisms as marine animals. The tadpoles of most amphibians and the larvae of most dragonflies, for example, have external gills that are similar in structure to the internal gills of aquatic invertebrates such as molluscs. Some animals in ponds and lakes have not evolved such structures and must have access to air from the surface. The larva of the drone fly (*Eristalis tenax*), which lives in detritus at the bottom of small ponds and puddles, has a long telescopic breathing tube, and is consequently known as the rat-tailed maggot.

Many organisms have evolved methods of maintaining their positions in fast-running water. Larvae of caddisflies spin silk, with which they create shelters by sticking together debris and fastening them to stones or plants. For example, larvae of the grannom (*Brachycentrus subnubilus*) fix their shelters to the stalks of reeds or other plants.

2.11 Chapter summary

The system of classifying the world's ecological systems into biomes is used to categorize similar communities on a broad, regional scale. Classifying biomes relies upon the outward appearance of the dominant vegetation types in an area. Biomes differ in their productivity and biodiversity. Equatorial regions have the highest productivity and biodiversity, and both tend to decrease at higher latitudes. As with terrestrial systems, we can divide marine and freshwater systems into broad categories that differ in biodiversity and productivity.

Recommended reading

Archibold, O.W. (1995) *Ecology of World Vegetation*. Chapman and Hall, New York.

Ennos, A.R. (1997) Wind as an ecological factor. *Trends in Ecology and Evolution* **12**: 108–111.

Finelli, C.M. (1999) Physical–biological coupling in streams: the pervasive effects of flow on benthic organisms. *Annual Review of Ecology and Systematics* **30**: 363–396.

Knapp, A.K., Blair, J.M., Briggs, J.M. *et al.* (1999) The keystone role of bison in North American tallgrass prairie. *BioScience* **49**: 39–50.

Chapter 3

Interpreting ecological information

3.1 Studying a broad subject requires many techniques

The definition of ecology that was developed in Chapter 1 recognized the enormous breadth of the subject. In essence, ecology is concerned with every individual of every kind of organism living in every part of the Earth, and is also concerned with every aspect of the physical environment in which those organisms live and die.

Chapters 1 and 2 built on an initial appreciation of this diversity by exploring how this variety of organisms and environments can also be viewed as a variety of genes, populations, communities and ecosystems, and how the broad pattern of diversity can be described in a small number of large-scale biomes. These concepts help ecologists to understand particular parts of their subject, but they also draw attention to the fact that the world's ecology is even more complex and difficult to understand than might be suggested by the existence of millions of species, living in millions of populations that are combined into millions of communities.

Any one technique for studying ecology will allow ecologists only a partial insight into the complexity of their subject, and the fullest understanding of a community will rely on a variety of methods and approaches. Like all scientists, ecologists can carry out experiments, draw inferences from observations, or calculate possibilities using theoretical concepts. All of these methods are valuable, and they can sometimes be used imaginatively in order to extend their usefulness, so that ecologists can maximize the amount of information and understanding that they can achieve.

However, this wide selection of techniques and applications carries potential problems and pitfalls for the ecologist who is not meticulously careful. Different kinds of scientific study provide different kinds of information. A student may spend a few months studying the physiology of a single species of grass, and learn something about the requirements of grass, while a team of full-time plant ecologists may spend a decade studying several species over many generations. The professional ecologists will learn more, and have a deeper understanding, but the type of information they collect will be broadly similar.

At other times, however, two separate studies might give qualitatively different kinds of information. One study might describe the **correlation** that when the number of individual fungus-eating insects in a community increases, the density of the fungi they eat tends to decrease, but may not be able to offer any proof of a causal link. Another study might carry out experiments, resulting in substantially different information, to prove that increase in one species actually causes the decrease in the other.

Issues about the quality and type of information are common to all scientific endeavours, but they are particularly important in ecology, partly because of the scope and breadth of the subject, and partly because some of the questions that ecologists choose to ask absolutely demand that they use certain techniques, while other problems cannot be solved using the same techniques. For example, questions about interactions between existing populations cannot be answered in the same way as questions about the evolutionary history of those same populations.

3.2 Experiment, observation and theory

Edward Jenner discovered that vaccination can prevent diseases when, in 1796, he performed an experiment that must have seemed obscure and complicated to his contemporaries, but which, with hindsight, was incredibly simple in its design. Jenner drew fluid from a sore on the hands of a woman with cowpox, and injected it into an 8-year-old boy called Edward Phipps. He then tried to infect the boy with smallpox. The child remained healthy, because his body had developed antibodies to the cowpox virus and those antibodies conferred immunity to smallpox, as well as combating cowpox.

As a scientific discipline, medicine has often advanced in this empirical way, although modern medical researchers are more rigorous in the way they test treatments, with larger scale experiments that contain proper control groups. Of course, there have been non-experimental advances in medicine, such as the mathematical modelling of epidemic diseases, which could help health managers decide on the most effective times to hold major vaccination campaigns. However, experimental trial and error have been a major factor in the progress of medical science. This approach has led to incredible results, with doctors eradicating smallpox, curing some cancers, successfully transplanting organs from dead bodies into living people, and performing remarkable keyhole surgery.

By contrast, the science of astronomy cannot be conducted experimentally. It is not possible to move the planet Mars in order to observe the effects of the change on the gravitational forces of the other planets, or to increase or decrease the number of black holes in the universe. Nevertheless, detailed astronomical observation has led to very powerful predictive abilities, foretelling the date and time of eclipses and the position of comets. More usefully, a knowledge of the gravitational forces exerted by the planets, the sun and the moon allows accurate predictions about the ocean's tides. More impressive still, a knowledge of how space works has led to people walking on the moon, and to the orbiting satellites that transmit television pictures and telephone messages around the world in seconds.

A third, related type of scientific method is the theoretical approach. In the world of quantum physics, for example, it is not really possible to observe tiny particles, such as electrons and quarks. Moreover, although experiments are possible, they are problematical and expensive, and the results are difficult to interpret. In this field, powerful results have been achieved through logical thought, divorced from the particles involved. Physicists have accurately predicted the existence of different types of subatomic particle long before those particles were actually proved to exist.

Some scientists believe in the strict experimental approach known as Popperian science, after the philosopher Karl Popper, who introduced the emphasis on falsification in the 1930s. In this system, a scientist asks a question about why something is the way it is, or why something happens (or does not happen). Then the investigator suggests a possible cause, and tests the hypothesis. Either the experimental results falsify the hypothesis, or they fail to rule it out as a possibility, in which case our confidence in it may be strengthened. This experimental test is achieved by interfering with the proposed cause and seeing if the intervention has the effect that was predicted by the hypothesis. For instance, Jenner tested his hypothesis about vaccination by injecting Edward Phipps and observing that his intervention had given the boy immunity to smallpox.

Nobody doubts that this is a powerful technique and that, ultimately, it is the only real way of demonstrating cause and effect. However, some features of ecology cannot be manipulated, just as planets cannot be moved as part of enormous astronomical experiments. Some ecologists believe that the science has suffered from a 'myopic infatuation with manipulation experiments' (Lawton 1995) and they may well be correct. It is difficult to imagine experiments that could fully explain why tropical rainforests contain more species than temperate woodlands. Nevertheless, this is such an important question that ecologists cannot simply disregard it as being impossible to answer (see Chapter 14). Ecologists sometimes need to swallow their pride in the face of criticisms that their science is fuzzy or less analytically rigorous than scientific perfection might demand. In fact, they should be proud that they can learn as much as they can without always being able to perform elegant manipulative experiments.

Ecology as a discipline uses a very diverse array of techniques. To understand ecological phenomena fully, ecologists must not rely too heavily on any one set of methods; they need to make use of experiment, observation and theory. They must then integrate the knowledge and understanding that is derived from the three techniques, taking care not to give too much weight to a theoretical framework that apparently disagrees with observations or experimental results.

It is impossible to make a precise determination of the relative importance of each of these three kinds of approach in ecology, but it would be broadly accurate to suggest that each is roughly as important as the others. As an example, take the quest to understand ecological competition (see Chapter 8), where insights have come through all three approaches. For example, the observation that on the islands of New Britain,

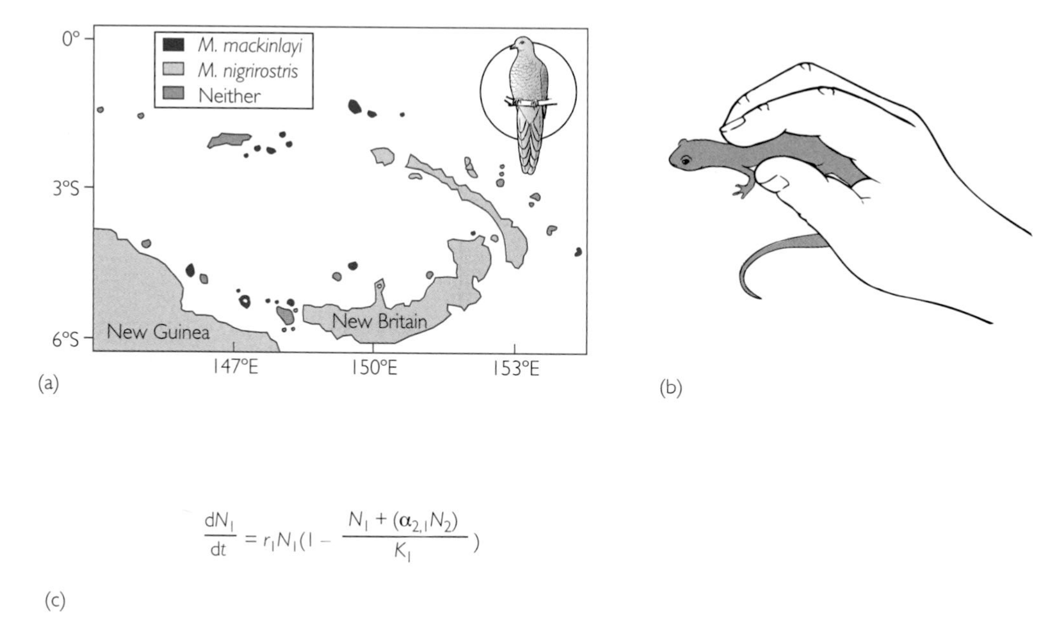

$$\frac{dN_1}{dt} = r_1 N_1 \left(1 - \frac{N_1 + (\alpha_{2,1} N_2)}{K_1}\right)$$

Fig. 3.1 Insights into ecological competition have come from (a) the observation that cuckoo dove species on the Bismarck Islands tend to have non-overlapping distributions—the islands shaded dark purple are home to Mackinlay's cuckoo dove (*Macropygia mackinlayi*) and those shaded grey have the black-billed cuckoo dove (*M. nigrirostris*) (Diamond 1975); (b) experiments involving the removal of Jordan's salamander (*Plethodon jordani*) from habitats in North America (Hairston 1980); and (c) the theoretical mathematics that produced a model of competing species (see Section 8.3.1).

between the Solomon Sea and the Bismarck Sea, there are two species of cuckoo dove (*Macropygia* spp.), but that each island only ever has one or the other, never both, leads to the suggestion that competition prevents them from coexisting. In North America, experimental removal of one species of salamander can lead to the increase in abundance of another species. In the world of theory, equations of competing populations, developed by A.J. Lotka and V. Volterra, are among the best known of all mathematical models in biology. Each of these approaches has given us different information in our overall picture of the importance of ecological competition (Fig. 3.1).

It can, of course, be a benefit to have such a versatile set of methods available to answer questions. But equally, it can prove to be a problem, especially if we misunderstand the limitations of the information that each type of approach can give us. People sometimes criticize non-experimental studies without acknowledging that the authors of those studies have taken great care to outline the limits of what can be inferred from their own results. The truth is that a careful correlation analysis based on data compiled from previous studies can tell us very much more than a poorly designed experiment ever could. Likewise, a non-experimental study that fails to recognize its limitations tells us almost nothing by comparison with a well-designed, carefully executed experiment.

3.3 Experiments that manipulate the world

Many scientific advances rely on experiments, in which people manipulate a small part of the universe, and then observe the effects of that manipulation. In order to be certain that it was their actions that caused any effects, they compare those things they have manipulated with similar things that were left alone. These unmanipulated parts of the experiment are called **controls**. The importance of proper controls in experiments cannot be underestimated because, without this comparison, experimental results can be easily misinterpreted.

An example of an ecological experiment might stem from a desire to know about the effects that high levels of carbon dioxide might have on organisms. Carbon dioxide presently constitutes 335 parts per million (ppm) of the Earth's atmosphere, and an increasing concentration of the gas is one of the features of the global climate change that, according to scientific consensus, is taking place at the moment. If humankind is to manage this change effectively, or to minimize the effects of this change on natural systems, ecologists will need to understand how the extra carbon dioxide might affect ecological communities.

Schäppi and Körner (1997) were interested in the effects of elevated levels of carbon dioxide on the physiology of alpine plants in the Alps in Switzerland, especially the sedge *Carex curvula*. They conducted an experiment in which they divided their study area into 48 circular plots. Sixteen of the plots were designated as control plots, and were left alone, in order to chart the normal physiology of the sedge plants over the period of the experiment. The other 32 plots were covered with open-topped chambers, attached to which were computer systems for controlling the level of carbon dioxide. In 16 of these chambers, the level of the gas was elevated to 680 ppm, roughly twice the natural level. In the other 16 chambers, the computer system maintained the level of carbon dioxide at the natural level of 335 ppm. Thus, this experiment used two controls. The plots without any chambers at all allowed the scientists to control for the presence of chambers on the plants. The plots with chambers maintained at natural levels of carbon dioxide provided the scientists with a control for the high carbon levels of dioxide.

Another principle of experimentation illustrated in this example is that of **replication**. A curious observer might ask why Shäppi and Körner decided to have 16 plots in each of the three treatments, rather than saving time and effort by using just one plot. Certainly the experiment would have been more manageable using only one plot in each treatment. The answer is that replication is necessary to ensure that the effects of treatments are not due to random factors. For example, if a single plot had been chosen for the high carbon dioxide treatment, and by chance that plot had been located in a small depression with deeper soils and more water available, then the growth of the plants in that plot may have been due more to its location than to high carbon dioxide levels. These types of random effects that can obscure treatment effects can be minimized by replicating the treatments several times. Random placement of the plots to minimize the effects of location would also be important in such experiments.

Scientists talk about results being **significant**, which is a way of saying that what they have found is unlikely to have occurred by chance alone. Replication is necessary to decide whether a result is significant. In the experiment on sedges, an unusual result in one chamber might be caused by unusual soil conditions in one particular spot, but unusual results in all the chambers with high levels of carbon dioxide is very unlikely to have occurred by chance, and is almost certainly caused by the experimental manipulation.

The ecologists were able to compare the plants in the three different kinds of plots. In some years, they showed that there was a lower concentration of nitrogen in the leaves of the plants that had been subjected to elevated levels of carbon dioxide. The plants in the control plots were essentially indistinguishable from those in the chambers that maintained natural levels of carbon dioxide.

Schäppi and Körner were able to infer cause and effect from their experiment. They were certain that the variation in nitrogen concentration in the plants was not merely due to the presence of the experimental apparatus, because there was no difference between the one set of control plots, which were left alone, and the other set of controls, which had chambers that maintained carbon dioxide levels at the natural average level of 335 ppm. However, they were equally certain that the level of carbon dioxide in the atmosphere was an important factor, because sedge plants in the chambers with elevated levels of the gas showed altered physiology, relative to the controls with no chamber and relative to the control chambers that maintained natural levels of carbon dioxide.

They were able to draw these conclusions because they were able to compare what had happened before and after the experiment, in both the control plots and the manipulated plots. As a general rule of thumb, these four terms—before, after, control, manipulation—describe the essence of a good experiment.

3.3.1 Experiments in ecological communities have special problems

Relative to many of the experiments that one might theoretically imagine carrying out in an ecological community, the alpine manipulation of carbon dioxide levels was very simple. The chambers were only 40 cm in diameter, there were only a few plant species involved, and the experiment could be carried out in a single growing season.

These features will rarely apply to ecological situations that ecologists wish to investigate. If they were to limit themselves to the study of communities with a relatively small, manageable number of species, ecological scientists might know a great deal about the tundra of northern Europe, about hot deserts in Africa or about small remote islands in the Atlantic Ocean, but they would know nothing about tropical forests, the Great Barrier Reef or the pampas of South America. They would certainly not be able to draw general conclusions or construct theories to explain large-scale ecological phenomena.

3.3.1.1 Sample sizes in ecological experiments

There was a significant difficulty when Jenner performed the experiment that proved the value of vaccination (Section 3.2). As the story is popularly told, he successfully inoculated just one small schoolboy, so it was possible that by chance, he had accidentally picked a boy who already had immunity to smallpox, perhaps because he had previously recovered from a cowpox infection (and hence produced appropriate antibodies), or because he had inherited natural immunity through his genetic make-up.

The truth of Jenner's conclusions only became certain as more and more people gained immunity to smallpox by being inoculated. The real experimental results were obtained from a larger sample of observations of these people. This is one of the most credible experiments of all history, because of the unquestionable effects it has had on human ecology. Smallpox has changed from having a significant effect on human populations to being irrelevant to all but a handful of the 6 billion living humans. Throughout most of history, the virus was a constant, serious and lifelong threat to billions of people, but the disease was completely eradicated from the world in 1979, and the virus only exists under tightly controlled conditions in two safe laboratories. The only people who are in any way affected by the virus are the few people who are responsible for looking after those laboratory samples.

In the case of most scientific experiments, statistical experts would strongly advise that the experimenter should make a large number of observations. Large sample sizes tend to provide more reliable evidence than smaller ones, which is why people are (or should be) often critical of surveys that appear in the media, purporting to reveal features of human nature, but which turn out to be based on interviews with a very small number of individuals.

The need for a sufficiently large sample of manipulations on which to base experimental observations can create particular problems for ecological researchers for two reasons.

The first reason is that there tends to be more experimental error and inexplicable (apparently random) variation in ecological experiments than in experiments in other subjects. This is because of the huge variety of unknown and unpredictable factors that affect the physiology, behaviour, growth and reproduction of organisms. In the alpine carbon dioxide experiment (Section 3.3) factors such as soil depth, slope, solar radiation, insect herbivores, water availability, temperature variation and differences in soil nutrients could all affect the plants growing in the chambers. These are just a small subset of potential variables that ecologists must deal with in field studies. These things are more difficult to account for than the small errors and unpredictable features of experiments conducted by physicists studying the action of gravity or by engineers who are interested in the strength of different materials. Even other kinds of biologist, such as physiologists and geneticists, working in laboratory conditions with mice, fruit flies or plants, tend to be able to carry out intricate experiments.

The other reason why ecologists might experience particular problems obtaining large enough samples of experimental observations is that each of their manipulations can require a much greater effort than manipulations in many other scientific disciplines. When Galileo tested his theories about gravity, it is said that he simply dropped objects of different weights from the top of the leaning tower of Pisa. The only manipulation he needed to make was in the substitution of the different objects that he dropped. The only thing that took any time was running up and down the tower's staircase to collect objects from the bottom and carry them back to the top.

Ecologists tend to need to perform far more complex manipulations. Imagine, for example, that a group of people wanted to test a hypothesis about the effect of a population of one species on a population of another. They might wish to learn about the competition between two species of bacteria, or the predation of a population of mice by a population of hawks. In order to know that the presence of one species was the cause of population changes in the other, they would need to conduct an experiment in which they removed the predator (or one of the competitors) and observed the effects on the remaining species. To perform such a removal on a scale that was likely to provide any meaningful results, they might need to remove predators over a very large area, which would itself be particularly difficult. But in order to conduct an experiment that could draw reliable conclusions, they would need to remove the predators from several areas, and also monitor the population of prey species in a number of control areas.

That may seem unfeasible, but in order to be certain that one species genuinely has an effect on the other, such an experiment is ultimately the only answer. When bird-lovers in Britain disagreed with farmers and landowners about the effects that birds of prey, such as hen harriers (*Circus cyaneus*) and peregrine falcons (*Falco peregrinus*), had on populations of gamebirds such as black and red grouse (*Tetrao tetrix* and *Lagopus lagopus*), a 4-year experimental study was the only way of settling the issue with any degree of conclusiveness.

3.3.1.2 Dealing with complexity in ecological experiments

Most experimental scientists attempt, as far as they can, to reduce the variation among the things that they manipulate, leaving only the important variation between experimental and control manipulations. In this way, they can be clear that their results are due to the variation they have allowed between control observations and experimental observations.

Ecologists can rarely do this, because of the complexity of ecological systems, and the sheer number of interactions between various populations of organisms and physical features of the environment. Since each of these interactions can vary in many subtle ways, it is not possible to remove much of the variation between manipulations. When Schäppi and Körner carried out their test of increasing carbon dioxide levels (Section 3.3), they did not attempt to account for tiny and subtle differences in the angle at which the ground faced the sun in the morning, or the number of micro-organisms in the soil.

They simply had to do their best to **distribute those differences randomly** among their plots of grassland. For example, they did not place all the control plots at the southern end of the field site and all the plots with elevated carbon dioxide at the north, so they were able to rule out any possibility that their results were caused by some extra, unmeasured, factor that was correlated with how far north the plot was placed.

3.3.2 The ecotron

In the early 1990s, an attempt was made to treat experiments on ecological communities more like the tightly controlled manipulations that characterize many other scientific disciplines. A piece of apparatus called the ecotron was designed, allowing a group of ecologists to conduct highly controlled experiments of small communities (Fig. 3.2).

The ecotron is an experimental set-up that contains 16 identical chambers, each of which is 4 m^2 in area, and in each of which the experimenters can create a small terrestrial community consisting of a few plant species with a few invertebrates. The physical features of each chamber—temperature, humidity and light levels—can be independently controlled, and the biological community can be initiated fairly precisely by introducing into each chamber as near an identical set of organisms as possible. The model communities contain herbaceous plants, soil fauna such as springtails, woodlice and snails, herbivorous insects such as aphids, and also parasitic insects that attack the soil fauna and herbivores.

One of the first experiments in the ecotron investigated the effects of biodiversity on the productivity of the community. Chambers that were set up with fewer species produced less organic material overall, and the researchers found that it was largely irrelevant which species were included—it was the number of species, not their identities, which mattered (Naeem *et al.* 1994). Subsequent experiments have studied the effects of changing carbon dioxide levels on the individual experimental communities.

Scientists often have difficulty in interpreting the results of ecological experiments, and even apparently simple experiments can prove problematical. In these cases, theoretical ecologists may suggest refinements that could be made to the design of the experiment; these refinements are used to improve the execution of the next generation of experiments.

One of the main lessons of the ecotron experiments is that, however hard ecologists try, they will never be able to carry out the kind of experiments that can answer large-scale questions about large ecosystems. The cost of the ecotron was enormous, but it still has only 16 replicate chambers, and the small size of the

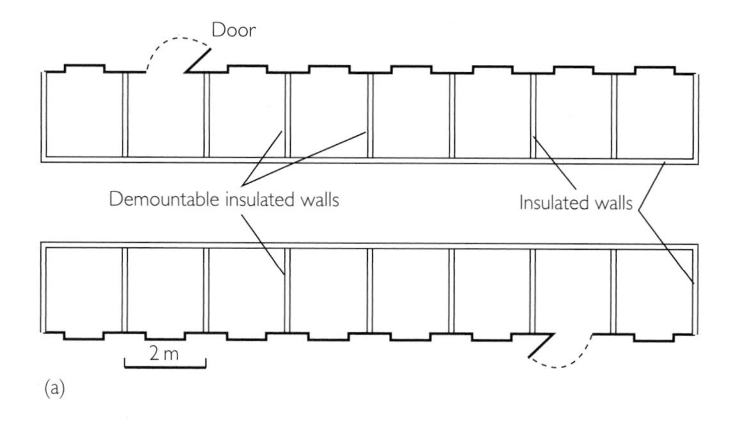

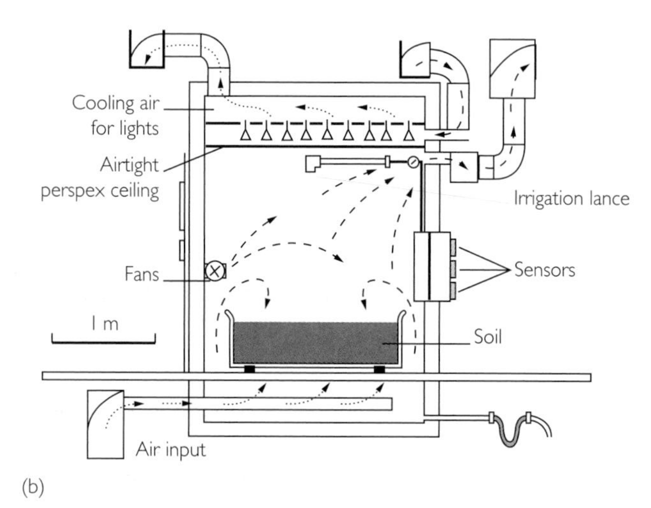

Fig. 3.2 The ecotron: (a) the floor plan, showing the 16 chambers, and (b) the cross-section of a single chamber, showing how the physical environment can be closely controlled.

system restricts the kinds of organisms that can be included in the experimental communities. Larger predators can never be part of the communities.

This does not detract from the results that have been produced from the experiments conducted in the ecotron, because they shed light on a whole variety of processes within ecological communities. But these small-scale experimental results will always need to be integrated with larger scale observations or theoretical inferences if we are to understand the real ecological communities that we see around us.

3.4 Events that have the appearance of experiments

Sometimes it is difficult or impossible for ecologists to carry out a well-designed experiment to demonstrate that some natural phenomenon has a particular cause, but, by good luck, a natural event or accident occurs that has some of the features of an experiment. If ecologists happen to realize at the right time that the natural or accidental experiment has happened, then they can study the effects of the events as if they had been part of a deliberate experimental set-up.

3.4.1 Natural experiments

An obvious example of such a natural experiment would be in the study of ecological succession (see Section 1.10.2) by which a sequence of different species invades the community as time progresses.

If ecologists wished to know how succession progressed (like that in the Netherlands described in Section 1.10.2), and to be absolutely certain in ascrib-

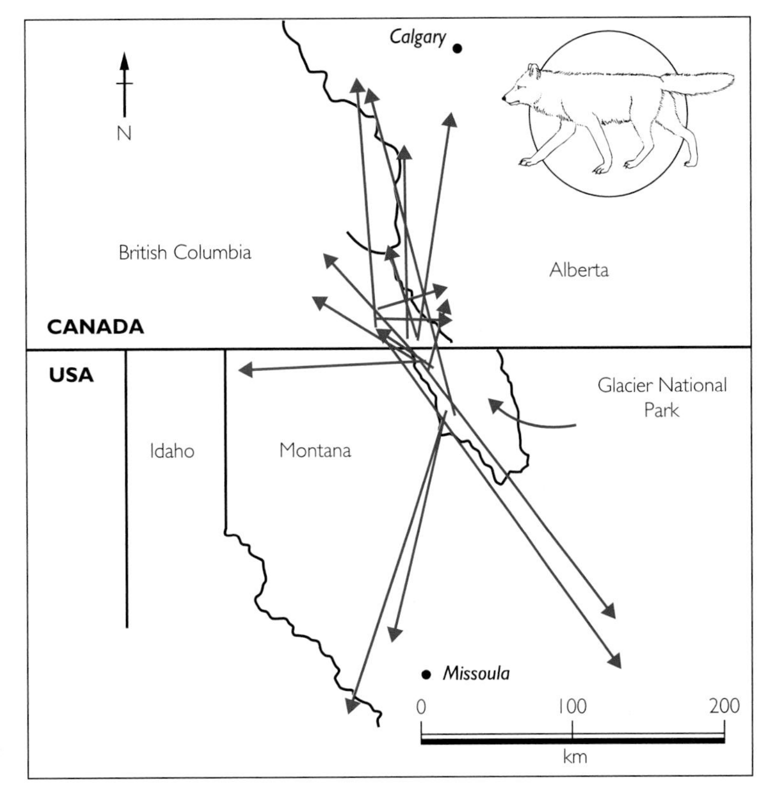

Fig. 3.3 As the wolf (*Canis lupus*) spread outwards from the Glacier National Park to surrounding areas in Canada and Montana, Boyd and Pletscher (1999) were able to treat the areas as if the wolf population had been experimentally manipulated, and observe the effects on the deer that formed the wolves' main prey. Each arrow represents a migration of a known individual wolf.

ing causes of the process to their particular effects, then strictly speaking they would need to carry out a large-scale experiment. As an example, they might plough up areas of land at different times, and watch how the sequence of species on each plot progressed. There are two reasons why such an experiment might be impracticable.

First, if it were concerned with one of the many parts of the world where the ultimate community is dominated by large, long-lived species (such as redwood trees in North America, elephants in African savannas or brain corals in Caribbean reefs) it would be an extremely long-term experiment, and would be almost impossible to carry out within the lifetime of a single ecologist. Long-term studies do take place, but they are exceptional. Moreover, much of the published work that relies on information referring to long periods of time actually uses observations that were reconstructed later, and not collected for the particular purpose for which they were eventually used (see Section 3.8).

The second reason why it would be difficult to carry out a huge experiment on how succession takes place is that it would cost too much money. The project would rely on large tracts of land, with no economic products, being committed to the study, for long periods. Such expense could not be justified when natural experiments, other kinds of observations, and theoretical calculations can give us most of the same information with various degrees of certainty.

To overcome these problems, ecologists could potentially make use of **natural experiments**. Studies of succession can be carried out by investigating the ecological communities in areas that happen to have suffered natural disasters, such as volcanoes, hurricanes, floods or fires. Zobel and Antos (1997) did precisely this by tracking changes in plant communities in the area of the volcano called Mount Saint Helens, after it erupted in 1980 and covered the surrounding area with a layer of ash. Their study is described in Section 13.7.

Other natural experiments include the recolonization of the Glacier National Park by the grey wolf (*Canis lupus*) in the 1990s, which allowed Boyd and Pletscher (1999) to test hypotheses about the effects of predators on populations of white-tailed deer (*Odocoileus virginianus*) and elk (*Cervus elaphus*) (Fig. 3.3).

3.4.2 Accidental experiments

On occasions, events occur because of human activity for reasons unconnected with a particular ecological problem, and these events later turn out to have the appearance of ecological experiments. In general, they will not be as perfectly designed as ecologists might have hoped in the case of a real experiment, but as long as their imperfections as experiments are recognized, a great deal can be learned from them. In effect, these events are similar to natural experiments, but differ in that they were generated, accidentally, by human activity.

A good example comes from studies of how populations of species can invade communities from which they were previously absent. Such invasions can occur naturally, as in the case of the inexplicable spread of the collared dove (*Streptopelia decaocto*) across Europe in the second half of the twentieth century. Until 1930, the species did not occur regularly in Europe outside the Balkans, but from that date onwards, populations began to invade areas of Northern Europe. The species first bred in Germany in 1946, in the Netherlands in 1950 and in Finland in 1966. Nobody really understands why this series of invasions was successful.

To understand the factors that allow some invasions to be successful, while others are not, the ideal solution would be to perform an experiment in which ecologists tried to introduce many different alien species into a variety of habitats, and varied the conditions. Ecologists might, for example, be interested in whether or not the size of the original founding population is important, so they could introduce different numbers of individual organisms into various sites, and observe which populations became established and which perished. To perform such an experiment would be technically difficult, illegal in many countries, and of dubious ethical value, since the ecology of the planet is too important for us to make major, irreversible changes merely to satisfy our curiosity.

However, it so happens that when human settlers from Europe established populations in the Americas, Australia, New Zealand, Africa and Asia, they imported large numbers of familiar songbirds to make them feel at home. The blackbird (*Turdus merula*) was introduced from Britain to Melbourne in Australia in 1862, and later introductions took place in Sydney and Adelaide. At least 10 other bird species introduced from Britain are now living wild in Australia as a direct result of human introductions, including finches (e.g. the greenfinch, *Carduelis chloris*), waterfowl (e.g. the mute swan, *Cygnus olor*) and gamebirds (e.g. the pheasant, *Phasianus colchicus*, which was itself introduced to Western Europe from Asia). Similar species, such as the chaffinch (*Fringilla coelebs*), Canada goose (*Branta canadensis*) and grey partridge (*Perdix perdix*) were also introduced to Australia from Europe, but they died out, and the artificial invasions were unsuccessful.

These introductions could be treated as if they were part of an experiment that was imperfectly designed. In the case of bird species introduced to New Zealand, Veltman *et al.* (1996) were able to ask questions about which features of attempted introductions were the most important in determining their success (Fig. 3.4). For example, they were able to test the hypothesis that migratory birds were less likely to be successful colonists than sedentary ones. The hypothesis appeared to be true, and is exemplified by the Cape Barren goose (*Cereopsis novaehollandiae*), which is sedentary and was successfully introduced, while the introduction of the migratory snow goose (*Anser caerulescens*) was a failure. This was despite the fact that slightly fewer individuals of the Cape Barren goose were involved in the initial introduction.

This was a perfect example of an accidental experiment, because it was almost exactly what would have happened in a well-designed experiment. Similar numbers of individuals of two closely related species were introduced, with the one major difference being the migratory habits of birds in each species.

Accidental experiments may allow us to test similar hypotheses to those that we study using natural experiments. Section 3.4.1 described how investigations into the way in which ecological succession occurs can be carried out by studying the effects of natural experiments caused by hurricanes and volcanoes. These investigations might instead be undertaken by studying areas that have been cleared by human activity for other reasons, and have then been abandoned, such as old fields or deserted settlements. For example, Section 13.6 discusses what ecologists might learn from observations of the ecological community at Tikal in Guatemala, where an enormous ancient city was abandoned centuries ago.

Fig. 3.4 Introduced birds of New Zealand. Veltman *et al.* (1999) were able to treat the introductions as if they were experiments, and compare the features that made some successful and others unsuccessful.
(a) Successful introductions: mallard, *Anas platyrhynchos*; rock partridge, *Alectoris graeca*; rook, *Corvus frugilegus*; house sparrow, *Passer domesticus*. (b) Unsuccessful introductions: wigeon, *Anas penelope*; red-legged partridge, *Alectoris rufa*; jackdaw, *Corvus monedula*; tree sparrow, *Passer montanus*.

3.5 Experiments do not expose mechanisms

The experiment that Edward Jenner performed in 1796 (Section 3.2) showed that immunity to diseases could be acquired through inoculation with appropriate matter, even though that matter was not itself infectious. We now know that this occurs because our immune systems create antibodies to attack the inert matter, and that those antibodies are also capable of attacking the living organisms that cause the infectious disease. We also know that in the case of smallpox, the infective agent is a virus, and that the antibodies work against the smallpox virus because it is structurally similar to the harmless cowpox virus. For a century after Jenner's first experiments, however, the mechanism of immunity remained a complete mystery—the word 'antibody' was not used in English until 1901 and the modern sense of the word 'virus' derives from the work of Pasteur in the 1880s.

Edward Jenner's experiment proved that fluids from cowpox pustules conferred immunity from smallpox, but it gave essentially no information about the mechanism by which that immunity was generated. This is also an important point in ecology. Just because an experiment demonstrates a link between two phenomena, it is not sensible to assume that any particular mechanism is at work, however plausible that mechanism seems, or however unlikely the alternatives might seem.

If we watch a farmer spraying some of his fields while leaving others unsprayed, and we then observe that corn in the sprayed fields grows faster and larger than the corn in fields that were not sprayed, we might infer that the spraying caused the extra growth. We cannot, however, know the mechanisms by which the spraying worked. It is possible that the spray was a pesticide that killed herbivorous insects that would otherwise have stunted the growth of the corn. But it is equally likely that it contained extra nutrients such as nitrogen and phosphorus. It might even have been a selective herbicide that killed weeds without harming the crops, and thus reduced ecological competition.

3.6 When experiments are not possible, observations must suffice

Sometimes, experiments are impossible. On occasions, they may simply be extremely impractical—it would be unworkable to perform a removal experiment to test the effects of a single species of bacterium in the soil, because it would be impossible to remove all of the individuals effectively without also causing other effects, such as destroying many individuals of other species. In other cases, it may be literally impossible to perform manipulation experiments. The reason ecologists cannot carry out experiments at the scale of the whole planet, for example, is not merely the impracticability, but also the fact that we only know of one planet that sustains life, so the experiment would have a sample size of just one planet, with no control planet to make a comparison.

The difference between true impossibility and extreme impracticability is sometimes an interesting distinction to make, but it is unimportant to the working ecologist. Ecologists must sometimes find ways of collating evidence about a theory without the need for experiments.

There are scientists who claim that, since experiments are essential to demonstrate cause and effect with certainty, it is pointless to postulate theories that cannot be tested experimentally. Ecologists do not have that luxury. It matters greatly whether global climate change is happening, and what its effects may be, so we cannot simply give up our interest in it because experimental tests of competing theories are impossible on a planetary scale.

3.6.1 Evolutionary theories

Many ecological theories that cannot be tested experimentally are those that depend heavily on the fact that ecological communities contain organisms that have evolved over long periods of time by natural selection (see Section 1.4). Sometimes humans set in process events that will rely on future generations for their completion—examples include building the pyramids in Egypt or interpreting the American constitution. But the initiators of even these know that they cannot foresee the potential results that might develop in the long term. Even with this degree of wisdom, the projects often have to be altered before very long (the odd shape of the 'bent' pyramid at Dashur and the amendments to the American constitution prove this). Looking further into the future is impossible.

Experiments to test theories about millions of years of evolution are on a different level altogether, and cannot seriously be contemplated. There are circumstances in which it is possible to conduct experiments to test hypotheses about evolution, but they tend to be restricted to species of organisms with short generation times (so that many generations can be studied in the realistic timescale of a human working life), and generally involve laboratory situations in which only a very small number of species coexist. Evolution in real natural communities, where there are many species, some of which are long-lived, cannot be so readily studied through simple experiments.

This does not mean that we have no information about what happened during the last few hundreds of millions of years. In fact, many sources provide evidence of the organisms that lived during this period. Fossils are probably the most obvious, but there are others, such as the construction of evolutionary trees based on genetic data (Fig. 3.5), and the geographical study of movements in the enormous tectonic plates that form the Earth's crust.

Moreover, this evidence is augmented by examples of detailed tests of theories of natural selection, speciation and evolution. For example, in areas where mining activity has occurred, the processed earth is deposited as mine spoils. These mine spoils are soils high in toxic metals. Ecologists, by finding out when the mine was active and when the mine spoils were deposited, can investigate the evolution of metal tolerance in grasses that revegetate the mine spoils. Reciprocal transplants are often then used to determine the amount of differentiation that has occurred between non-metal-tolerant populations and metal-tolerant populations. Reciprocal transplants are manipulative experiments where indi-

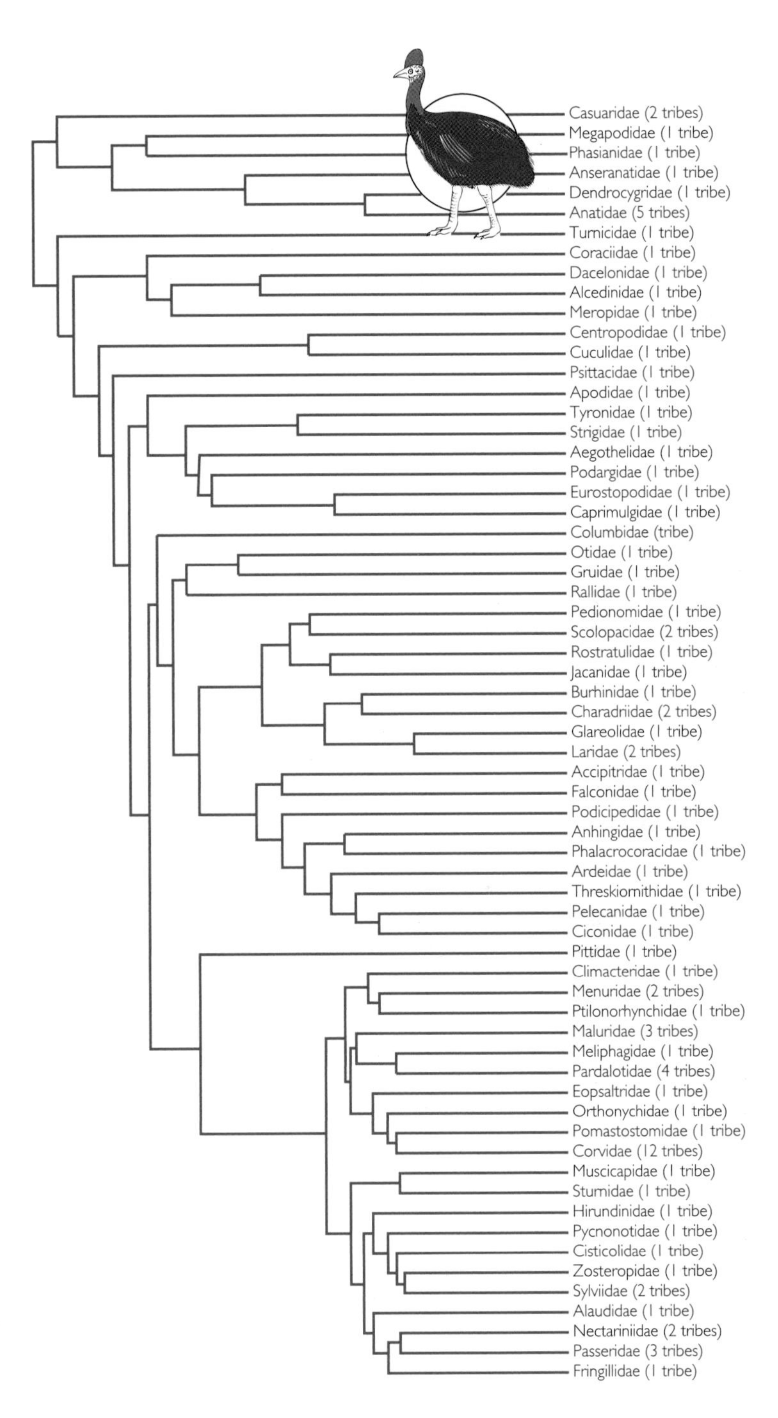

Fig. 3.5 Evolutionary tree of the birds of Australia reconstructed from experiments that directly compared the DNA of the species. (Adapted from Sibley & Ahlquist 1990.)

viduals from one habitat are transplanted to another, and vice versa. These experiments represent a powerful approach that ecologists can use to investigate questions about natural selection and adaptation to the environment (Bradshaw 1971).

3.6.2 Predictions

One way of evaluating theories for which there is no strict experimental test is to make predictions about what would happen if the theory were true, and to see how close actual events are to the predictions. That is essentially how meteorologists test their hypotheses about the causes of weather patterns. They may construct a theory that predicts that it will rain tomorrow, and another theory that forecasts sunshine. If the weather the following day is observed to be sunny, then the first theory cannot be correct. Because there is no experimental evidence of what caused the sunshine, it is not possible to know whether the second theory is correct, but it remains a possibility, and if it accurately predicts the weather every day for 10 years, then we can have a high degree of confidence that it is indeed accurate.

Not all of these theoretical 'predictions' are strictly about what will happen in the future; they might relate to how the world already is, but be concerned with features that have not yet been measured or observed. In such cases, the 'prediction' is about what scientists might find when they make the necessary measurements or observations. For example, someone might predict that there is a statistical correlation between the average height of individual trees and the number of herbivore species that eat parts of the trees—taller trees support more diverse herbivore communities. The supposition about the ecological phenomenon is not strictly a prediction, because it is not about the future—the trees and herbivores are already there. But a 'prediction' has nevertheless been made concerning what might be found when the statistical test is performed.

3.7 Macroecology

Macroecology is the name given to one of the ways in which ecologists overcome the problem that some theories are not experimentally tractable. It is the study of large-scale patterns, and involves comparisons between different communities or different species. It relies heavily on the kinds of predictions discussed in Section 3.6.2. At its simplest, macroecology may simply involve measuring two ecologically important features of many different species and then carrying out a simple statistical test to investigate a prediction. An extremely simple example may be to compare carnivorous animals with herbivorous ones, and to note that carnivores tend to be larger than herbivores. Lions, eagles and spiders are generally larger than the antelopes, pigeons and insects that they eat.

More complex comparisons of a variety of species may seek to look for factors that determine the average population density at which each species lives. In such an analysis, ecologists might include factors such as the average size of an individual organism of the species and various properties of their life histories, such as whether individuals mature late in life and are long-lived (such as lichens, mahogany trees or elephants), or if they generally reproduce at an early age and die young (such as annual herbs or shrews). In such an analysis, it would not be surprising to find that populations of small organisms with a short generation time lived at higher densities than populations of large, long-lived ones. Nor would it be difficult to think of plausible explanations for such a pattern. It would, however, not be possible to determine whether those explanations were true without a more complex, and problematic, study involving an experiment. Since it is not easily possible to manipulate the size of an organism (at least not without having undesirable side effects), such an experiment would be so unrealistic that ecologists who were interested in the true reasons for variation in population density would have no option but to think of further predictions that can be tested by broad comparisons among many species.

A well-known example of such a macroecological comparison is the observation that the population density at which each species lives is strongly correlated with the size of individual organisms belonging to the species. It is generally exemplified by animals, partly because of the relative ease with which one can define the average size of individuals within the population. The pattern is not dissimilar in plants, however, and probably also applies to fungi and micro-organisms.

Other macroecological comparisons involve comparing different communities rather than different kinds of organisms. Chapter 14 will investigate the observation, noted in Section 2.2.3, that tropical communities tend to contain more species than temperate or polar ones. Such comparisons are inevitably somewhat general, and should never be overinterpreted. Ecologists are forced into them because the possibility of experimental study is denied them, and they learn a great deal from macroecology. But the careful ecologist must always keep its limitations in mind.

3.8 Reconstructing observations

Sometimes it is necessary for ecologists to reconstruct observations that could have been made at some point

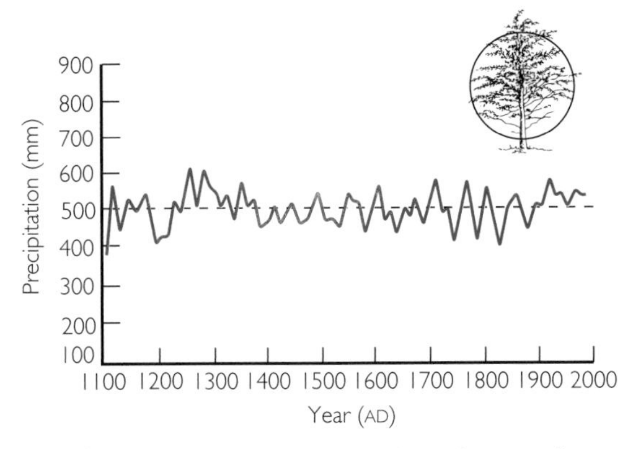

Fig. 3.6 Annual precipitation in a semi-arid area of Morocco from AD 1100 to the present day, as reconstructed from the growth rings of cedar trees. (From Till & Guiot 1990.)

in the past, but which were not made at the time, either because nobody thought to make them, or because the events occurred before humans evolved. In the extreme case, such observations might be reconstructed from fossils that are hundreds of millions of years old, while information about the abundance of an animal species might be reconstructed from records collected by human hunters over a few years or decades. Between these two extremes, information about the climate over the past few centuries or millennia can be reconstructed from other sources of evidence, such as the growth rings of trees (Fig. 3.6).

In many cases, reconstructing observations will rely on inferences about the way in which organisms and the environment have reacted to particular conditions. This is true, for example, of growth rings on trees, which are assumed to be smallest in years when the weather is dark, cold, or in some other way unsuitable for trees to grow well. Ecologists assume that because present-day trees react in this way to weather conditions, trees reacted in the same way in the past. Then, when they find particularly small rings in tree trunks, they assume that a particularly cold or dry year occurred, and resulted in this response by the tree (Fig. 3.6). By examining corroborative evidence, such as silt deposition in nearby streams, or chemical records found in ice cores of Greenland, Antarctica and elsewhere, ecologists gain more and more confidence in their interpretation of past events.

To obtain information from ice cores, deep cylinders are drawn from the ice, and information about the atmosphere in the past is obtained by chemical analysis, working with the assumption that the deeper ice contains chemicals that were trapped in the freezing water at an earlier date than those in the ice that is near the surface of a glacier. The detailed chemical nature of the gases trapped and dissolved in the ice allows ecologists to make inferences about the Earth's temperature, because it is known that temperature correlates with the concentration of atmospheric gases—this is the basis of the 'greenhouse effect' by which increasing levels of carbon dioxide are believed to cause increases in average temperature. The ice core investigated in a study known as the Greenland Ice Sheet Project gives information about the Earth's temperature more than 60 000 years ago.

In other cases, the requirement for such inferences and assumptions may be less, but it may still be necessary to remember that data were collated for purposes other than those for which ecologists choose to use them. In 1995, Fitter and his colleagues relied on two such datasets when they asked about the effects of temperature on the flowering patterns among plant species in southern England. They used their own information about the date at which the flowers of each species first appeared each year, collected over a period of 36 years. Although these data were collected in a consistent manner by the same person each year, they were not collected for the purpose for which they were eventually used in the scientific study. The ecologists realized that they needed to take account of this problem because the records of flowers were not evenly distributed across the geographical area. They solved this issue by making the arbitrary decision to discard observations that were recorded more than 80 km from the village of Chinnor.

Fitter *et al.* (1995) used information about the weather from a long-term series of temperature records collected by meteorologists. The ecologists were able to use these data because the information was all collected in a comparable manner, and there was no reason to believe that it contained any biases. The ecologists were able to demonstrate that higher temperatures in spring advanced the flowering of 219 species of plants by an average of 4 days for every degree Celsius above the seasonal average. This result was possible because the botanical ecologists had managed to reconstruct a set of observations and use them for this study even though that had not been their own intention when they started making observations about the local flowers; nor had this use been the intention of the meteorologists who had recorded the temperature every day for decades.

3.9 Theory helps to interpret experiments and observations

The experiments and observations that ecologists carry out are generally aimed at testing theories about why the world has the ecological features that we see around us. The concepts of theories and hypotheses are common to all sciences, so rational science has developed techniques for improving and developing those theories.

In the case of ecology, the principal technique that dominates theoretical research is the construction of mathematical models. Mathematical models are ways of helping people to reduce the complexity of the real world to a more manageable set of ideas, and involves encapsulating real-life phenomena in numbers and symbols. The use of mathematical models will be explored more fully in Chapters 6–10, which will look at the details of interactions between coexisting populations of different species. Such interactions form one of the subjects in ecology where models have proved extremely useful, and in consequence, the relevant models are both well-defined and well-known.

Modelling has, however, proved of interest in many other areas of ecological investigation, particularly in parts of the subject that are especially difficult to study by experimentation or observation, such as global climate change. Many environmental scientists believe that the average temperature of the Earth's atmosphere will rise by somewhere between 2°C and 5°C by the end of the twenty-first century. The methods by which they arrive at this estimation are themselves based on mathematical modelling, but assuming that this estimate is accepted, it is possible to use models to predict what changes will occur as the temperature rises.

Dunbar (1998) modelled the extent of grassland habitats in the uplands of Ethiopia, which are the habitat of the gelada baboon (*Theropithecus gelada*). He estimated that for every 2°C rise in temperature, the lower limit of the crucial grassland habitat will be forced up the mountains by about 500 m, as the more equable climate allows lush lowland vegetation to spread to higher elevations. Because the upper limit of the grassland cannot expand (it already encompasses the summits of the mountains), the result would be a reduced amount of the baboons' grassland habitat. He then modelled the extent of habitat that would be suitable for the baboons as the climate changes. He found that, if the predicted rise of more than a few degrees were to occur, the habitat would become extremely fragmented, and the population of gelada baboons would suffer because their habitat would be broken into many small areas that may not be sufficient to support a viable population (Fig. 3.7).

3.9.1 Grounding models in the truth

An important feature of mathematical models in ecology is that they must be grounded in the real world. When Dunbar modelled the effects of rising temperature on grassland habitats, he did it from a knowledge of the physiology of the grass species and other plants of the region, and having studied the vegetation of a range of African habitats that currently experience a range of different climatic conditions (recall that Section 3.3 described the kind of detailed experiment that tells us about how plants react to changes in atmospheric gases). He did not make wildly unrealistic assumptions about the way in which grass populations might respond to increased temperatures.

This need to fix the terms of a mathematical model with real observations often causes confusion between two groups of people, namely those who think they lack mathematical aptitude and are frightened of models, and those for whom mathematical truths are so obvious that they sometimes forget about the real world altogether.

For those who distrust mathematical models, it is necessary to understand that the symbols and mathematical quantities stand for broad, average patterns, not the precise details of the situation. Dunbar did not seek to imply that he could predict precisely which square metre of grassland would disappear, nor which individual baboons would be unable to survive.

For those whose mathematical prowess inclines them to rely heavily on models, it is essential to understand that the model might be wrong, either because the ecologists who created it have overlooked some important factor, or because circumstances change and the model is not designed to take account of those changes.

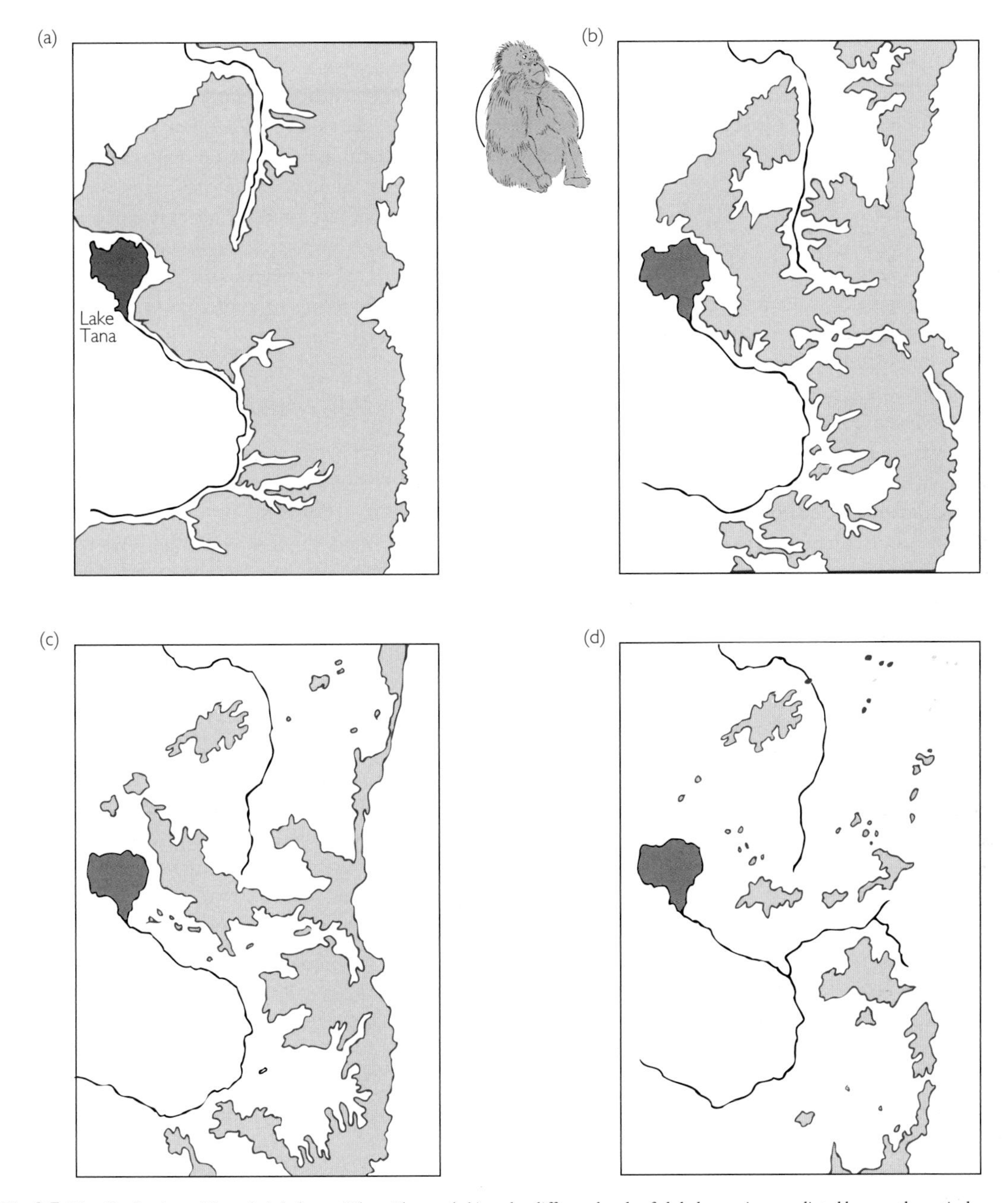

Fig. 3.7 The distribution of the gelada baboon (*Theropithecus gelada*) under different levels of global warming, predicted by a mathematical model: (a) current distribution, (b) with global warming of 2°C, (c) with warming of 5°C, and (d) with warming of 7°C. (From Dunbar 1998.)

3.10 Linking different kinds of information

Faced with different kinds of information, gathered in different ways, and offering different degrees of certainty, the ecologist must put together an explanation of the natural phenomena that surround us. Integrating all these different pieces of information into a more comprehensive understanding of ecology is never easy. Some conclusions that are apparently secure will ultimately rest on information about which we cannot be certain.

In the example of how climate change affects gelada baboons (Section 3.9), ecologists could have conducted extremely careful mathematical modelling of Ethiopian grassland habitats and calculated the details from manipulation experiments designed to give very precise information about the effects of temperature changes on individual plants of all the local species. Nevertheless, the conclusions about what would happen to the habitats, and hence to the baboons, would still rest on the assumption that the temperature is likely to rise by two or more degrees over the next century. However expertly, carefully and precisely the meteorologists have calculated this figure, it remains uncertain, and will continue to be unproven until the century has elapsed.

Scientists are human, and as such, they cannot help being influenced by their own personal opinions. The truth is that, in any debate, individual ecologists will tend to favour one set of theories and explanations over other possible hypotheses. They will have a habit of seeing the evidence that supports their theory, and of marshalling a case to support it, in the way a lawyer prepares a case by stressing the factors that support one side of an argument and placing less emphasis on the weaker parts of the case.

There is nothing wrong with this, and it would be extremely foolish to presume that any kind of scientist was any less human than other people. All science moves forward as researchers challenge one another's ideas, with theories winning support when they fit all the available evidence. Ecology is no different, and our knowledge of organisms can only progress if we sometimes challenge received wisdom.

Ecology is, however, different from many other scientific disciplines in one significant respect. It is a very young science and we have few real rules, unlike the physicists who have the immutable laws of motion that were set out by Newton over 300 years ago. Of course, Einstein showed that these laws turned out to be less absolute than they had seemed—they do not apply at the very small scales of subatomic particles. But they are still applicable for all practical everyday purposes, and they formed the foundation on which Einstein built. Ecologists can sometimes become frustrated because they do not have such a firm foundation on which to build their studies.

This lack of a firm framework can make it difficult to interpret a collection of information, and to link it into a coherent explanation of natural phenomena. But, if we remember the principles that were established in Chapter 1, and take care to interpret each ecological study on its merits, taking into account the degree of experimental, observational and theoretical evidence, then we can build up an impressive picture of the way the Earth's organisms live.

3.11 Chapter summary

Ecologists use a range of different information, derived from observations, experiments and theories. Sometimes, they reconstruct information from sources such as fossils or the growth rings of trees. In order to analyse information for statistical significance, scientists must have a large enough sample size, and their results must be repeatable.

Theories and models must be grounded in observations and experiments, but they also suggest new experiments and opportunities for observation. Because experiments are often difficult to perform on the large scale of ecological studies, ecologists often rely on accidental and natural experiments, which share many of the important features of well-designed scientific experiments. In some cases, this involves making large-scale comparisons, and this approach is given the name of macroecology.

Recommended reading

Arnqvist, G. & Wooster, D. (1995) Meta-analysis: synthesizing research findings in ecology and evolution. *Trends in Ecology and Evolution* **10**: 236–240.

Condit, R. (1995) Research in large, long-term tropical forest plots. *Trends in Ecology and Evolution* **10**: 18–22.

Lebreton, J.-D., Pradel, R. & Clobert, J. (1993) The statistical analysis of survival in animal populations. *Trends in Ecology and Evolution* **8**: 91–95.

Resetarits Jr, W.J. & Bernardo, J. (eds) (1998) *Experimental Ecology. Issues and Perspectives*. Oxford University Press, New York.

Chapter 4

Climate and life on Earth

4.1 Similar climates lead to similar life forms

In Section 1.5, ecology was defined as the study of how the distribution and abundance of organisms are determined by the interactions of individuals with their physical and biological environments. The physical (abiotic) environment with which organisms interact is influenced to a great deal by climate—the long-term solar radiation, rainfall and temperature conditions in a region. This is true for both aquatic and terrestrial systems. In oceans, the penetration of sunlight into the water column and the mixing of water layers due to wind and temperature changes have major impacts on the distribution of living organisms. On land, the climate strongly influences the animal and plant life found in different regions of the world. The abiotic factors that help determine organismal distribution in the world's major terrestrial and aquatic biomes (see Chapter 2) form the focus of this and the next chapter.

The long-term climate in an area determines when and how much water is available, the development of the seasons, the availability of solar radiation, and the temperature conditions to which organisms are exposed. Climate also affects the type of soil that develops, the amount of ice cover on the surface of oceans, lakes and streams, the seasonal mixing of lake and ocean layers, and the water flow in streams and rivers (Fig. 4.1). Because plants anchor themselves in soil and obtain their water and nutrients from soil, the relationship between the type of plant community found in a particular area and the climate and soil of that area is especially pronounced (compare Figs 2.1 and 4.2).

It is not the same species of plants and animals that are found in similar climates throughout the world, but similar **life forms**. A life form is the outward, external appearance of an organism, sometimes combined with descriptions of its lifespan or habitat. Examples of different life forms would include evergreen trees, microtine rodents (small-toothed, mouse-like mammals) and benthic marine invertebrates (bottom-dwelling ocean animals without a vertebral column, e.g. crabs and mussels).

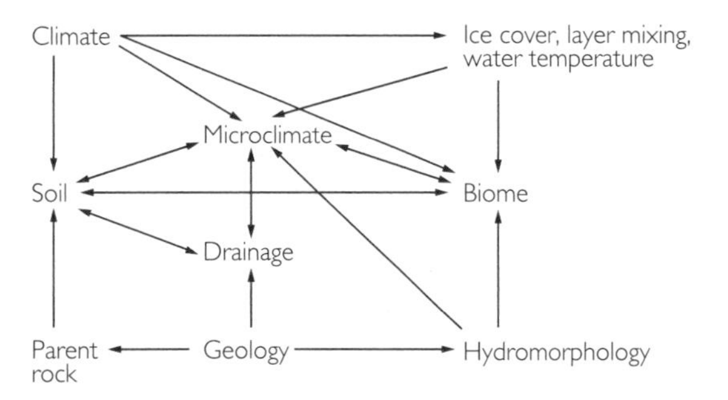

Fig. 4.1 The inter-relationships between climate, geology, substrate and biomes. The microclimate is the climate immediately surrounding an organism, and can be affected by the organism itself. For example, shading of the ground by plant leaves will create a microclimate with lower temperatures and lower solar radiation levels than the surrounding climate. Burrowing by small mammals creates microclimates buffered from the surrounding temperature extremes.

4.1.1 Convergent evolution

One of the most striking examples of comparable climates leading to similar life forms on land occurs in areas of the world with Mediterranean-type climates. These climates have hot, dry summers and cool, moist winters. Such climates occur in the coastal areas of western Australia, South Africa, Chile, California, Portugal, Spain, France, North Africa and Italy (see Fig. 2.9). In these areas, shrubs about 0.5–1.5 m tall with tough, leathery, evergreen leaves (see Section 2.7) often dominate the plant community. This is an example of **convergent evolution**, the evolution of similar adaptive traits in unrelated organisms in response to similar selective pressures (see Section 1.4.3). However, the plants found in each of these areas are often unrelated to each other and have unique, separate evolutionary histories.

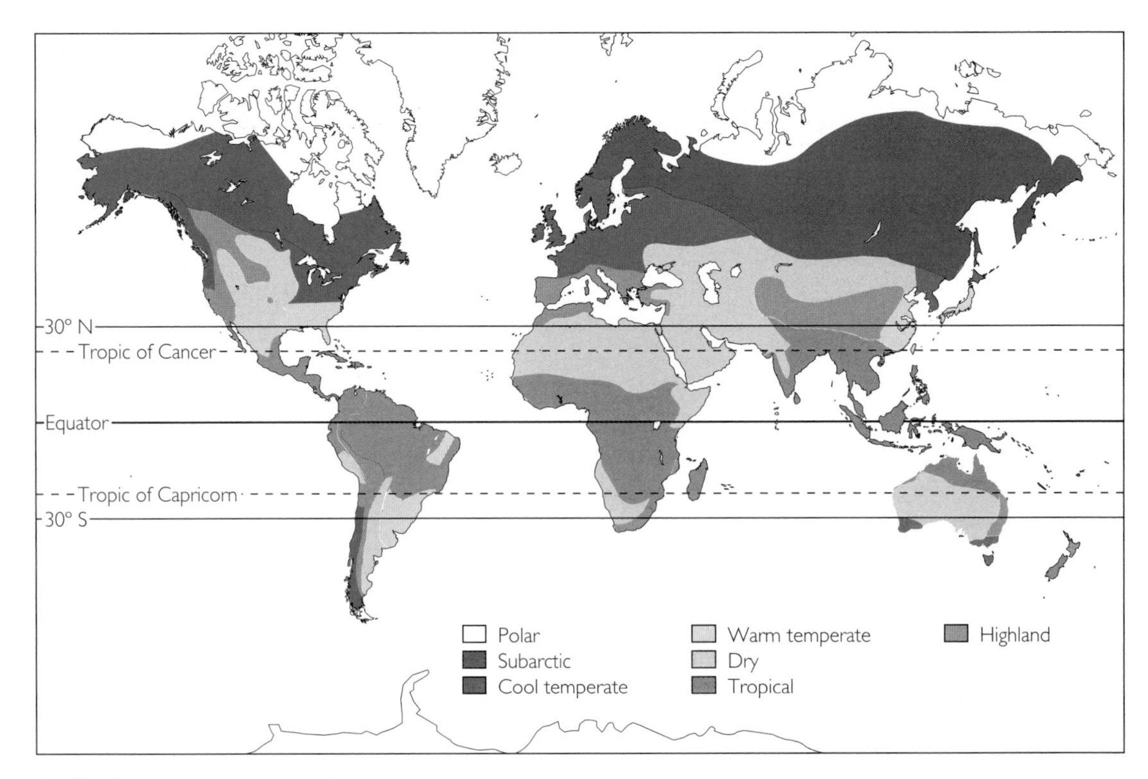

Fig 4.2 The major climates of the Earth.

Convergence in regions with a Mediterranean climate goes beyond the vegetation. Streams and rivers are affected greatly by flow rates, flooding and drying. Streams in Mediterranean-type climates are very similar to each other. Most of these areas are characterized by rapid changes in elevation near the coast, with streams originating in cooler, wetter high-elevation sites. Over 60–80% of seasonal rainfall occurs in the autumn and winter, with spring floods that scour streambeds. Drying in the spring and summer is gradual, but severe, with a low probability of relief until autumn rains. Anoxic conditions in isolated pools and intermittent stream flow are common in the summer. Animal and plant life in these streams have adapted to these conditions, and have many similarities across continents. For example, a higher percentage of fish species that live in Mediterranean-climate streams spawn in spring, before the anoxic conditions develop, than those that inhabit other temperate streams.

Oceanic reefs provide a marine example of convergent evolution arising from similar physical conditions. Reefs have been formed repeatedly in warm ocean waters by a variety of organisms over the last 3.5 billion years, starting with cyanobacteria. The present-day distribution of coral reefs depends strongly on sea level, geochemistry and climate. Tropical reefs throughout the world are found in high-light, low-nutrient, shallow, warm waters, saturated with calcium carbonate. Coral reefs, in turn, generate specialized microhabitats, protecting coastlines from erosion and creating sheltered harbours. Reefs are important storehouses of organic carbon, and help buffer atmospheric carbon dioxide concentrations. They also are sites of substantial tropical marine biodiversity.

The climate and biotic community in an area will also affect the activity of **decomposers**, organisms that feed on dead organisms and recycle nutrients back into the living components of an ecosystem (see Section 2.2.2). The rate at which nutrients are recycled has a large effect on the overall productivity of an ecosystem. This, in turn, affects the biodiversity of the system (see Section 2.2.3). For example, evergreen leaves are

Fig. 4.3 Examples of grassland-savanna grazing animals from different continents. (a) The North American pronghorn antelope (*Antilocapra americana*, Antilocapridae family) (Courtesy of P. Kempenich); (b) the Indian continent blackbuck (*Antilope cervicapra*, Bovidae family) (Copyright the Zoological Society of London); (c) the impala (*Aepyceros melampus*, Bovidae family) (Courtesy of P. Cotgreave); and (d) the pudu (*Pudu pudy*, Cervidae family) (Copyright the Zoological Society of London).

generally thicker and tougher for bacteria or fungi to break down than are deciduous leaves. Hence, nutrient cycling may take longer in communities with tough, evergreen leaves, and the availability of nutrients may limit plant and animal productivity. The activity of phytoplankton and decomposers in aquatic systems is affected by the rates of mixing between different water layers. When mixing is prevented, due to temperature differences between surface and deeper water layers (see Section 2.10.3), nutrients can quickly become depleted in upper layers. For example, spring phytoplankton blooms in the North Atlantic occur in response to increasing light and temperature. However, as surface waters warm, temperature gradients will prevent mixing below about 30–50 m. The bloom then decreases as phytoplankton become limited by nutrient depletion. Productivity that is sustained is due to rapid decomposition of dead phytoplankton and zooplankton by bacteria in these upper layers, before it falls to deeper layers.

There are strong associations between animals and climate, despite greater mobility and behavioural responses compared with plants. For example, major grassland and savanna areas of the world all have large, mobile, grazing animals adapted to feeding on grasses (Fig. 4.3). Nevertheless, mobility and behavioural responses do allow animals some flexibility in choosing their local habitat and moving to other, more favourable climates. For example, the migratory swallow (*Hirundo rustica*) flies south to warmer areas during northern winters. Thus, individual swallows do not have to deal with the food shortages and cold temperatures that occur during winter in the northern hemisphere.

4.2 Solar radiation controls climate

The primary factor that determines the climate of an area is the amount of radiation that is received from the sun. Regions near the equator receive the greatest amounts of solar radiation over the course of a year (Fig. 4.4). There are two main reasons for this. The first is that the direct radiation from the sun strikes the equator almost at right angles (Fig. 4.5). When a beam of light strikes a surface at an angle other than perpendicular, the beam is spread out, diminishing its intensity. You can test this simple principle by shining a flashlight directly on the surface of a desk or wall, and then shining it at an angle on the same surface. Notice how shining the beam at an angle causes it to spread out over a larger area. Since the total energy in the beam of light remains constant, spreading it over a larger area reduces its intensity at any one point on the surface. Because of the spherical shape of the Earth, radiation from the sun is spread over a larger area of the Earth's surface at higher latitudes than it is near the equator.

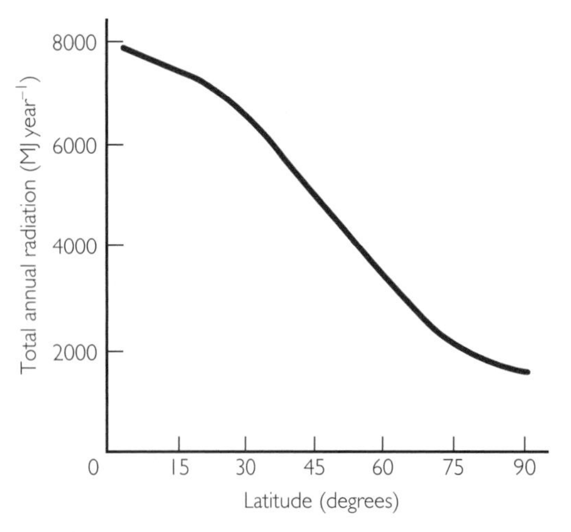

Fig. 4.4 Average annual solar radiation received at different latitudes on the Earth's surface.

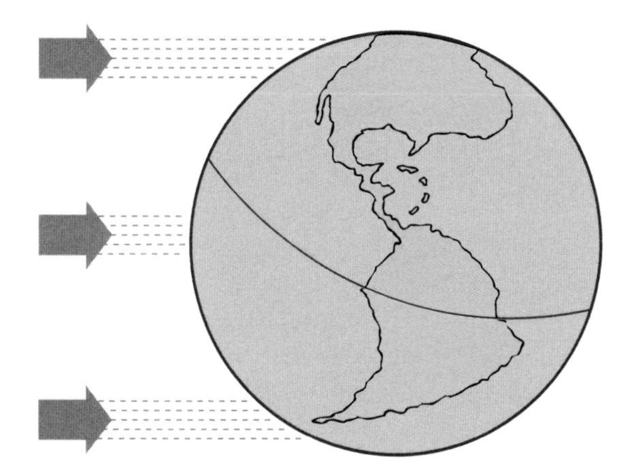

Fig. 4.5 The increased angle between the Earth's surface and the direct rays of the sun at higher latitudes results in lower radiation at higher latitudes.

The second factor in the reduction of solar radiation with increasing latitude is that the Earth is blanketed by an atmosphere. The atmosphere is made up of various gases, water vapour and particulates. All of these absorb a small fraction of the sun's radiation, reducing the amount of radiation reaching the ground. Solar radiation has to pass through larger amounts of the atmosphere at high latitudes, or when the sun is low in the sky (during the morning and evening hours). As the solar radiation travels a greater distance through the atmosphere, the amount of radiation reaching the Earth's surface is reduced.

Latitudinal differences in solar radiation result in higher average temperatures near the equator (Fig. 4.6). The variation in temperature is also greater nearer the poles than it is at the equator, where there is often less than a single degree separating the average June and January temperatures. At higher latitudes there are large seasonal differences in temperature, with distinct summer and winter seasons. Above the Arctic circle, differences between January and June average temperatures can approach 40°C, and average temperature for the whole year can be below 0°C.

As winters become cooler, and freezing temperatures occur, many organisms find it harder to find food and maintain activity. Organisms in seasonal climates show many adaptations to this variation. For example, brown bears (*Ursus arctos*) in northern latitudes go into torpor (see Section 2.8.3), lowering their metabolism to minimal levels as they spend the winter in caves or

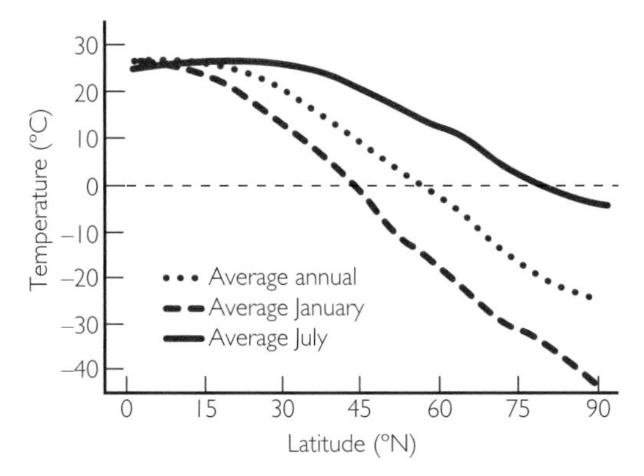

Fig. 4.6 Average annual temperatures for different latitudes north of the equator. Also shown are the average temperatures during January and July for different latitudes. Temperatures decrease and the difference between January and July temperatures increases as latitude increases.

burrows. The monarch butterfly (*Danaus plexippus*) migrates, spending summers at high latitudes when temperatures are favourable and food is available, and then moving to tropical latitudes during the winter months. Many temperate bird species follow a similar pattern of migration. Plants, such as the North American-white oak (*Quercus alba*), drop their leaves during the winter and accumulate antifreezing compounds in their trunks and branches (see Section 1.4.3). Natural freeze-tolerance supports the winter survival of many animals, including terrestrial insects, intertidal marine invertebrates, and amphibians and reptiles that hibernate on land. Freeze-tolerant animals can endure the conversion of 50% or more of their body water into extracellular ice, and employ a suite of adaptations to counter the negative consequences of freezing, including controlling the sites and rate of ice formation, regulating cell volume and accumulating cryoprotectant compounds. Some fish live year round in polar seas at a virtually constant water temperature of −1.86°C. The freezing point of their blood is maintained below −2°C by antifreeze peptides or proteins.

4.3 The Earth's revolution about the sun creates the seasons

A lower solar radiation level at higher latitudes does not completely explain the differences in winter and summer seasons. For example, why is there such a large difference in daylength between seasons at higher latitudes, but not at the equator? The answer lies in the way that the Earth revolves around the sun. The Earth is tilted on its axis in relation to the sun (Fig. 4.5). In other words, an axis drawn through the Earth from the north to the south pole is not perpendicular to the sun. It is tilted by 23.5°. Because of this tilt, the northern hemisphere is pointed towards the sun for part of the year and the southern hemisphere is pointed towards the sun during another part of the year (Fig. 4.7).

The tilt of the Earth relative to the sun is defined as the **declination**. During the spring and autumnal **equinoxes** (around 21 March and 23 September), the sun is directly over the equator, and the declination is 0°. On 22 June and 22 December, the summer and winter **solstices**, the declination is 23.5° and −23.5°, respectively. Changes in declination through the year will cause changes in daylength that correspond closely with radiation and temperature changes associated with the seasons (Fig. 4.8). Thus, organisms often use daylength cues to aid them in preparing for seasonal temperature and radiation changes. Among the responses to daylength that organisms have developed are the timing of flowering and leaf drop in plants, and food gathering, migration and hibernation in animals.

4.4 Global air circulation

Differences in the amount of solar radiation falling on the Earth's surface produce temperature patterns that, coupled with the Earth's rotation on its axis, generate wind patterns and ocean currents. These wind and ocean currents in turn affect the distribution of rainfall. They also play a major role in distributing heat from the equator to higher latitudes. Without dispersal of heat by the wind and ocean currents, the higher latitudes would be much colder and equatorial regions much warmer than they are presently.

The high levels of solar radiation near the equator heat the Earth's surface, and some of this heat is transferred to the surrounding air. As this warmed air rises, it cools again (Fig. 4.9). As the air cools, water vapour condenses, because warm air can hold more water than cool air. The result is a zone of rising air over the equator and high rainfall amounts. This is where tropical rainforests occur in South America, Africa and South East Asia (see Figs 2.1 and 4.2). It is also the region where permanent thermoclines reduce the productivity of the open oceans (see Section 2.10.1.2) and

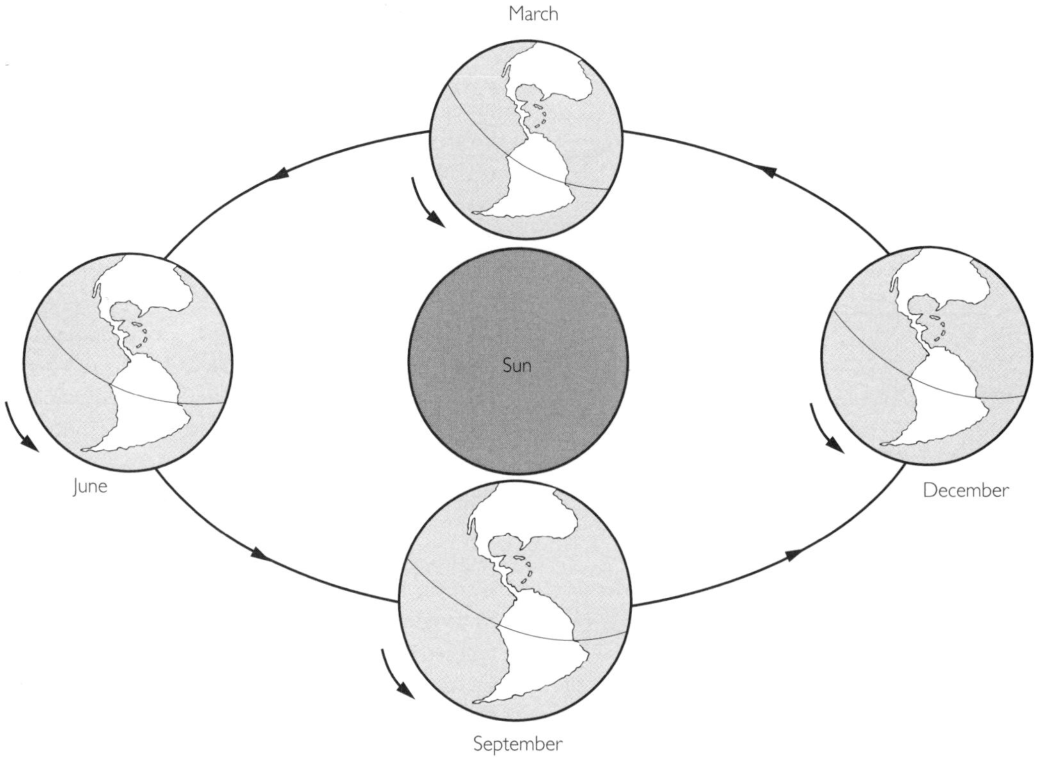

March

June

Sun

December

September

Fig. 4.7 The tilt of the Earth as it revolves around the sun results in either the northern or southern hemisphere being pointed towards the sun at different times of the year. This results in seasonal differences in temperature, radiation and daylength at higher latitudes.

where coral reefs occur in a band around the Earth (see Fig. 2.13).

As air masses rise over the equator, they spread north and south, travelling to higher latitudes. At about 30° north and south latitude, these air masses start to sink and warm up. Because warm air holds more water than cool air, the descending air masses at 30° N and S latitude are dry. These are the latitudes where the world's great deserts are found, including the Sahara (see Fig. 2.1). These air masses will then spread out north and south along the Earth's surface, part will go back to the equator, completing circulation cell 1 (Fig. 4.9). Other air masses spread to higher latitudes, where they will meet cold, dense air coming from the poles at about 60° N and S latitude. This forces some of the air up, where at higher levels in the atmosphere it spreads out north and south, completing two more cells of air circulation (Fig. 4.9). There are three air circulation cells in each hemisphere, one located approximately between 0° and 30° latitude, one between 30° and 60°, and one between 60° and 90°. As the Earth's declination changes these cells move north and south accordingly.

4.5 Prevailing winds and ocean currents

The Earth's rotation west to east on its axis complicates the simple atmospheric circulation pattern shown in Fig. 4.9. Since the Earth is a sphere, its diameter varies with latitude. Yet, every point on its surface travels once around in 24 h. This means that relative to a stationary point suspended above the Earth's surface, the velocity of the land varies. Tokyo travels a distance of 32 000 km as it spins round once in 24 h, while Rio de Janeiro travels 37 000 km in the same time. As air masses spread out along the surface of the Earth, friction with the surface will change their velocity, but not instantaneously, because energy transfer is less than 100% efficient. For example, air at 30° S latitude moving north towards the equator will be moving to an area of wider diameter and faster relative land velocity.

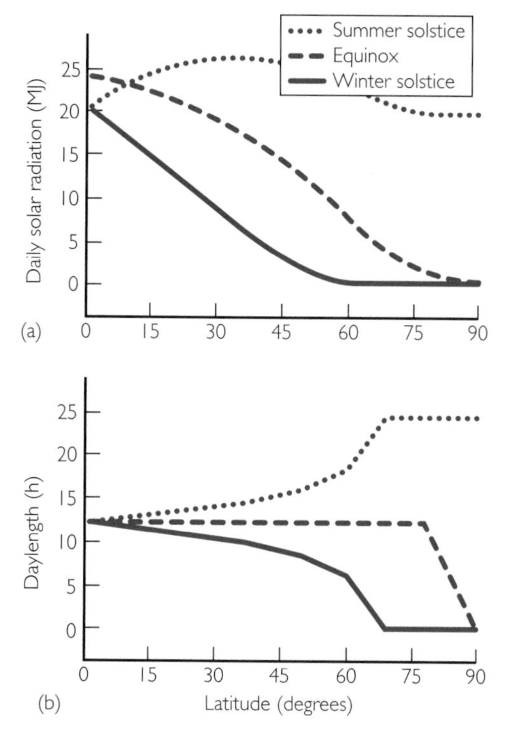

Fig. 4.8 (a) Daily solar radiation on a horizontal surface as a function of latitude for different declinations. During winter there is a steep decline in radiation as latitude increases. During the summer high latitudes receive most of their yearly radiation. (b) Variation in daylength as a function of latitude for different declinations.

This air will be going progressively slower compared to the Earth's surface below. This will result in a deflection of the air mass to its left, relative to the land surface (Fig. 4.10).

This phenomenon is called the **Coriolis** force, and it results in this case in the southeast tradewinds. In the northern hemisphere, air masses are deflected to their right, hence the northeast tradewinds occur between 30° N latitude and the equator, and the westerlies between 30° and 60° N and S latitude (Fig. 4.10).

Wind patterns, combined with the Earth's rotation create the major ocean currents. In the northern hemisphere, these currents travel clockwise, while in the southern hemisphere they travel counterclockwise. Water has a higher specific heat capacity (the amount of energy needed to raise 1 g of a substance by 1°C) than does land or air, so these ocean currents are a very effective means of dispersing heat from the equator towards the poles. Without the ocean currents, temperature differences between the high latitudes and the equator would be much larger than they are today.

4.6 Regional climates

Regional factors (on the scale of tens or hundreds of kilometres) also strongly affect the climate of an area. For example, over the course of a year, the sun will be directly over different latitudes between 23.5° N and S. Wherever the sun is located, there will be a zone of maximum heating of the Earth's surface beneath it. This zone of maximum heating is called the intertropical convergence zone. Its location determines the monsoon season in many subtropical areas. Thus, in India, the presence of the intertropical convergence zone creates a low-pressure area (warm air rising) over the land that will draw in moist air from the Indian Ocean and start the wet season. Conversely, when the intertropical convergence zone moves away from this area the dry season follows.

A common adaptation of vegetation in seasonally dry conditions is the timing of growth to correspond to periods of available water. Thus, north and south of the tropical rainforests seasonally dry tropical forests occur (see Figs 2.1 and 4.2). These biomes have many drought-deciduous trees and shrubs (leaves are dropped during the dry period). Eventually, savannas (areas of widely spaced trees and grassland) ensue, with marked wet and dry seasons and domination by grass life forms rather than trees.

The presence of large bodies of water, such as oceans, results in a **maritime** climate. These areas (such as coasts of continents and islands) have smaller ranges in annual temperatures than do areas far from large bodies of water. This is due to the moderating influence of the nearby water, which requires large amounts of energy to change temperature (high specific heat capacity). In contrast, the interiors of continents, far from oceans, experience **continental** climates. Here, temperature extremes are more pronounced and the differences between seasons are larger. This is due to the lower specific heat capacity of land, which causes it to heat and cool more rapidly than water in response to changing solar radiation levels. These interior continental regions are characterized by grasslands, or steppes, with hot summers and cold winters (see Figs 2.1 and 4.2).

Regional topography, especially the occurrence of mountain ranges, also has a large effect on temperature

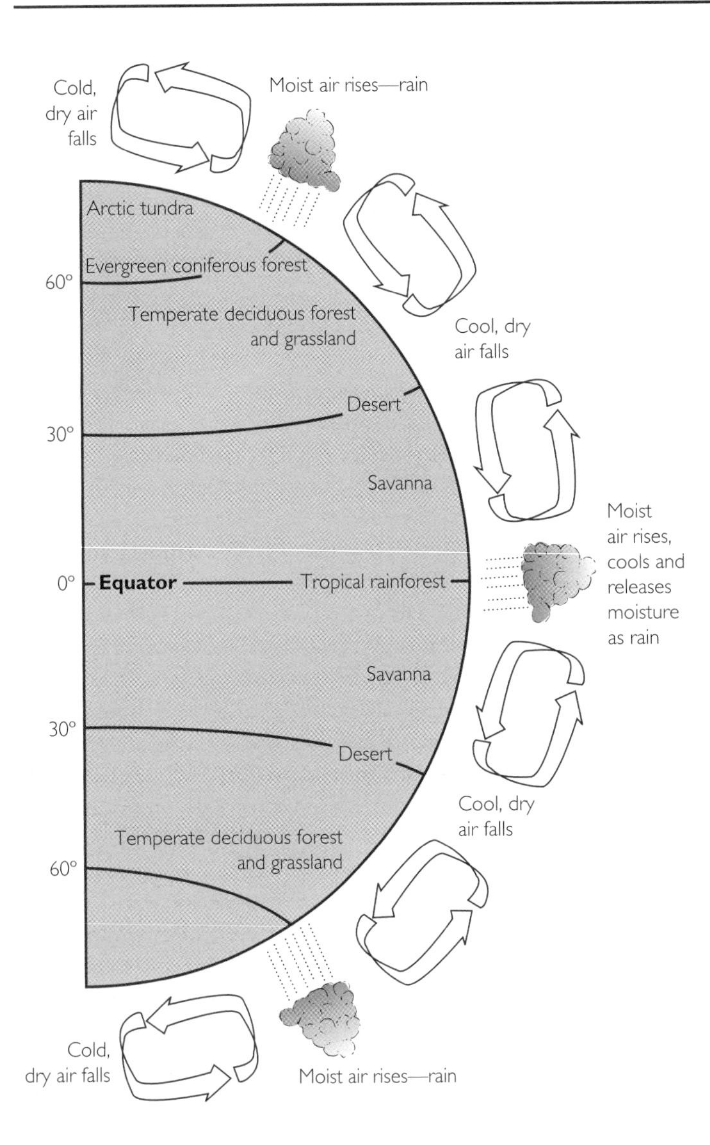

Fig. 4.9 Air circulation patterns that develop in the Earth's atmosphere due to differential heating of the Earth's surface.

and precipitation patterns. As an air mass encounters the windward side of a mountain range, it will be deflected up the side of the range. As the air rises, it will cool at the rate of about 1°C for every 100 m of elevation (Fig. 4.11). Eventually, the air will cool to the point where it cannot hold any more water, and condensation (precipitation or fog) will occur. The act of water condensation will actually slow the cooling of the air mass to a rate of about 0.6°C per 100 m rise in elevation.

On the lee side of the mountain range (away from the prevailing wind), the air will descend and start warming. Since it has dropped moisture on the windward side, and it is descending, the temperature will rise at the dry rate of about 1°C per 100 m. This warm, dry air often leads to rainshadow deserts on the leeward side of mountains. Parts of the southwestern deserts of North America and the Gobi Desert in Asia are due to this rainshadow effect (see Figs 2.1 and 4.2).

Higher precipitation levels on the windward sides of mountains often lead to dramatic differences in the biotic communities on different sides of the mountain range. For example, the spruce and fir forests of the western sides of the Sierra Nevada mountain range in

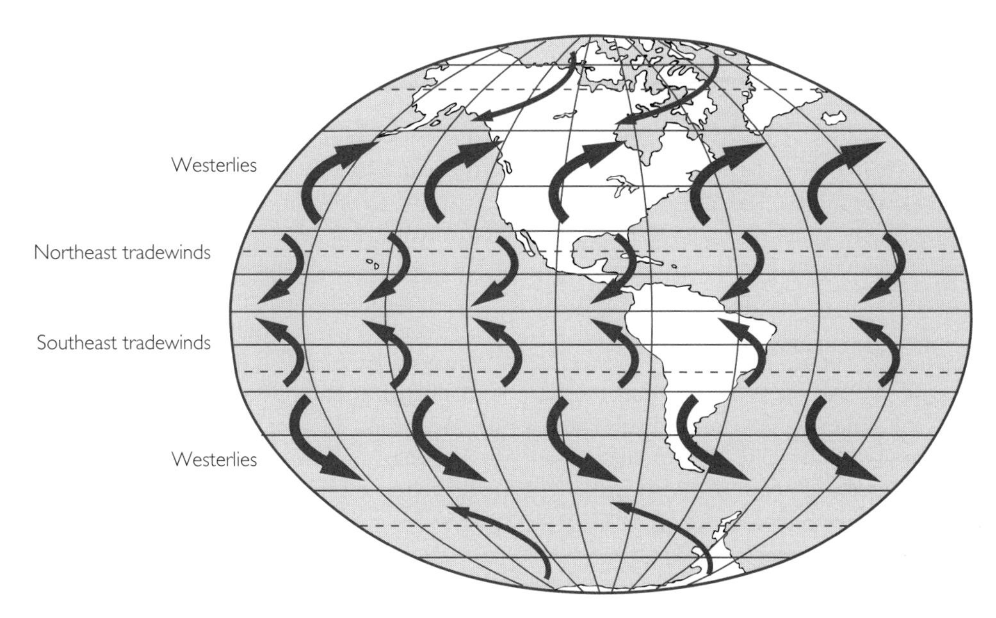

Fig. 4.10 Prevailing wind patterns on Earth. The rotation of the Earth, west to east, results in major air currents being deflected to their right in the northern hemisphere, and to their left in the southern hemisphere. Major wind patterns include the mid-latitude westerlies and the tropical tradewinds.

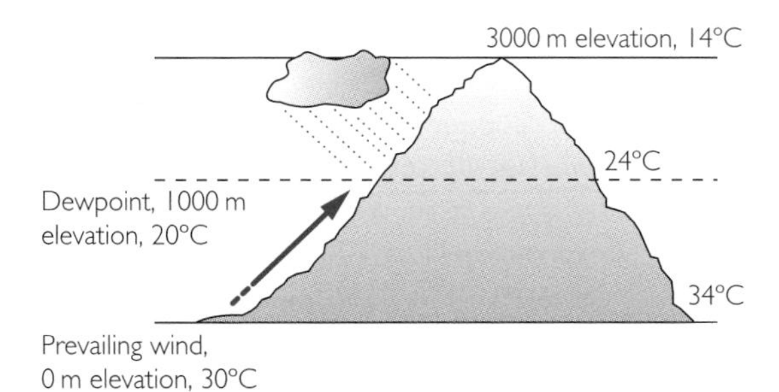

Fig. 4.11 Topographic effects and rainshadows. As moist air rises on the windward side of mountains, it cools. Eventually, water condenses in the form of precipitation. On the leeward side of mountains, cool, dry air descends, warming as it does. This warm, dry air results in rainshadow deserts on the lee side of mountain ranges.

western North America are replaced by drought-resistant shrubs and grasses on the eastern sides. Mountain ranges in the tropics also show this phenomenon, with rainforest on the windward side of mountains and deciduous tropical forest on the leeward side.

Another phenomenon found on mountains is that, at high altitudes, cooler temperatures allow vegetation that would normally be found at high latitudes to be found near the equator. Thus, tundra and coniferous forest communities, which normally occur at high latitudes, are found on mountain ranges at lower latitudes. The degree of exposure can exaggerate this phenomenon, and in the northern hemisphere, the exposed north-facing slopes of many tropical mountains are populated with species that are usually more typical of high latitudes. The elevation at which deciduous forest gives way to coniferous forest is lower on north-facing slopes than it is on south-facing slopes on mountains in the northern hemisphere. This is because solar radiation levels are higher on south-facing slopes, leading to higher temperatures. The situation is reversed in the southern hemisphere.

Slopes facing opposite directions may contain the same vegetation, but at different phenological stages. **Phenology** is the timing of life cycle events, such as germination and flowering. A population of plants on a slope facing the sun may emerge and flower 1–2 weeks ahead of a population of the same species on the opposite facing slope.

For example, the timing of snowmelt has a large impact on reproduction in the alpine buttercup (*Ranunculus adoneus*) in the Rocky Mountains of North America. Earlier snowmelt leads to a longer growing season, earlier flowering and bigger seeds. Similar situations apply to animal species. Earlier snowmelt and higher ground temperatures on ridges and slopes facing the sun result in earlier emergence from hibernation and a longer reproductive season for mountain-dwelling mammals such as the Belding's ground squirrel discussed in Section 7.1.1.2. Bee and butterfly species may emerge earlier and extend foraging seasons by travelling between slopes with different exposures. Insect activity may be concentrated on warmer slopes during cooler periods of the year.

4.7 Global climate change

The human population on Earth has grown to the point that it is having an effect on the Earth's atmosphere and ecosystems. Burning of fossil fuels, deforestation, urbanization, cultivation of rice and cattle, and the manufacture of chlorofluorocarbons (CFCs) for propellants and refrigerants are increasing the concentration of carbon dioxide, methane, nitrogen oxides, sulphur oxides, dust and CFCs in the atmosphere. About 70% of the sun's energy passes through the atmosphere and strikes the Earth's surface. This radiation heats the surface of the land and ocean, and these surfaces then re-radiate infrared radiation back into space. This allows the Earth to avoid heating up too much. However, not all of the infrared radiation makes it into space, some is absorbed by gases in the atmosphere and is re-radiated back to the Earth's surface. A greenhouse gas is one that absorbs infrared radiation, and then re-radiates some of this radiation back to the Earth. Carbon dioxide, CFCs, methane and nitrogen oxides are greenhouse gases. The natural greenhouse effect of our atmosphere is well-established. In fact, without greenhouse gases in the atmosphere, scientists calculate that the Earth would be about 33°C cooler than it currently is.

The current concentration of carbon dioxide in the atmosphere is about 360 ppm. Human activities are having a major influence on atmospheric carbon dioxide concentrations (see Section 12.9.2.1), which are rising so fast that current predictions are that atmospheric concentrations of carbon dioxide will double in the next 50–100 years. The International Panel on Climate Change (IPCC) report in 1992, which represents a consensus of most atmospheric scientists, predicts that a doubling of carbon dioxide concentration would raise average global temperatures anywhere between 1.4°C and 4.5°C. The IPCC report issued in 2001 raised the temperature prediction almost two-fold. The suggested rise in temperature is greater than the changes that occured in the past between ice ages. The increase in temperatures would not be uniform, with the smallest changes at the equator and changes two or three times as great at the poles.

The local effects of these global changes are difficult to predict, but it is generally agreed that they may include alterations in ocean currents, increased winter flooding in some areas of the northern hemisphere, a higher incidence of summer drought in some areas, and rising sea levels, which may flood low-lying countries like Bangladesh.

Scientists are actively investigating the feedback mechanisms within the physical, chemical and biological components of the Earth's climate system in order to make accurate predictions of the effects the rise in greenhouse gases will have on future global climates. Global circulation models are important tools in this process. These models incorporate current knowledge on atmospheric circulation patterns, ocean currents, the effect of land masses, etc. to predict climate under changed conditions. There are several models, and all show agreement on a global scale. For example, all models show substantial changes in climate when carbon dioxide concentration is doubled. However, there are significant differences in the regional climates predicted by different models. Most models project greater temperature increases in mid-latitude regions and in mid-continental regions, relative to the global average. Additionally, changes in precipitation patterns are predicted, with decreases in mid-latitude areas and increased rainfall in some tropical areas. Finally, most models predict that there will be increased occurrences of extreme events, such as extended periods without rain (drought), extreme heat waves, greater seasonal variation in temperatures, and increases in the frequency and magnitude of severe storms. Plants and animals have strong responses to virtually every aspect of these projected global changes.

The relative roles that abiotic and biotic factors play in the distribution of organisms is especially important now, when the world is confronted with the consequences of a growing human population. Changes in climate, land use and habitat destruction are currently causing dramatic decreases in biodiversity throughout the world. An understanding of climate–organism relationships is essential to efforts to preserve and manage the Earth's biodiversity.

4.7.1 Lessons from the past

The challenge of predicting organismal responses to global climate change is difficult. Partly, this is due to the fact that there are more studies of short-term, individual organism responses than there are of long-term, system-wide studies. It is extremely difficult, both monetarily and physically, for scientists to conduct field studies at spatial and temporal scales that are large enough to include all the components of real-world systems, especially ecosystems with large, freely ranging organisms. One way around this limitation is to use lessons from the geological past. Palaeobiologists attempt to reconstruct past climates by examining fossil life (see Section 3.8). Much of this work depends on the assumption that life forms adapted to a particular climate in the present were adapted to the same type of climate in the past. For example, reef-forming marine organisms in the past are assumed to have occurred in warm, shallow oceans, as they do now.

Studies focusing on the period from the last glacial era (18 000 years BP) to the current time are especially useful because this period was characterized by global warming, major evolutionary differences in organisms between that era and this are small, and radioactive isotopes can be used to get accurate dates. A radioactive isotope is an unstable form of an element that gradually decays to other elements, or to stable isotopes of the same element. The rate of decay is measured in terms of **half-lives**. Thus, an element with a half-life of 4000 years would decay to half its present concentration in 4000 years. By comparing the current concentration of radioactive isotopes with the concentration of their decay products, the age of a rock, sediment or fossil can be estimated.

The first pattern that emerges from studies of the geological past is that many organismal communities are loosely organized collections of species, whose coexistence is dependent upon a variety of physical, biological and historical variables. Past environmental changes have largely caused species to respond individually. Different species migrate in different directions, at different rates and with different lag times in response to changing climate, resulting in a variety of species associations forming in one locale at different times. Davis (1981, 1989) and Webb (1987) found that over the last 18 000 years, the species composition of forests in North America has changed considerably, with some forest communities reaching their present species composition only within the last 2000 years. The same pattern has been found in studies of European forests.

Several factors can lead to time lags between climatic changes and the responses of species. For example, limited mobility and dispersal of propagules can slow rates of movement. Asexual reproduction and long lifespans can also contribute to delayed responses to climatic change. Arseneault and Payette (1997) examined tree rings from black spruce (*Picea mariana*) trunks preserved in peat bogs near the treeline in northern Canada, about 50 km east of Hudson Bay. Treeline in this area is a habitat where trees give way to lichen–tundra vegetation. The border between these areas is dynamic, and is affected by changes in climate of only a degree or two. One thousand years ago, a slight climatic cooling favoured the spread of tundra. Despite this cooling, spruce forests maintained themselves in the area through vegetative reproduction for over 500 years. Arseneault and Payette found a layer of carbon in the substrate dating to 1567–1568 AD, indicating the occurrence of forest fires. These fires destroyed trees on dry ridge tops. However, in wetter, low-lying areas, most trees survived the fires. Nevertheless, the loss of ridge top trees caused a series of changes that eliminated the remaining spruce forests within the next 25 years. The loss of ridge top trees reduced the trapping of snow, with a corresponding reduction in snow cover in low areas. Reduced snow cover exposed spruce branches to winter winds and ice damage. As exposed branches died, there was even less interception and accumulation of snow. Eventually, snow cover was reduced to the point that it did not fully insulate the soil below the trees. This caused lower soil temperatures and the spread of permafrost. As permafrost spread, drainage was impeded in low-lying areas, eventually resulting in flooding and the replacement of trees with peat bogs. Nevertheless, the spruce forest survived for a long time after the climatic cooling.

A second pattern of response to climate change learned from the geological past is that levels and types of disturbance change as climate changes. Over 7000 years ago, fire was the major source of disturbance in low elevation forests in the White Mountains of New Hampshire, USA (Spear 1989). However, as precipitation increased about 7000 years ago, windstorms replaced fire as the major source of disturbance (Davis 1985). Because plants have different adaptations to different disturbances, changes in disturbance regime may produce a larger change in vegetation than would have resulted from the effects of climate change alone (Davis 1989). For example, fire-adapted species may not regenerate after a disturbance caused by wind.

Significant variation in global climate has occurred during the last 3000 years (the period known as the late Holocene), often on timescales as short as decades or as long as centuries. This period is especially valuable for examining the response of organisms to climate because there is a variety of independent records documenting changes during this time. Tree ring analyses were used by Lloyd and Graumlich (1997) to reconstruct the history of tree abundance in the southern Sierra Nevada of California. Regions such as the alpine treeline are called **ecotones**, because they are areas where two different habitats meet. These areas are very dynamic and sensitive to climate change, since they represent the limits to the distribution of species from each habitat type.

Lloyd and Graumlich examined five different sites that are currently treeless, but have the remains of tree trunks from past forests. Tree trunks were dated using tree rings and a process of cross-matching. In this method, an event that affects tree ring deposition strongly (like a very cold year that results in a narrow ring), and whose date is known is used to age tree trunks from different sites (see Section 3.8). The results showed that across all five sites, synchronous increases in the abundance of trees and an extension of the treeline to higher altitudes have occurred several times in the past 3500 years (Fig. 4.12). Synchronous declines in tree abundance across the five sites also occurred. The synchronous nature of these changes in tree abundance among five separate sites suggests that a common, widespread mechanism caused the changes. Using other sources for climate data, Lloyd and Graumlich found that increases in abundance were associated with increases in temperature. The correlation between increased temperatures and higher treeline has often been used to explain treeline elevation in the Sierra Nevada. However, the decrease in the abundance of trees between 950 and 550 years ago was accompanied by a warming trend, the opposite of what was expected. This apparent contradiction was resolved by evidence from a large nearby lake. Mono lake experienced a large reduction in area during this period, indicating a severe drought. Lloyd and Graumlich concluded that drought

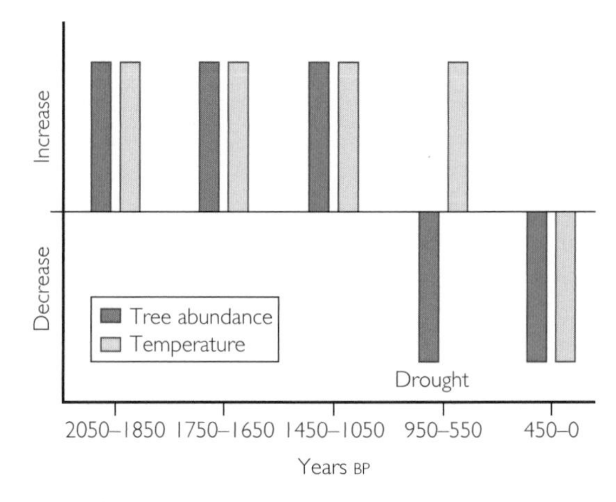

Fig. 4.12 The location of the treeline in the Sierra Nevada mountains of California is associated with temperature. Increased temperature leads to increased tree abundance and increases in the elevation of treeline. Synchronous increases and decreases in tree abundance in five different sites over the last 2000 years supported this correlation, except for the change between 950 and 550 BP. Here, despite an increase in temperature, trees decreased in abundance at treeline, most probably due to a severe drought throughout the area.

plays a larger role than previously thought in the control of the treeline in the Sierra Nevada, and that this factor should be considered in future predictions of how the treeline will respond to climate change.

4.8 Climate space

Another approach that scientists use to predict the responses of organisms to future climate change is to combine energy balance approaches (see Chapter 5) with global circulation model predictions. First, the **climate space** of an organism is determined by plotting the extreme values of temperature and radiation that the organism has been observed to tolerate (Fig. 4.13). Then, the distribution of habitats that the species currently occupies is plotted on the climate space diagram. Finally, predictions of temperature and radiation change are used to predict the habitat conditions the species will experience in the future. This graphical approach allows ecologists to determine if organisms are at or near the limit of their tolerances, and whether they will exceed tolerances in future climates.

Johnston and Schmitz (1997) used computer simulations of climate space for four North American mammal species (elk, *Cervus canadensis*; white-tailed

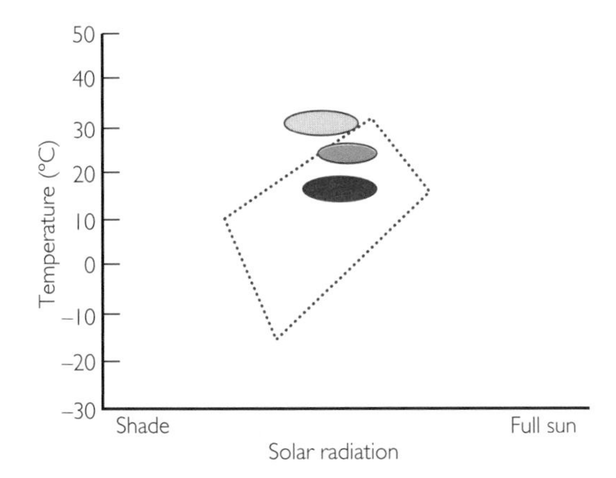

Fig. 4.13 Idealized climate space diagram. The climate space of a species is delimited by the polygon connecting extreme temperature and radiation conditions that the species may tolerate. The dark purple ellipse represents the current distribution of the species within this climate space. The species is well within its tolerances. The grey ellipse represents a projection of the conditions the species would experience with a moderate future warming of global climate. The species is now nearer its limits in parts of its range. The light purple ellipse represents projected conditions with a more severe warming of future climates. The species is now completely outside its thermal and radiation tolerances. This species would either go extinct, have to migrate to more favourable habitats or change its thermoregulatory behaviour in its current habitat.

deer, *Odocoileus virginianus*; Columbian ground squirrel, *Spermophilus columbianus*; and chipmunk, *Tamias striatus*) to predict species response to global warming. These species were chosen because there is a great deal of information on their thermal physiology, and because they represent a range of body sizes. Size has a large effect on an animal's ability to cope with extreme or unusual thermal conditions, because small animals have a very large surface area, relative to their total body mass, and thus lose heat energy relatively rapidly. Johnston and Schmitz looked at the three warmest months of the year, assuming that heat loads during these months would be the most stressful. The results showed that with predicted changes in climate, these species would still be within their climate space, although nearer their limits. The scientists concluded that predicted changes in temperature should have little effect on current distribution for these species.

Since these four mammal species are tightly associated with certain vegetation types, Johnston and Schmitz also looked at how predicted changes in vegetation distribution might affect the animals. Here, their results suggested that changes in vegetation distribution would have little effect on white-tailed deer or the eastern chipmunk. However, there would be major changes for elk and ground squirrels, with smaller future distributions and little chance for these species to spread to new localities. Johnston and Schmitz concluded that animal–vegetation links are very important to understanding potential changes in distribution of animals due to climate change.

Organisms may not be distributed evenly across climate space. For example, the middle of a species' geographical range may be the area best suited to its growth and reproduction, and its individuals may be found in greatest numbers in this core area. A prediction from this hypothesis would be that the lowest abundances would be at the edges of the organism's range. Lower abundance may make the organism more vulnerable to climate changes, and the most dramatic effects of climate change would occur at the limits of the organism's range (Fig. 4.14a). An alternative hypothesis is that the greatest abundance of an organism may have no relationship to its geographical range, but that high densities occur in suitable patches anywhere within its range (Fig. 4.14b). These patterns are important, because they may point out regions where the species is most vulnerable to local extinction.

Mehlman (1997) looked at the response of three different North American passerine bird species to a series of harsh winters in the late 1970s. Mehlman was concerned not only with how distribution may change with climate, but also with how the birds' abundance changed in different parts of their range. All three species in this study (the Carolina wren, *Thryothorus ludovicianus*; eastern bluebird, *Sialia sialis*; and field sparrow, *Spizella pusilla*) showed declines in abundance in response to harsh winter conditions. However, the proportional abundance change was greatest at sites closest to the edge of the range, both for declines in response to the severe winters, and for increases during recovery periods. Additionally, local extinctions of populations were greatest at the boundaries of species' ranges. This was countered by greater rates of colonization in these areas during recovery. For these species, population response to climatic perturbations resulted in a contraction of ranges to a central core area that had possessed the highest abundances before the climate change.

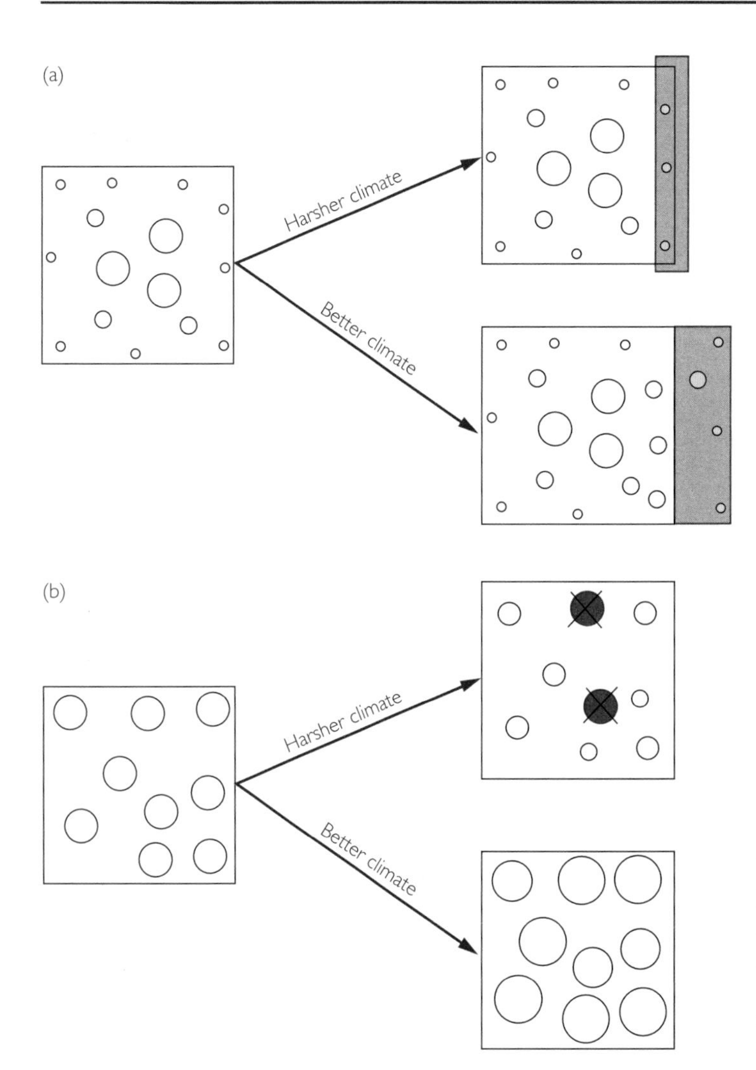

Fig. 4.14 Idealized distributions and their response to climate change. (a) Population size decreases at the limits of an organism's range. Climate changes, indicated in the hatched areas, would have the greatest effect, either positive or negative, at the limits of distribution. (b) Distribution within the range does not affect population size. Here, climate changes may affect populations throughout the range.

4.9 Climate–species interactions

Climate affects the structure of natural communities, and climate change will affect the interactions between species, and this may lead to further changes in communities. On rocky, intertidal habitats, marine invertebrates, such as barnacles, anchor themselves to rock surfaces. The elevation of a barnacle on the rocks will determine its exposure to heat and drying during low tides in the summer and to freezing during low tides in the winter. On rocky marine shores, temperature and desiccation are critical community structuring forces. Positive species interactions, where neighbours benefit one another by buffering physical stresses, sometimes occur in physically harsh environments. Bertness *et al.* (1999) hypothesized that high barnacle densities might enhance individual survivorship at high tidal heights because of a thermal buffering effect from neighbours. Alternatively, high densities may lead to higher rates of predation and/or competition. Bertness *et al.* (1999) tested their hypothesis using the acorn barnacle, *Semibalanus balanoides*. They took advantage of a natural, geologically driven change in climate over a short distance on the rocky, intertidal coasts of eastern North America. Cape Cod, by deflecting warm, southern ocean currents, serves as a dividing line between warmer, southern sites and cooler, northern sites. Bertness and co-workers manipulated barnacle density and constructed artificial shades over barnacles in sites north and south of Cape Cod. They found that barnacle

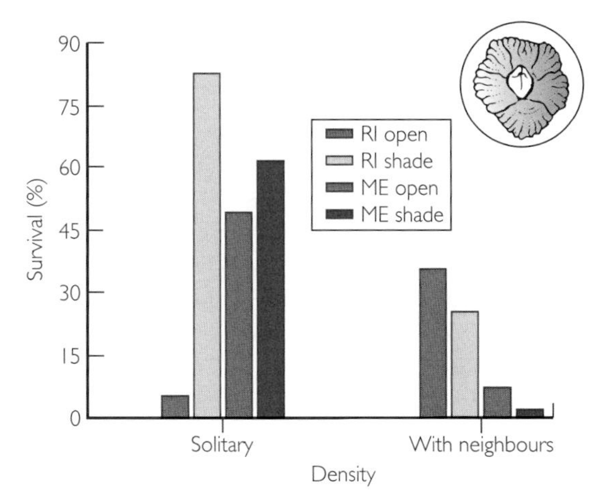

Fig. 4.15 Survival of solitary and grouped barnacles in warm sites south of Cape Cod (Rhode Island, RI) and cool sites north of Cape Cod (Maine, ME). Shading and crowding increase survival in warmer areas, but crowding decreases survival in northern areas where predators control barnacle distribution.

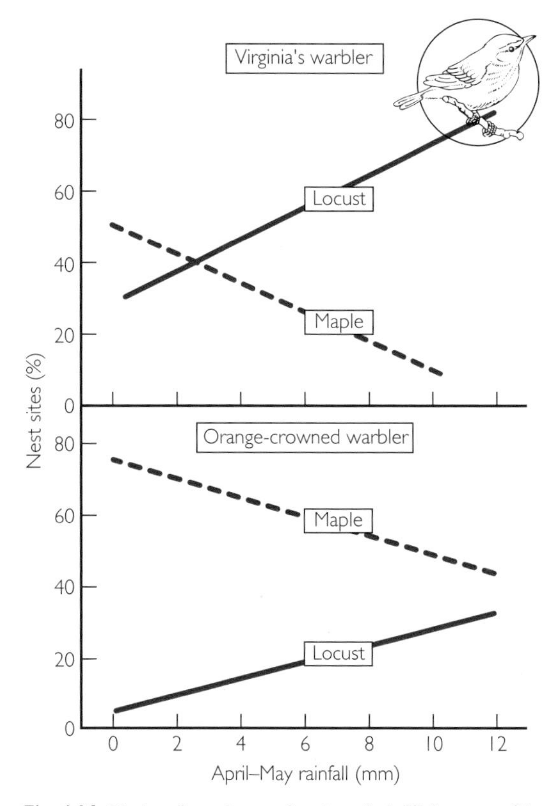

Fig. 4.16 Choice of nest sites as a function of rainfall for two warbler species in the high elevation forests of Arizona. Orange-crowned warblers prefer moist, maple-dominated sites, but use more drier, locust-dominated sites in wet years. Virginia's warbler will switch nest preferences, choosing drier, locust-dominated sites in wet years, and moist, maple-dominated sites in dry years.

survivorship at higher intertidal elevations was enhanced by artificial shading or by neighbours in southern sites. This was because shading reduces heat and thus lowers the chance of desiccation stress (Fig. 4.15). However, at cooler, northern sites, neither shading nor crowding increased barnacle survival. In fact, mortality at northern sites was increased at high barnacle densities, primarily through predation from predators not found south of Cape Cod. These results show how the relative importance of abiotic and biotic factors can change in different parts of the range of a species. Such complexity makes predicting organismal responses to climate change a challenge.

The relative importance and interaction of biotic and abiotic factors in determining habitat use has important implications for the ecological consequences of global climate change. The eggs and immature stages of an organism's life may be especially sensitive to these factors. For example, in birds, the choice of a nest site is often made in response to abiotic factors such as temperature, but may be modified due to vegetation preferences or the risk of predation or parasitism. Martin (2001) studied four ground-nesting birds that segregate along climatic gradients in the mountains of Arizona. For these ground-nesting birds, selection of nest sites is especially critical to offspring success. One species, the orange-crowned warbler (*Vermivora celata*), commonly occurs in cold, damp northern regions. In contrast, Virginia's warblers (*V. virginiae*) occur in warm, dry southern regions. These two species occur together in the high altitude (2400 m) forest valleys studied by Martin. There is a strong microclimatic gradient in these valleys from warm, dry upper slopes to cooler, moist bottoms that receive heavy snowmelt. Often, only 100 m separates dry forest communities with abundant New Mexican locust trees (*Robinia neomexicana*) from moist stands dominated by canyon maple (*Acer grandidentatum*).

Martin (2001) found that birds shifted their nest sites toward higher positions in wet years and lower positions in dry years (Fig. 4.16). Virginia's warblers actually reversed their habitat preference for nest sites between the wet 1994 and dry 1996 years. This change in choice of upper and lower elevation nest sites corresponded to changes in the use of vegetation dominated by locust trees versus maple trees. However, when orange-crowned warblers nested under locust trees, nest predation increased and nesting success decreased compared to nesting under maples. The opposite pattern occurred for Virginia's warblers. These results imply that some birds shifted habitat choices independent of plant type and in direct response to changing weather. These shifts occurred out of the species' preferred vegetation, increasing predation costs and lowering nesting success.

4.10 Year-to-year variability

Many organisms show large year-to-year variation in population recruitment, numbers and survival. Often, these population fluctuations are related directly and indirectly back to climate. In the seasonally dry tropics of northern Australia, the aquatic Arafura filesnake (*Acrochordus arafurae*) shows large year-to-year variation in population dynamics in response to rainfall at the end of the wet season in February and March. Madsen and Shine (2000) examined filesnake population dynamics in order to understand the mechanisms responsible. They found that high late-season rainfall slows the fall in water levels that occurs in lakes during the dry season. This promotes fish population growth and fish feeding by the filesnakes. Higher fish availability increases reproduction by female snakes and increases yearling snake survival.

In contrast to the filesnake, the water python (*Liasis fuscus*) shows the opposite response to high late-season rainfalls. The python occupies similar floodplain habitats to the filesnake, but it feeds on land animals. Years with low late-season rain experience drops in water level, exposing more land and increasing the population of dusky rats (*Rattus colletti*), the primary prey of the python. In response, feeding rates, growth rates and reproductive rates of the python are highest during dry years. Thus, a 'good' year for filesnakes is a 'bad' year for pythons and vice versa. Long-term changes in rainfall during the critical period (February and March) would have lasting consequences for the distribution and population size of snakes, rats and fish in this ecosystem.

Key species interactions that are sensitive to temperature may act as leverage points through which small changes in climate could generate large changes in natural communities (Sanford 1999). For example, the sea star *Pisaster ochraceus* feeds on the rocky intertidal mussel *Mytilus californianus* in the coastal seas off northwestern North America. Without sea star predation on mussels, a monoculture of *M. californianus* develops, displacing a diverse assemblage of intertidal algae and invertebrates. The algae provide a habitat for young fish, and without its protection, fish and fish predators (other fish, hawks and eagles) suffer declines. Sanford (1999) found that densities of the sea star responded strongly to changes in ocean temperature caused by weather events such as the El Niño southern oscillation (ENSO). The ENSO is a pattern that lasts for about 2 years, and repeats every 7–9 years. During El Niño years, warm water accumulates in the eastern Pacific, and ocean temperatures in the Pacific, upwelling patterns and rainfall in large areas of North and South America are altered. Sea star activity is highest in spring and summer, but drops when water temperatures drop. During low temperature periods, many sea stars become inactive in low zone channels or shallow subtidal waters. Changes in upwelling patterns and ocean temperatures due to climate change may have large effects on the species interactions and community structure of these regions. Section 7.3.4 will detail the effects of ENSO on Darwin's finches in the Galapagos islands.

The North Atlantic oscillation (NAO) is similar to the ENSO in that it involves both ocean currents and air masses, and it can affect temperatures and precipitation patterns over large areas of the Earth. The NAO influences weather patterns in the northern hemisphere north of 20° latitude. This oscillation is characterized by subtropical air masses moving further north, causing warm winters over much of Europe, reducing ice cover on lakes and snow cover on land. The phytoplankton and zooplankton dynamics in Lake Constance, a large lake approximately 1000 km from the Atlantic Ocean on the northern fringe of the Alps, have been studied for over 15 years. Such long-term studies are invaluable in revealing long-term trends and correlations between climate and organisms. Straile (2000) found that warm water temperatures in April and May associated with a high NAO index led to increased growth of the water flea (*Daphnia* spp.) with peak biomasses occurring in May. *Daphnia* feed on phytoplankton, hence high populations of *Daphnia* in May lead to depressions in phytoplankton growth and early onset of the clear phase in the lake. High water clarity is associated with low phytoplankton populations. The *Daphnia*–phytoplankton connection is important in many freshwater lake systems, so this climatic forcing is an important source of variation in many temperate aquatic ecosystems.

Year-to-year extremes in temperature, precipitation, storm occurrence, flooding and drought can all have major impacts on local populations. For rare species that have limited distributions, these chance events can have severe consequences, up to and including extinction. However, more subtle weather occurrences can also have large impacts on local population dynamics and interactions. Inouye (2000) has examined the impact of late spring frosts on alpine communities of the Rocky Mountains. These frosts are relatively common in alpine habitats and can result in severe damage to plant tissues. Especially vulnerable are the reproductive structures, and flowering in some plants in some localities can be completely wiped out by a particularly severe frost. Observations over several years pointed out some interesting complications due to reduced flowering. Many alpine animals rely on flowering plants for food. This is especially true for pollinators that use nectar and pollen as food sources, and for insect seed predators. Reproductive failure in important plant species can have large effects on populations of these pollinators and predators. While perennial plants may recover by the next year, many seed predators and their parasitoids do not recover for several years, and may suffer local extinction. However, low seed predator populations can actually improve plant seed production for several years immediately after a damaging frost.

4.11 Climate–human interactions

Anthropogenic factors also interact with climate to influence the distribution and population abundance of many species. An example that illustrates the complexity of these interactions involves Townsend's ground squirrels (*Spermophilus townsendii*), a species that occurs in shrub–steppe habitats in the western United States (Van Horne *et al.* 1997). The typical habitat occupied by the ground squirrel is dominated by shrubs like sagebrush (*Artemisia tridentata*) and rabbit brush (*Chrysothamnus nauseosus*) interspersed with native perennial bunchgrasses such as Sandberg's bluegrass (*Poa secunda*). The ground squirrel can occupy different habitats, grassland or shrub, but usually needs succulent perennial grasses to support its reproduction. Over the last century, this region of North America has been invaded by several annual grasses, such as cheatgrass (*Bromus tectorum*). These invaders were introduced by humans, and have three main effects on the habitat. First, they are not as good a food resource as the native plants. Second, they out-compete and eliminate the native perennial bunchgrasses. Third, they are highly flammable and serve to increase the incidence of fires in these regions. The native shrubs like sagebrush cannot survive fire, hence the introduced grasses spread at the expense of native vegetation.

Townsend's ground squirrels have a unique life history in that they have very short active periods (approximately 4 months) and high reproductive efforts. The adults emerge from hibernation in early February and breed almost immediately. The young are born in early March and emerge from their natal burrows in early April. The adult males, after breeding, build fat reserves and go back to their burrows in late April. This reduces competition for food for females, which take more time to recover from giving birth, eventually hibernating in early to mid May. The juveniles keep feeding until late May or early June, at which time they join the adults in hibernation. Thus, juvenile squirrels have only 3 or 4 months in which to acquire sufficient fat to survive the inactive period, which lasts for 8 or 9 months.

An important resource for fat storage is a high-quality diet of succulent grasses (mainly Sandberg's bluegrass) and forbs. By displacing native grasses, exotic annual grasses make the habitat less suitable for the ground squirrel. This was evidenced in late spring of 1992, when a severe drought occurred in the western United States. The drought caused early drying of Sandberg's bluegrass, resulting in low adult and juvenile body masses in the ground squirrels prior to hibernation. The following winter was prolonged by low temperatures and late snowmelt, resulting in late emergence of females in spring of 1993. The combination of a dry spring and long winter resulted in near zero survival for juveniles born in 1992, and a marked reduction in survival of adults, especially females. The adult females that did survive each produced fewer young than normal and the females born in 1993 were smaller than in previous years. Fewer, smaller females led to reduced population growth for 2–3 years after the drought, in

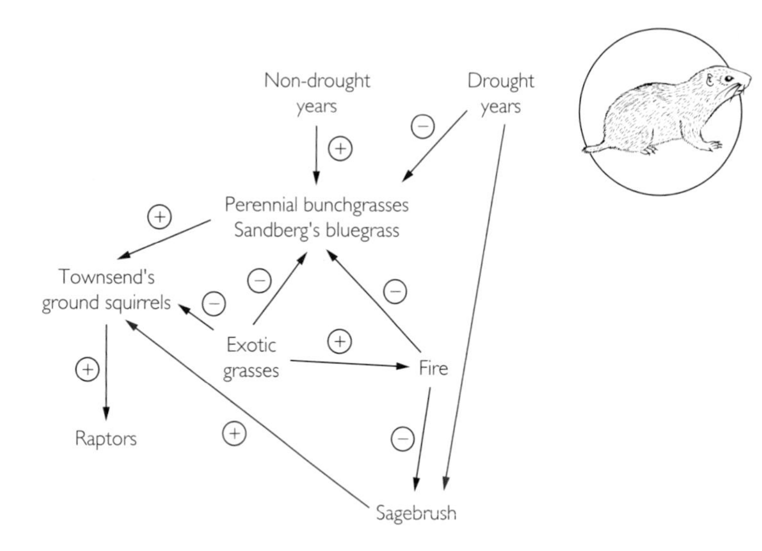

Fig. 4.17 Interactions between introduced grasses, native vegetation, drought and ground squirrel populations in the western United States. (+) indicates positive interactions, for example growth of Sandberg's bluegrass increases ground squirrel populations. (–) indicates negative interactions, for example fire eliminates sagebrush from the community. The positive effect of sagebrush on ground squirrels occurs primarily during severe droughts, indicated by purple arrows.

addition to the loss of almost the entire 1992 cohort. Loss of an entire cohort is especially important for the population dynamics of Townsend's ground squirrel, since adults usually live only for 1–3 years.

Interestingly, ground squirrel colonies in sagebrush communities had better survival and reproduction than those in grasslands. In years with average levels of precipitation, densities of ground squirrels were higher in grasslands than in shrublands, and these habitats provided the greatest number of young to the population. However, shading by shrubs allowed grasses to stay green longer during the 1992 drought, increasing forage for the squirrel. The shrubs also provided cooler temperatures for burrows and the animals aestivating in them. Thus, during an extreme drought event, habitat quality was reversed, and shrub habitats become an important refuge for Townsend's ground squirrels. As fire frequencies increase in this area due to the invasion of exotic grasses, shrubs are rapidly disappearing, reducing the availability of this refuge. Repercussions from decreases in shrub and squirrel numbers are widespread in this ecosystem, involving a number of other organisms (Fig. 4.17). The ground squirrel is an important prey item in the community because it is relatively large and is active during the day. It is especially important for most of the raptor (hawk) species in the area. In fact, the timing of egg laying in the raptors is such that their young emerge at the same time as juvenile ground squirrels. Decreased ground squirrel populations have a large, negative effect on raptor populations.

4.12 Substrate, geology and geochemistry

The distribution of individuals can also be influenced by non-climatic factors such as substrate, i.e. the type of ocean floor (sand or rock) or type of soil. An example is found in areas where soils are poorly drained, forming marshes or swamps. If flooding is combined with salinity, then the local community may be quite different from surrounding areas. Saltmarshes dominated by salt-tolerant grasses develop in temperate coastal areas that are climatically suited for temperate deciduous forest. Rocky intertidal substrates support an entirely different ecosystem than do sandy shores. In northern latitudes, spruce forests may be displaced by sphagnum moss bogs on poorly drained substrates. In areas where soil is thin, due to rocky outcrops, forests will fail to develop, although the climate could easily support them. On serpentine soils (soils with low nitrogen and phosphorus levels, a pH far from neutral and high levels of toxic metals such as manganese and nickel) herbaceous plant communities quite different from surrounding communities develop.

The dynamics of plankton communities in the North Atlantic and Pacific oceans are quite different, despite similar latitudes. In the North Atlantic, there is a strong spring bloom of phytoplankton, which does not occur in the North Pacific. During the winter in the North Atlantic, wind shear and surface cooling deepen the mixed layer to about 1000 m. The average irradiance experienced by phytoplankton in this mixed layer is too low to support photosynthesis and growth. Consequently, the dominant zooplankton overwinters without feeding. In spring, increased solar radiation heats the ocean surface, raising the mixed layer to about 30 m or 50 m. The higher light in this zone supports rapid phytoplankton growth until nutrient depletion prevents further accumulation. Low zooplankton population numbers due to lack of winter feeding and slow population growth rates that lag behind the phytoplankton results in a spring bloom. However, in the North Pacific, a permanent halocline (salinity gradient) at 100 m prevents deep mixing of waters during the winter. Some phytoplankton growth occurs throughout the winter because of higher light in this top 100 m, allowing a population of zooplankton to be maintained. These zooplankton have growth rates similar to those of phytoplankton, hence their population is closely coupled to that of the phytoplankton. During spring, increasing solar radiation moves the mixed layer to 30–50 m and primary production increases, but phytoplankton biomass remains low because of grazing by the zooplankton. The dichotomy between the North Pacific and North Atlantic is due to responses of phytoplankton to different vertical structures generated from differences in the halocline.

In addition to these physical factors, biotic and historical factors such as predation, disturbance (fires, storms), competition, diseases, continental drift and chance also play a large part in the distribution of species within climatic regimes. The effects of humans, mediated through land use changes, disturbance, hunting, introduction of exotic species and habitat destruction are becoming more and more a dominant force in species distribution and biodiversity on local, regional and global scales. Subsequent chapters will outline the physiological bases of species' response to environment, the nature of interactions among species and the impact of humans on global climate change and biodiversity.

4.13 Chapter summary

Climate is a major determinant of the physical environment of an organism. Climate determines the availability of water and the degree of heat. Similar climates throughout the world often have similar life forms. These associations are due to the convergent evolution of life forms in response to selective pressures exerted by their environment. Although life forms may be similar in like climates, widely separated areas have unique histories that limit the degree of taxonomic similarity.

On a global scale, the most important determinant of climate is solar radiation. Latitudinal differences in radiation result in temperature differences that drive global air mass circulation patterns. Air mass movements coupled with the Earth's tilt on its axis and rotation around the sun lead to the development of distinct seasons at mid to high latitudes and the prevailing wind and ocean currents. Regionally, the presence of large bodies of water and mountains can dramatically affect climate.

Climate can have major effects on the distribution and abundance of plant and animal species. However, it is more common that a variety of interacting factors, including climate, substrate, geochemistry and biotic interactions combine to determine a species' distribution. A greater understanding of climate–organism interactions will help ecologists to predict the consequences of future, human-caused global changes. This is especially important given the large effects the growing human population is having on the world's ecosystems.

Recommended reading

Coma, R., Ribes, M., Gili, J.-M. & Zabala, M. (2000) Seasonality in coastal benthic ecosystems. *Trends in Ecology and Evolution* **15**: 448–453.

Hayward, T.L. (1997) Pacific ocean climate change: atmospheric forcing, ocean circulation and ecosystem response. *Trends in Ecology and Evolution* **12**: 150–153.

Johnston, K.M. & Schmitz, O.J. (1997) Wildlife and climate change: assessing the sensitivity of selected species to simulated doubling of atmospheric CO_2. *Global Change Biology* **3**: 531–544.

Oechel, W.C. & Vourlitis, G.L. (1994) The effects of climate change on land–atmosphere feedbacks in arctic tundra regions. *Trends in Ecology and Evolution* **9**: 324–329.

Thebaud, C. & Johnston, A. (1997) Plant responses to global changes in CO_2: unfinished business? *Trends in Ecology and Evolution* **12**: 425–426.

Chapter 5

Interactions between individuals and the physical environment

5.1 The microenvironment that organisms encounter

Chapter 4 focused on global and regional climate patterns and the relationship between climate and the distribution of life forms. The abiotic environment of an organism, although broadly determined by climate, also consists of local factors including the substrate on which it lives (such as sand, rock or mud), the velocity of wind or water passing over it (known as the flow regime), geochemistry (pH, oxygen concentration, nutrient supply) and electromagnetic radiation. The abiotic environment immediately surrounding an organism exerts strong selection on its morphological and physiological traits. To distinguish between factors that occur on a regional scale, such as climate, and factors that occur on a scale immediately surrounding an organism, ecologists use the terms **macroenvironment** and **microenvironment**. The macroenvironment includes climate, soil and biological communities on a scale of several metres to thousands of kilometres and is most often associated with the biomes described in Chapter 2. The microenvironment is partially determined by the macroenvironment, but is defined as the environment immediately surrounding an organism.

The term 'microenvironment' is somewhat of a misnomer, because it occurs on a scale similar to the size of the organism. Thus, the microenvironment of an elephant would encompass a much larger area than the microenvironment of an ant. A rule of thumb to help define microenvironment is that it is the part of the physical environment immediately surrounding an organism, and which could potentially be modified by the organism itself. To illustrate this, imagine the interior of a honeybee hive on a cool day. The temperature may be several degrees higher than the outside air, owing to the metabolic heat from the flight muscles of the bees. Alternatively, on a hot, sunny day, the interior of a beehive may be several degrees cooler, due to the bees waving their wings over evaporating water droplets brought to the hive by workers. The wind speed inside the hive would be lower than outside the hive, and the humidity inside may be much higher than outside the hive. The bees, through their activities, would be altering their immediate microenvironment.

There can be large differences between macro- and microenvironments. For example, in Arctic regions, some fly and bee species will be active during periods colder than might be expected. Their ability to remain active during cold periods is due to a behavioural response—they bask inside flowers shaped like shallow cups. Some of these flowers actually follow the sun's arc in the sky during the day, focusing the sun's rays on the centre of the flower. This raises the flower's temperature several degrees above the surrounding air temperature, resulting in faster development of the reproductive parts of the flower, as well as providing a warmer microenvironment for the insects.

The potentially large differences between macro- and microenvironment of an organism have led to another common term used by ecologists—microhabitat. The microhabitat of an organism is the localized area that that organism exploits for resources to support its growth and reproduction. For example, Section 4.9 described how the orange-crowned warbler and the Virginia's warbler, despite occurring together in the high-altitude forests in Arizona, differed in the microhabitats in which they chose to build their nests. The microhabitat forms part of the niche of a species (see Section 1.6). The definition of an organism's microhabitat might vary according to which aspect of the individual's ecology or behaviour is being studied. For the warblers, the microhabitat used for gathering food would not be the same as that for building nests.

5.2 Physiological ecology

Ecologists that focus their studies on the interaction between organisms and the abiotic microenvironment are called physiological ecologists. Physiology is the study of the function of organismal processes (such as kidney function or photosynthesis). Physiological ecologists are interested in how the environment interacts with the metabolic and structural functions of organisms. Examples of the sort of investigation that they carry out might include a study of which plants can grow in an environment polluted by industrial waste products, or how animals on the seashore cope with being inundated by the tide twice a day as well as remaining dry for long periods when the tide is out. Many other disciplines may be involved because of the need to quantify the environment and organismal response. Expertise in physics, meteorology, physiology, biochemistry and geology, as well as a good grasp of other aspects of ecology, may all play a part in investigations of physiological ecology.

Why are ecologists so concerned with the interaction between organisms and their physical environment? There are basic ranges of temperature, moisture, oxygen, nutrients, salinity, pH and other physical factors that are known to support life. These ranges are constantly being extended as scientists find microorganisms that can tolerate high salinity, extreme temperatures and high pressures, such as those around volcanic vents and hot springs. Nevertheless, the range of physical factors exploited by the majority of plant and animal species is relatively narrow. Any one particular species will be found in an even narrower subset of these ranges. The abiotic conditions tolerated by an organism will define its fundamental niche (see Section 1.6.1). It is important to examine the response of the individuals in a population to their physical environment in order to determine the fundamental niche of the population. Ecologists can then determine the relative roles of abiotic and biotic factors and their interactions in determining a species' realized niche, which is the part of its fundamental niche that it actually exploits (see Section 1.6.1). One of the things that makes ecology a fascinating subject is the challenge of defining the various limits and interactions a species experiences in nature. The rest of this chapter will focus on the response of organisms to the abiotic environment. The types of biotic interactions that help to determine the realized niche of a species (including competition, predation or parasitism) will be the focus of Chapters 6–9.

Some important abiotic factors and their primary biological effects are outlined in Table 5.1. Many of these factors act **synergistically**, that is, the effect of a combination of factors may be more than the sum of their individual effects. For example, a plant that can tolerate air temperatures of 40°C when shaded may suffer severe damage to its photosynthetic machinery at 35°C under high light levels. If soils are dry, reducing the ability of the plant to cool its leaves through transpiration, then photosynthesis may be damaged at even lower air temperatures.

5.3 Electromagnetic radiation

Microenvironments in terrestrial systems and shallow marine and freshwater systems will be dominated by solar radiation that includes ultraviolet, visible and near-infrared wavebands (Fig. 5.1). They will also receive other sources of electromagnetic radiation, including long-wave **infrared** or terrestrial radiation (Fig. 5.1). Infrared radiation received by an organism will heat it (think of heat lamps), while infrared radiation given off by an organism will reduce its temperature.

Electromagnetic radiation in the environment stems from the fact that any matter above the temperature at which all molecular motion stops, 0 kelvin (–273°C), will emit radiation. The total amount and wavelength of radiation emitted will depend on the temperature of the matter. Higher temperatures result in greater amounts of radiation emitted. (For example, the average temperature of the sun's surface is 6000 K). Additionally, the higher the temperature, the shorter the wavelength of the radiation emitted. The colour of a very hot fire is yellow or blue (shorter wavelengths), while cool fires burn red (longer wavelengths). Infrared radiation is a major avenue for heat exchange for plants, animals and the Earth's surface. In fact, the Earth would be much warmer than it is presently if infrared radiation was not radiated out into space (see Section 4.7).

Different wavelengths of electromagnetic radiation will have different effects on biological systems (Fig. 5.1). In order to understand the microenvironment of an organism, scientists measure both the solar radiation and infrared radiation that the organism experiences. These

Table 5.1 Physical factors in the environment and their primary biological effects.

Physical variable	Biological effects
Temperature	**Damage** (freezing, overheating, exceeding physiological tolerances) **Metabolism** (rates of growth, development, respiration, photosynthesis) **Environmental cue** (cold-hardiness, thermal acclimation) **Water loss** (rates of evaporation and transpiration)
Wind	**Damage** (storms, wind-throw, pruning) **Water loss** (rates of evaporation and transpiration) **Temperature** **Humidity** **Dispersal** (seeds and small animals) **Mixing** (turnover of lakes and ocean layers)
Solar radiation	**Temperature** **Metabolism** (rates of photosynthesis for both terrestrial and aquatic autotrophs) **Water loss** (rates of evaporation and transpiration) **Environmental cue** (photoperiod, red/far-red ratio) **Damage** (skin cancer, cataracts, photosynthetic inhibition, coral bleaching)
Infrared radiation	**Heat exchange** (warming and cooling, night-time temperatures, frost, dew formation)
Salinity	**Uptake and release of water** (osmoregulation in plants and animals) **Ion imbalance** **Toxicity** **Nutrient availability**
Substrate	**Water availability** (flooding, drought) **Germination sites** **Egg-laying** **Nutrient availability** **Anchoring** (roots, barnacles) **Burrowing activity** **pH**
Humidity	**Water loss** (rates of evaporation and transpiration) **Pathogens** (growth of bacteria and fungi)
Precipitation	**Water availability** (for growth of both plants and animals) **Drought stress** **Flooding**
Oxygen	**Metabolism** (marine and freshwater organisms) **Toxicity** (anaerobic organisms)
pH	**Availability of metal ions** and **nutrients** **Toxicity** **Membrane integrity**

radiation inputs are combined with other aspects of the microenvironment such as air temperature, humidity and wind speed to build an 'energy budget' for organisms (Box 5.1). This energy budget will account for all of the sources of radiant energy an organism is exposed to, as well as the radiant energy leaving an organism. The overall energy budget will determine the temperature of an organism. Excess energy coming into the organism will cause its temperature to rise, while excess energy going out of an organism will cause its temperature to decrease.

5.4 Organisms can alter their radiant environment

Many organisms can manipulate their radiant energy budgets by changing either the amount of direct solar

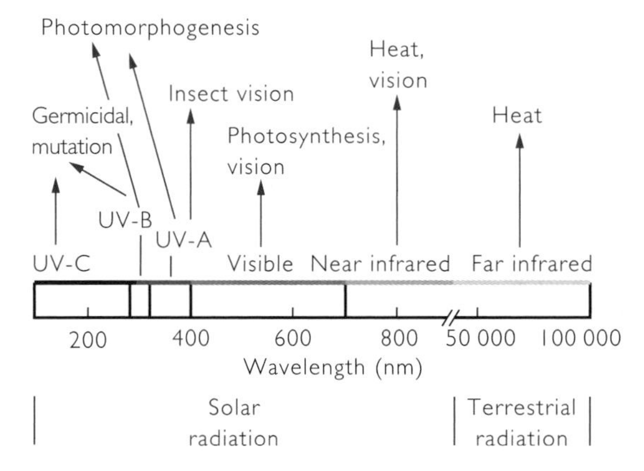

Fig. 5.1 The electromagnetic spectrum, and some of the effects of specific wavebands upon biological systems.

radiation striking their surfaces, or the amount of incident radiation that they absorb. Some striking examples are seen in hot, dry deserts, where many plants have white leaves (see Plate 5.1, facing p. 24). This is due to the growth of reflective hairs on the leaf surface that increase the amount of sunlight reflected from leaves, thereby reducing leaf temperature. Plants may also orientate their leaves to a more vertical angle from the horizontal. This increases the angle between the sun's direct rays and the leaf during hot, midday periods (similar to the latitudinal effect described in Section 4.2).

Smith *et al.* (1998b) examined the association between leaf properties and the amount of sun exposure and water availability in five different plant communities in Western Australia. They examined only shrubs with evergreen leaves, because these leaves would be exposed to environmental influences year round. The communities they examined ranged from a moist forest to a dry, open community with isolated shrubs. They found that the proportions of species in a community with leaf angles greater than 45° from the horizontal increased as precipitation decreased and sun exposure increased (Fig. 5.2). This change in leaf angle lessens light interception during hot, midday periods (Fig. 5.3).

Leaves on plants in these communities also became smaller, thicker and cylindrical as the environment

Box 5.1 Radiant energy balance for plants and animals

The balance between incoming and outgoing energy determines the temperature of an organism, which in turn affects metabolism, activity and growth. In the case of plants that use solar energy to drive photosynthesis, documenting radiation inputs also provides information on light energy used for photosynthesis. For animals, the intake of food energy and the use of this energy for metabolism, respiration, growth and reproduction could be measured to form a complete energy budget.

The incoming solar and infrared radiation, along with other forms of energy exchange for a plant is illustrated in Fig. 1. The leaf intercepts direct solar radiation, solar radiation scattered by dust and water vapour in the atmosphere (diffuse solar radiation), and solar radiation reflected off clouds and surrounding objects. In addition, the leaf receives infrared radiation from the sky, clouds and surrounding objects. Not all of this incoming radiation will be absorbed by the leaf, some will be reflected off its surface and some will be transmitted through the leaf. Some of the incoming solar radiation that is absorbed will be used to drive photosynthesis, the rest of the solar and infrared radiation absorbed by the leaf will be converted to heat.

The leaf also exchanges heat with its environment through radiating infrared radiation itself, **convective** heat exchange, **conductive** heat exchange and evaporation of water. Convective heat exchange is heat exchange between the leaf and the surrounding air. Think of how your skin is cooled by a breeze on a hot day—that is convective heat exchange. The smaller the leaf and the higher the wind speed, the greater the rates of convective heat exchange.

Conduction is heat exchange through direct contact between molecules, and is usually unimportant for a leaf suspended in air. It can be important for things like leaves pressed to the ground, tree trunks conducting heat to internal tissues, trunks of cacti, or animals in contact with other surfaces. Conduction is the process that cools your bare feet when you step onto a cool cement or tile surface, or burns your feet when you step onto hot sand or asphalt.

Evaporative heat exchange occurs because the change in the state of water from a liquid to gas requires energy, hence plants and animals can cool off by evaporating water from their surfaces. Humans and other animals do this through panting or sweating from sweat glands in the skin. Plants achieve this through stomata, which are pores in the leaf epidermis that allow water to escape from the interior of the leaf.

Figure 2 shows a picture of energy exchange for an animal, analogous to that in Fig. 1.

(cont'd on p. 80)

Box 5.1 (cont'd)

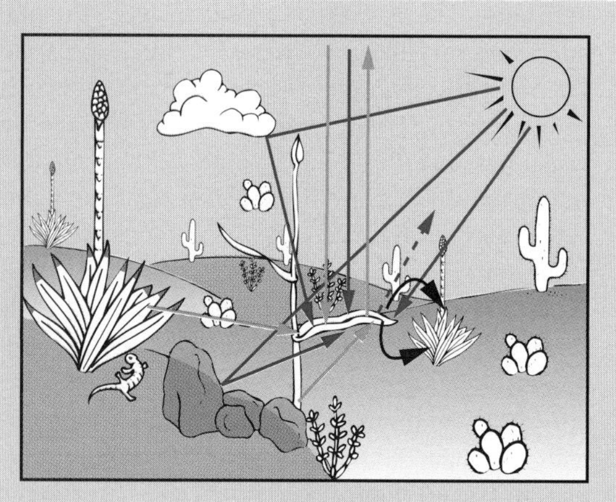

Fig. 1 Energy balance for a leaf. The leaf receives solar radiation, indicated by dark purple arrows, through direct radiation from the sun, diffuse radiation from the sky and reflected radiation from clouds and surrounding objects. The leaf receives long-wave radiation, indicated by light purple arrows, from the sky, clouds and surrounding objects. The leaf also emits long-wave radiation, exchanges heat energy with the surrounding air (convection, indicated by black arrows) and exchanges heat energy through evaporation of water through stomatal pores in the epidermis (transpiration, indicated by the dashed purple arrow).

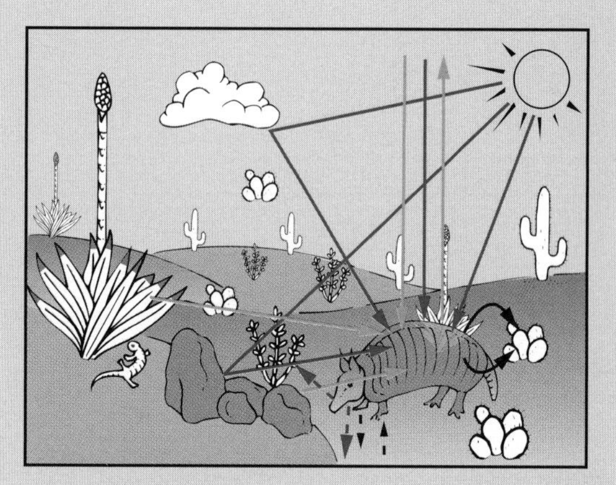

Fig. 2 Energy balance for an animal. The armadillo receives solar radiation, indicated by dark purple arrows, through direct radiation from the sun, diffuse radiation from the sky and reflected radiation from clouds and surrounding objects. The animal also receives long-wave radiation, indicated by light purple arrows, from the sky, clouds and surrounding objects. The animal emits long-wave radiation, exchanges heat energy with the surrounding air (convection, indicated by black arrows), exchanges heat energy through water evaporation (panting, indicated by the dashed purple arrow) and exchanges heat energy through contact with surfaces (conduction, indicated by dashed black arrows).

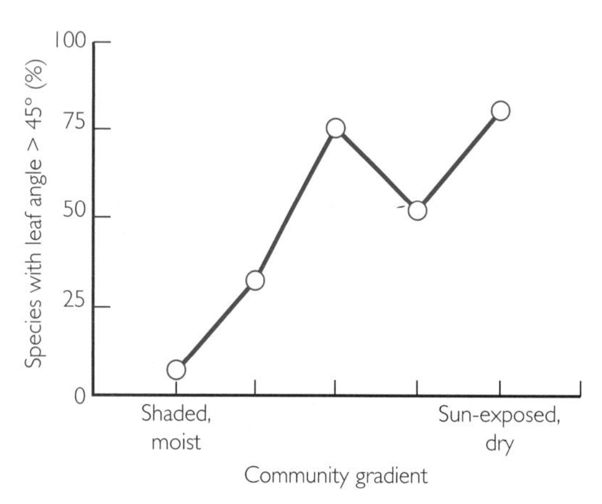

Fig. 5.2 The percentage of species having leaves orientated at an angle greater than 45° from the horizontal in five different communities in Western Australia. (Data from Smith *et al.* 1998b.)

became drier and sunnier. Cylindrical leaves reduce direct light interception, because the light strikes the curved leaf surface at an angle. A smaller leaf in high-light, dry habitats also increases convective heat exchange (Box 5.1). With low water availability, stomata close and reduce the amount of transpirational water loss. This in turn reduces evaporative cooling. This makes convective heat exchange an important means of reducing the chance of high temperatures damaging the leaf. There have been numerous studies showing how leaf size, shape, colour and angle all change along similar environmental gradients. In fact, paleoecologists (who study the ecology of fossil organisms) use these relationships to provide clues about past climates by looking at the shape and size of fossil plant leaves.

5.5 Thermoregulation

Body temperature is important to most aspects of ani-

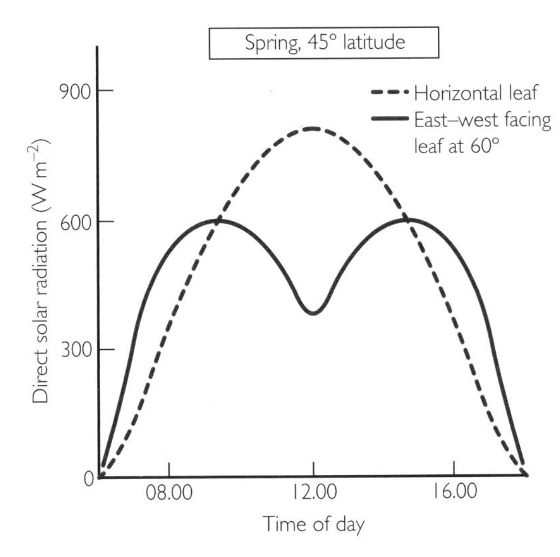

Fig. 5.3 Patterns of sunlight falling on leaves of different orientation throughout the day. A horizontal leaf gets its highest radiation at midday, while a leaf orientated 60° from the horizontal gets its highest radiation levels during morning and afternoon hours.

mal ecology and behaviour. Animals that generate much heat internally and have good thermal insulation are called **endotherms**; examples include mammals and birds. Other animals, such as fish and insects, that are poorly insulated and produce little internal heat are known as **ectotherms**. In everyday speech, these terms are often replaced by the phrases 'warm-blooded' and 'cold-blooded'. Many insect species manipulate their exposure to solar radiation in order to achieve favourable body temperatures. The inornate ringlet butterfly, *Ceononympha inornata*, a small butterfly of the northeastern United States, mixes two different basking postures with flight activity to maintain favourable body temperatures over a wide range of ambient temperatures. Heinrich (1986) found that when ambient temperatures were below 24°C, the butterfly spends most of its time in a lateral basking posture. This posture consists of the butterfly folding its wings dorsally and tilting to one side until the longitudinal axis of its body is orientated perpendicularly to direct solar radiation. Within 2 min on a sunny day, body temperatures can rise more than 20°C above air temperature in this position. When ambient temperatures are above 28°C, the butterfly modifies its basking position, with its head pointed away from the sun, wings closed dorsally, and wings orientated with their edges towards the sun. This posture can reduce body temperatures by 10°C within 1 min.

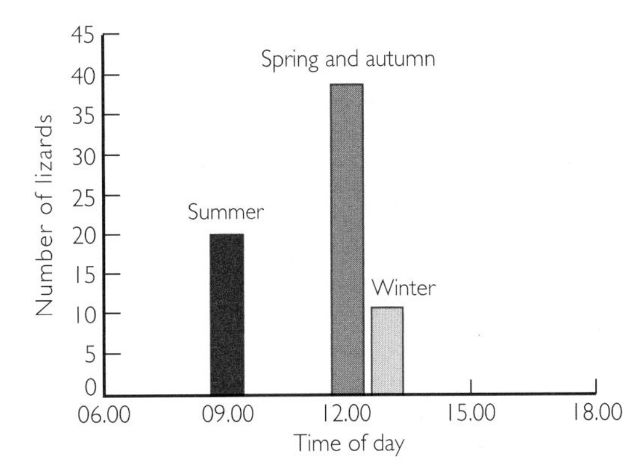

Fig. 5.4 Number of active lizards seen as a function of season and time of day. The greatest numbers of lizards were active at midday during the spring and autumn. Fewer lizards were active in the summer, with greatest daily activity during morning hours. The lowest number of lizards were active in winter, with activity restricted to midday periods. The study looked at 12 lizard species in the Kalahari Desert of Africa. (Data from Huey *et al.* 1977.)

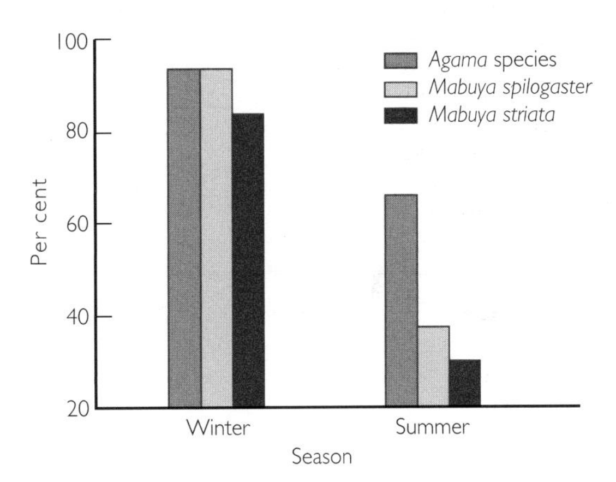

Fig. 5.5 Percentage of lizards perched in full sun when first observed. In winter, over 80% of first sightings were in full sun. In summer, fewer lizards were seen in the sun. This study was done on three semi-arboreal species from the Kalahari Desert in Africa. (Data from Huey *et al.* 1977.)

The behaviour of the inornate ringlet butterfly is an example of **thermoregulation**. Reptiles, especially lizards, have received a great deal of attention from ecologists because they have distinctive thermoregulatory mechanisms, including restriction of activity times (Fig. 5.4), selection of microhabitats (Fig. 5.5) and alteration of body posture to change heat exchange or sunlight exposure. Thermoregulation allows lizards to avoid extreme body temperatures and gain more control over metabolic processes.

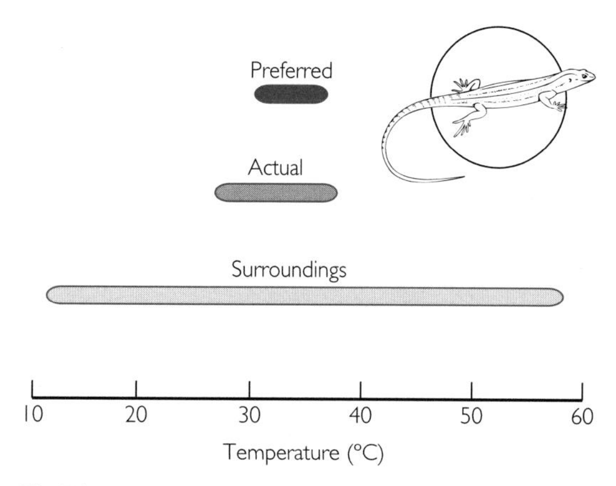

Fig. 5.6 The preferred body temperature of *Podarcis* lizards in comparison to actual body temperature and the range of temperatures in the immediate surroundings in early autumn. (Data from Bauwens *et al.* 1996.)

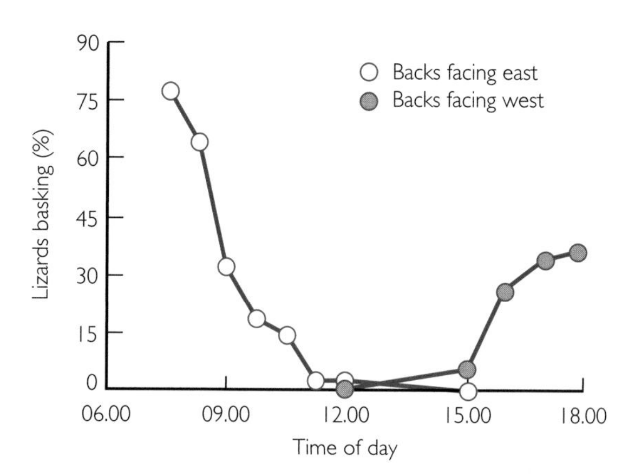

Fig. 5.7 In the morning, lizards bask with their backs towards the east, maximizing sun exposure. At midday, they avoid the sun. In the afternoon, they orientate their backs towards the west. (Data from Bauwens *et al.* 1996.)

5.5.1 Active versus passive thermoregulation

How do ecologists distinguish between body temperatures resulting from active thermoregulation and those due to other causes? For example, random movements among different favourable sites may result in a low range in daily body temperatures. Thermoregulation may also result from animals following thermoregulating prey into different temperature zones. A complete understanding of thermoregulation requires knowledge of: (i) temperatures that animals prefer or would select if given a choice; (ii) distributions of available microhabitat temperatures; (iii) behaviours that alter rates of heat exchange; and (iv) actual body temperatures experienced by animals.

A small lizard, *Podarcis hispanica atrata*, that occurs on the Columbretes Archipelago, small islets in the Mediterranean off the coast of Spain, demonstrates many of the aspects of active thermoregulation (Bauwens *et al.* 1996). *Podarcis* lives on the ground and actively searches for insect prey in surrounding shrubs and rocks. In early autumn, lizards maintain remarkably constant body temperatures throughout the day (Fig 5.6). In addition, these body temperatures are much closer to the species' preferred temperature range than a random sample of surrounding temperatures (Fig. 5.6). In order to achieve this narrow range of body temperatures, *Podarcis* employs different behaviours at different times of the day. In the cool morning hours, lizards bask in sun patches, with their backs facing east to maximize sun exposure (Fig. 5.7). In the afternoon, lizards bask in sunny patches, this time with their backs facing west. During warm midday periods, lizards shuttle between sun and shade patches without stopping to bask.

Animals are also capable of creating their own microenvironments that aid thermoregulation. For example, North American eastern tent caterpillars, *Malacosoma americanum*, have behavioural traits that allow them to regulate body temperature. Development time in this caterpillar is strongly related to temperature. Caterpillars raised at 15°C in the laboratory were unable to complete development past the larval stage (Knapp & Casey 1986). However, average springtime air temperatures over much of the tent caterpillar's range are below this critical temperature. The larval stage of insects often serves the primary function of gathering enough energy (food) to complete the pupal and adult stages and reproduce. Temperature is positively related to feeding rate and growth rate and negatively related to development time of these larval stages. Small, solitary caterpillars are generally unable to elevate their body temperatures much above ambient by metabolic means. Caterpillars that are capable of maintaining metabolically favourable temperatures may gain significant benefits in terms of acquiring energy and shortening the time it takes for them to complete their developmental stages.

Tent caterpillar larvae hatch from large egg masses in late April and together begin construction of a silk tent. They then remain in large aggregations either on

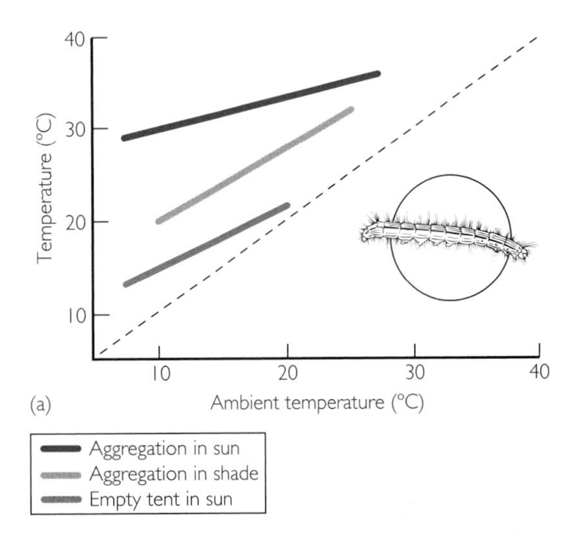

Fig. 5.8 (a) Maximal temperatures measured in the field for aggregations of eastern tent caterpillars, *Malacosoma americanum*. Thermoregulation was most pronounced in sunlit tents, as indicated by the largest deviation from ambient temperatures and the smallest seasonal range of temperatures. Sunlit, empty tents were slightly warmer than ambient, but aggregation by caterpillars and placement within the tent were the primary means of thermoregulation. (Data from Knapp & Casey 1986.) (b) The silk tent of the eastern tent caterpillar, along with a close up of the holes used by larvae to move in and out of the tents.

the surface of this silk tent, or at various depths within the tent, depending upon the time of day and season. In early season periods, the caterpillars aggregate either on the surface or immediately below the top surface of the tent. Here, wind is reduced, preventing cooling by convection (Box 5.1), but high sunlight levels allow elevated body temperatures. As the season progresses, and air temperatures rise, the tent becomes larger through the addition of multiple silk layers, usually on the sunny sides of the tent. In hot midday periods, caterpillars move deep within the tent, and are shaded by several silk layers. Thus, the tent construction, aggregations of many individuals, and movement of the caterpillars within the tent combine to make this species an effective thermoregulator (Fig. 5.8). Elevated body temperatures allowed growth rates in the field to be three times that expected, given average air temperatures during growth (Knapp & Casey 1986).

5.5.2 Thermoregulation, food and water

Animals eat a variety of foods, including plants, bacteria, fungi and animals. Plant food is especially difficult to digest for most animals due to its high fibre content (lignin, cellulose and other plant cell wall materials) and low nitrogen content. Digestion is often positively affected by increased temperature. Digestive efficiency and passage time through the digestive system can be affected by temperature. Ectothermic herbivores may choose higher temperature microenvironments in order to increase digestive efficiency. Alternatively, even if digestive efficiency was not increased, more rapid passage of food through the animal's body may allow it to eat more food and take in more nutrients and energy in a shorter time. This may have benefits in reducing the amount of time feeding animals are exposed to predators, or may lead to faster growth and reproductive rates.

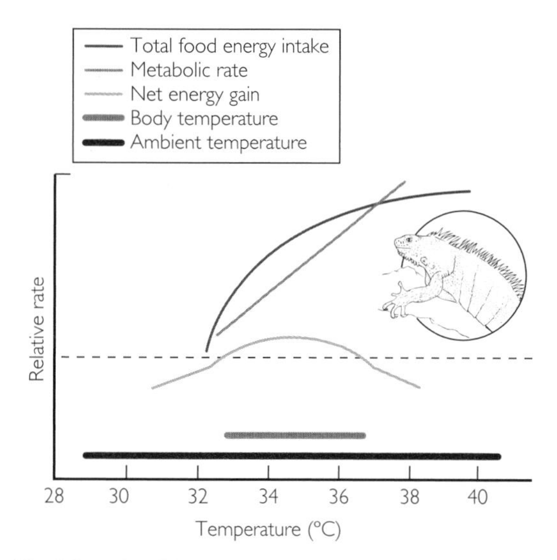

Fig. 5.9 Trade-offs between body temperature, metabolic rate and food intake in the green iguana. (Data from van Marken Lichtenbelt *et al.* 1997.)

Thermoregulatory behaviour by the green iguana, *Iguana iguana*, in Curacao (in the Netherlands Antilles) is thought to be influenced by its plant diet (van Marken Lichtenbelt *et al.* 1997). Active thermoregulation by the green iguana maintains midday body temperatures within a narrow range of temperature compared to surrounding temperatures (Fig. 5.9). Transit time for plant food in the green iguana's digestive system decreases as temperature increases. As transit time decreases, lizards increase their rate of intake of fresh food. The combination of decreased transit time and increased food intake results in an increase in total energy uptake as body temperature increases. However, the rate of metabolism also increases with increasing temperature. The point at which increased metabolism balances increased energy intake is about 36.5°C. This temperature is very close to the maximal body temperature for green iguanas observed in the field.

Water balance is an important part of the ecology of terrestrial insects. Water gain can occur through the uptake of liquid water, absorption of water vapour or production of metabolic water. Water loss occurs through spiracles (pores in the exoskeleton connected to the internal respiratory system) or through the exoskeleton, made up of chitin, a substance that restricts water loss. Despite the presence of an exoskeleton, insects are vulnerable to dehydration when temperatures are high and water availability is low. Because water is limiting in deserts, and temperatures often high, the manner in which desert insects obtain their water is of primary importance to their success.

The Namib Desert, on the southwestern coast of Africa, is unusual because of its cool temperatures and regular fogs. A cool ocean current (the Benguela) flows just offshore from the African coast in this area. The cold water reduces evaporation and rainfall, but it increases the incidence of fog clouds that travel inland. There are extensive sand dunes along the coast that contain a distinct insect fauna, including over 20 species of beetle. Despite low temperatures, many beetles in the sand dunes come out at night. They are sluggish at these temperatures and their slow movement makes them vulnerable to predators. Why are these beetles active at such thermally unfavourable times?

The answer lies in the beetle's dependence upon fog as a major source of water. Seely (1979) described three methods used by the beetles for the uptake of water. They can drink fog water droplets that condense on living or dead vegetation and stone surfaces. They can display a behaviour called fog basking, where they assume a head-down stance on or near a dune crest. Fog condenses on the beetle's dorsum and water droplets coalesce and glide down to the mouth. Finally, the beetles can take up water directly from fog-moistened sand. To do this, beetles construct shallow trenches 2–4 mm in depth. Beetles then drink the water that condenses on the ridges of these trenches. Thus, the beetles of the Namib Desert sand dunes not only display activity during risky periods of low temperature, but they also show unique behaviour patterns designed to exploit the limited water supplies of their habitat.

5.5.3 Costs of thermoregulation

Despite the benefits of maintaining a smaller range of favourable body temperatures, there are costs to thermoregulation. These costs include energy for locomotion, movements that make animals conspicuous to predators and ideal temperature sites that may be unsuitable for acquiring food, water or meeting mates. Because of these trade-offs, body temperatures in nature may not match the optimal physiological body temperature. For example, tree-dwelling lizards (*Anolis cristatellus*) in Puerto Rico that occur in open parks with scattered trees behave differently to lizards in a nearby forested habitat. These lizards bask in sunny sites in order to raise their body temperatures, and move to shady sites when air temperatures are too high. The forest has fewer open sites than the park. The average distance a lizard in the forest has to travel to a sun patch exceeds 7.9 m, while park lizards have to move only 1.2 m. This difference leads to lizards in the park shuttling between shaded and sunlit patches more than forest lizards. This enables the park lizards to regulate body temperatures within 1.6°C during the day, while forest lizards experience temperature variations of 4.9°C. Apparently, the higher cost of thermoregulation in the forest forces lizards to tolerate larger variations in body temperature.

Many small animals may suspend thermoregulatory behaviour and employ daily torpor in response to extreme temperatures (high or low) and seasonal food shortages. This can be an important mechanism to minimize energetic costs in unfavourable environments. In the cool, dry season in the deciduous forests of western Madagascar, food and water availability decrease along with the temperature. The grey mouse lemur (*Microcebus murinus*) normally feeds on fruits, small animals, gum and insect secretions—foods that are limited in the cool, dry season. Temperatures in this season can range anywhere from a midday high of 32°C down to night-time lows of 4°C. With limited food and large temperature ranges, the mouse lemur is faced with a high energetic cost for thermoregulation. In response, the mouse lemur undergoes bouts of daily torpor in the dry season, lasting anywhere from 3.6 h to 17.6 h (Schmid 2000). Body temperature during torpor is significantly cooler than for active animals, averaging slightly over 17°C compared to active body temperatures of 37°C. The lowest body temperature recorded by Schmid was 7.8°C. Arousal from torpor involved a passive warming stage, exploiting the rise in daytime temperatures, followed by a second stage with endogenous heat production that raised body temperature to normal. Metabolic rate during torpor was reduced by 76% compared to a resting metabolic rate, resulting in average daily energy savings of 38%. These energy savings are important to individual fitness. Daily savings reduce the time devoted to foraging, during which lemurs are exposed to predation. Since reproduction is controlled by daylength to start at the end of the dry season, energy savings over the entire dry season will contribute significantly to the breeding condition of the animals. Timing of reproduction at the end of the dry season and beginning of the wet season enables the offspring to exploit the higher food availability as the wet season progresses and to have as long a period to grow as possible before the onset of the next dry season.

5.5.4 Thermoregulation in endothermic animals

Although endothermic animals are more buffered from their physical environment than ectotherms, there remains a strong connection between the surrounding microenvironment and an endotherm's activity and behaviour. Activity during hot periods of the day or season may lead to overheating or excessive water loss, while activity during cold periods may cost too much energy to support the metabolic activity needed to maintain high body temperatures. Activity in thermally favourable microenvironments must also be balanced with predator avoidance, foraging for food and social interactions, just as it is for ectotherms.

The semi-arid, Mediterranean environment of northern and central Chile (see Section 2.7) is characterized by hot, dry summers. The plant community consists of scattered evergreen shrubs, small herbs and grasses. The degu (*Octodon degus*), a small, herbivorous rodent native to this area, is active during the day, lowering body temperature by panting and evaporating water from the tongue. However, these rodents are easily overheated if exposed to prolonged periods of sunlight and temperatures over 32°C (Torres-Contreras & Bozinovic 1997). This temperature is routinely exceeded in open microenvironments during spring, summer and autumn. When given a choice of plants to eat, degus prefer tender grasses and forbs with low fibre content. However, when the surrounding temperatures are too high, they will settle for low-quality plant food, as long as that food is in a cooler, shaded microenvironment. Thus, on hot days, degus restrict their activity to cooler areas under shrubs, or in burrows and rock piles, despite lower quality plants in these areas.

The small mass and high surface area to volume ratios of small, endothermic animals make them particularly sensitive to air temperature, solar radiation and wind.

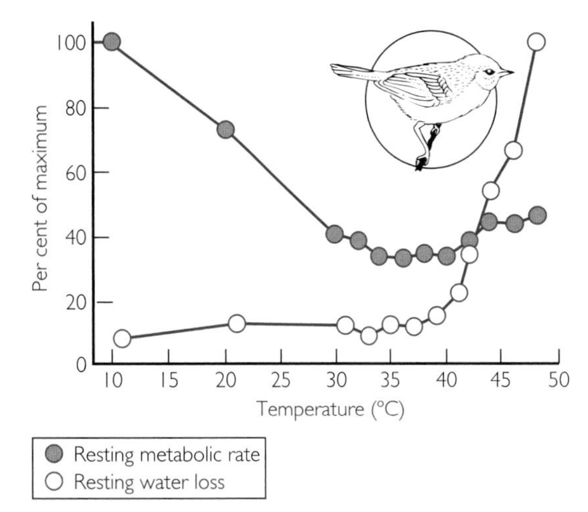

Fig. 5.10 The effects of temperature on resting metabolic rate and evaporative water loss from the desert songbird *Auriparus flaviceps*. The resting metabolic rate increases rapidly as temperatures drop under 28°C, but is relatively constant between 28°C and 48°C. Evaporative water losses increase rapidly at temperatures exceeding 38°C. (Data from Wolf & Walsberg 1996.)

Small birds are good examples of this. Maintenance of high body temperatures in a small bird requires high rates of energy expenditure. Basal metabolic and thermoregulatory costs in small birds may account for 40–60% of their total daily energy expenditure. In addition, their small mass means that they are not buffered against environmental change, and they have a limited capacity to store water. By selecting microenvironments that minimize thermoregulatory costs, birds may reduce their energy requirements and rates of water loss.

The microhabitat selection and activity patterns of the verdin (*Auriparus flaviceps*), a small, insectivorous songbird that resides in the deserts of the southwestern United States and northern Mexico, varies with season and environmental conditions (Wolf & Walsberg 1996). The surrounding temperature, solar radiation and wind all have large effects on metabolic rates and water loss of the verdin. Resting metabolic rates increase rapidly as temperatures drop below 28°C (Fig. 5.10).

In the cool temperatures of winter, verdins must spend 75–95% of the day searching for food to meet their energy needs. Short periods of bad weather can therefore have serious effects on the bird's ability to survive. Energy-saving strategies become of major importance, and natural selection may favour those individuals that select favourable microenvironments. Verdins can reduce their metabolic rate by up to 50% by shifting from shaded, windy sites to sunny, protected sites in the winter. At low wind speeds, changing a perch from the shade to one in the sun may be equivalent to raising the surrounding air temperature by as much as 12°C.

The summer season is characterized by high temperatures and low water availability. In response, verdins reduce their level of activity and seek shade. When the air temperature is below 35°C, verdins spend an equal time in shaded and sunlit sites, foraging about 75% of the time. Above 40°C, rates of foraging decline to between 9% and 21% of the time, and birds spend 95% of their time in the shade. There is an almost exponential increase in water loss from evaporation as temperatures increase above 40°C (Fig. 5.10). In the shade at 40°C, a verdin may lose 1.7% of its body mass per hour through evaporation. Shifting to a sunny site would increase water loss to 7% of its body mass per hour. A verdin can only withstand a water loss of 11% of its body mass, a value that would be reached in a little over 1.5 h in the sun. During the summer, selection of a protected, shaded site can reduce water loss by over four times compared with sunny sites. Because of the smaller size of juveniles compared with adults, high summer temperatures and water loss represent severe challenges to juvenile survival in the verdin.

5.6 Non-thermal components of microenvironments

This chapter has focused on the role of radiation and temperature in microenvironments. Nevertheless, as Section 5.1 outlined, an organism's microenvironment comprises many different parameters, including substrate, flow regime and geochemistry. Gray and Spies (1997) examined forest floor factors that affected seedling growth of trees in the Pacific Northwest of the United States. Despite a cool, wet winter, the climate of the Pacific Northwest includes warm, dry summers. Young tree seedlings often die when faced with summer water stress and high temperatures in large forest canopy gaps. Gray and Spies found that the type of forest floor on which a seedling germinated had a large effect on seedling success. For example, seedling establishment for Pacific silver fir (*Abies amabilis*) and western hemlock (*Tsuga heterophylla*) was greater on decayed wood than on bare forest floor. Decaying logs

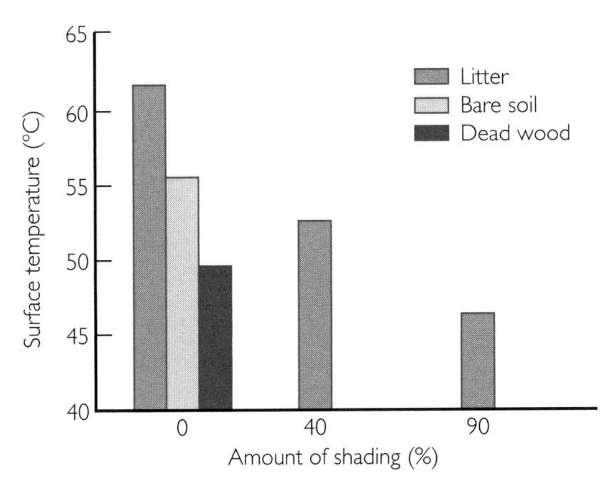

Fig. 5.11 The effect of forest floor type and amount of shading on midday surface temperatures in a northwestern United States coniferous forest. Litter surfaces are hottest, bare soils are several degrees cooler, and wood surfaces are coolest. As shading increases within the gap, surface temperatures can decrease by over 15°C.

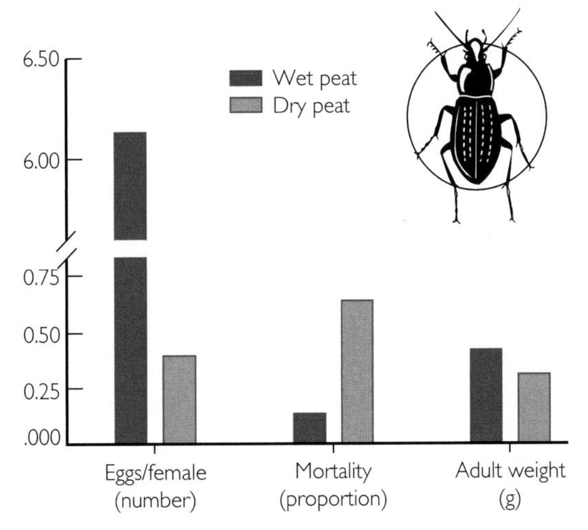

Fig. 5.12 Number of eggs per female laid by ground beetles in wet peat compared to dry peat when adult females were allowed to choose substrates. The mortality of larvae and pupae was significantly lower on wet peat substrate, and adults from this substrate were larger.

reduced competition from understorey plants and provided cooler substrates within large canopy gaps. The presence of shade (for example, from a neighbouring plant) within a canopy gap combine with substrate to affect seedling survival. Shading and substrate were both strong determinants of surface temperatures (Fig. 5.11).

Ground beetle larvae are highly vulnerable to abiotic conditions. Carabid ground beetles prefer high humidity and wet soils, and are often found in wetlands such as bogs. However, carabid beetle larvae have limited mobility, which restricts them to sites close to where adult females laid their eggs. Thus, habitat choice by females is a major factor in the success of larvae. Huk and Kuhne (1999) gave adult beetles a choice of wet peat, wet sandy peat, dry peat and dry sandy peat. Prior to mating, there was little substrate preference shown by adults. After mating, females preferentially moved to wet peat substrates to lay their eggs, a behaviour not shown by males. This choice of substrate had large effects on the survival of larvae and pupae, although egg survival did not differ among substrates (Fig. 5.12). Adults that came from eggs laid in wet peat were also larger than adults from other substrates.

5.6.1 A case study of caddisfly microhabitats in streams

Physical factors dominate the microenvironments of streams and rivers. Flow rates will affect resource availability, water mixing, oxygenation, sediment load and deposition, dispersal of adults and offspring, and the type and amount of substrate. Anoxic conditions and intermittent flow, along with isolation in independent pools are common in many temperate-zone streams and rivers, creating large longitudinal variation in habitat type and availability. It is not surprising then, that aquatic organisms have evolved numerous adaptations to these conditions. A wide array of these adaptations is seen within the Trichoptera order of arthropods. There are about 7000 described species of trichopterans, or caddisflies, which occur throughout the world in freshwater lakes and streams. Caddisflies are important to the ecology of many freshwater habitats because they occur in large numbers and are food for many fish and waterbirds. Additionally, they fill many trophic roles themselves, feeding on algae, stream litter, bacteria and other small animals.

Caddisflies are closely related to butterflies, and the adults resemble small, dull-coloured moths. With only a few exceptions, the aquatic larvae of caddisflies resemble caterpillars (Fig. 5.13). Caddisfly eggs are usually deposited in gelatinous matrices in or near water, and the larvae go through several instars before pupating. Female caddisflies can enter the water either by walking or by diving, and cement their eggs to stones or submerged plants. Females can stay under water for

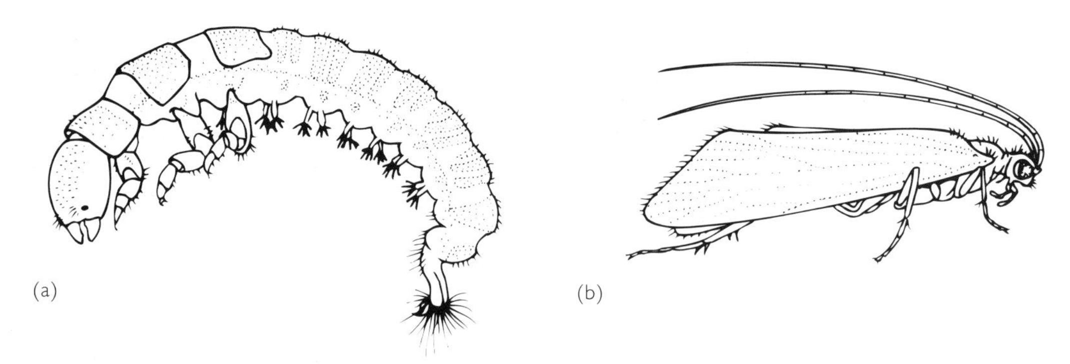

Fig. 5.13 (a) Larval and (b) adult forms of caddisflies. (From Gullan & Cranston 2000.)

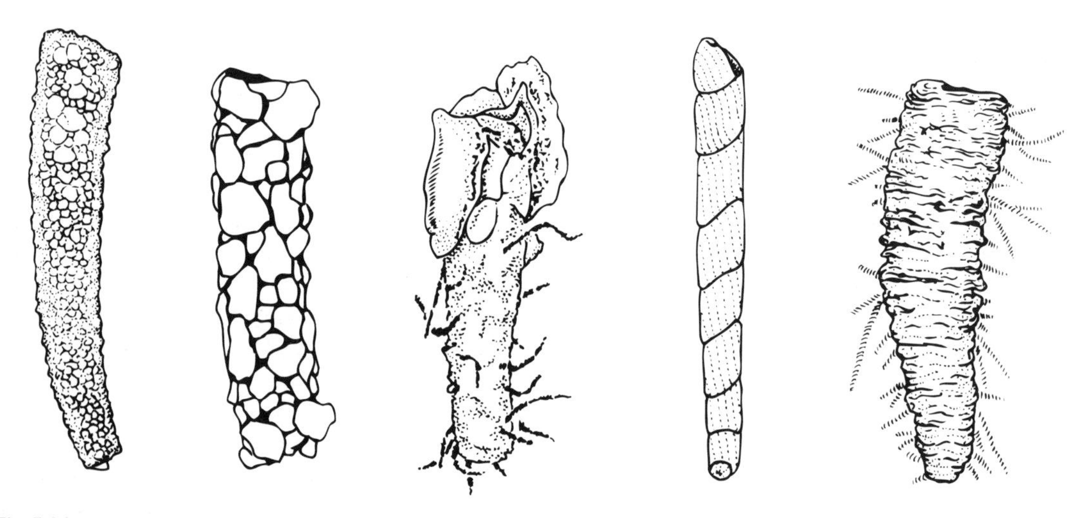

Fig. 5.14 Examples of caddisfly cases. (From Wiggins 1977.)

periods exceeding 30 min because small hairs on their body hold a film of air around the body that acts as an oxygen reservoir. The spiracles (breathing holes) are never in contact with water, so the insect is actually breathing air while under water. Adults are primarily nocturnal, short-lived and sometimes do not feed at all, serving a reproductive function only. Therefore, larvae form an important food-gathering and energy storage stage of the life cycle. Caddisfly larvae demonstrate a wide array of adaptations to flow rates, substrate, geochemistry and predators found in their aquatic habitat. Caddisfly species show a pronounced microenvironment segregation and specificity. Some are adapted to live on the tops or sides of rocks in high flow regions; others specialize in the undersurfaces of rocks in lower flow regimes (Harding 1997). A difference of only a few millimetres can change flow dramatically, allowing high species numbers to be supported within the same stream.

Two of the most striking adaptations seen in caddisfly larvae are case building and the ability to secrete silk. Silk is used to bind larval cases together and to form nets used to capture food. Caddisfly cases come in a variety of shapes and sizes, many of which are unique to individual species (Fig. 5.14). Some cases are open at one end only (purse cases), and others are open at both ends (saddle cases). Larvae must build a new case each time they moult.

Cases built by caddisfly larvae can be composed of plant material, gravel, sand or all three. They are usually constructed from surrounding materials, thus the case resembles its surroundings and provides camouflage from predators. Some caddisfly species have evolved the ability to build mobile cases that they carry with

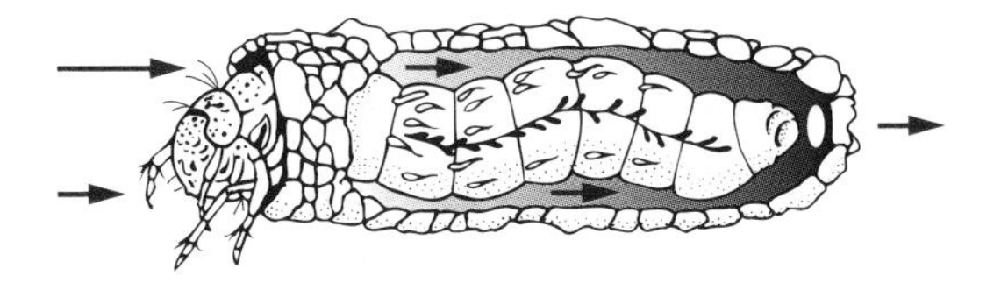

Fig. 5.15 Flow of water through caddisfly cases due to abdominal flexing. (From Wiggins 1977.)

them. Because they are constructed from surrounding materials, these mobile cases have the advantage of allowing individuals to occupy feeding patches and habitats (e.g. patches of rotting leaves) while disguising the larvae as part of the substrate. Thus, caddisflies are able to utilize optimal microenvironments that other aquatic insects cannot because of predation pressure.

Cases also aid in supplying oxygen to caddisfly larvae. By flexing their abdomen inside the case, larvae maintain a current of moving water carrying fresh oxygen past their gills (Fig. 5.15). Larvae in cases can survive in areas of lower oxygen concentration compared with larvae without cases. Feldmeth (1970) found that caddisflies increased the rate of case ventilation at lower current speeds, providing them with greater amounts of oxygen.

There are three families of silk net-building caddisflies, and they differ in net construction and appearance (Fig. 5.16). Hydropsychid larvae construct nets in fast-flowing water, usually on the upper surface of rocks. They use these nets not only to capture food, but also to hide from predators. Philopotamid larvae build nets in areas protected from strong currents, fastening them to the underside of rocks. These nets stretch when the current flows through them, and act as a sieve filtering food as water passes through. Polycentropid larvae build trumpet-shaped nets, usually in low current areas. Caddisflies may adjust the size of their silk nets in response to flow (smaller nets in higher flows) and food availability (larger nets in lower resource areas), allowing individual species to exploit a wider range of microenvironments (Kondratieff *et al.* 1997).

5.7 Human alteration of microenvironments

Human activities such as land clearing, road building and logging can change local microenvironments, and have a large impact on an organism's distribution, growth and reproduction. A case in point involves riparian zones, areas located next to streams or rivers. Riparian zones provide habitats for organisms that are adapted to high water availability, even in dry desert environments. The stream conditions themselves are dependent on the surrounding environment. Shading by overhanging vegetation can have a large influence on stream temperature. Water temperature affects the stream biotic community and water quality. Dissolved oxygen decreases with increased temperature. Photosynthesis, insect migration, soil microbial activity and fish all have specific thermal ranges. Vegetation around the stream also serves to protect water quality, reducing the amount of nutrients, sediments and pollutants that reach sensitive streams and rivers. Amphibians rely on the cooler temperature, higher humidity and reduced wind of near-stream environments to prevent dehydration and control respiration (Box 5.2). Small mammals and birds are also dependent on the riparian microenvironment.

However, riparian zones are popular areas for human uses such as recreation, housing and logging. Thus, these areas are often destroyed or disturbed by human activities. Brosofske *et al.* (1997) looked at the

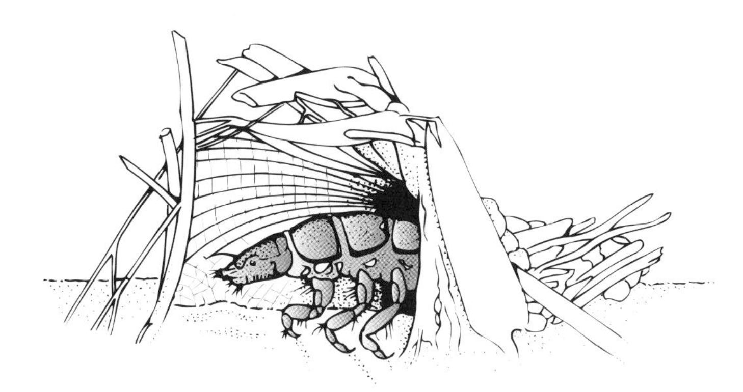

Fig. 5.16 A caddisfly hydropsychid larva in its retreat; the silk net is used to catch food. (From Gullan & Cranston 2000; after Wiggins 1978.)

Box 5.2 The amphibian decline phenomenon

The potentially dramatic effects of environmental change on populations of organisms can be illustrated by examining the rapid declines in populations of many frogs, toads and salamanders. Over the last 15 years, large declines in the range and number of more than 200 species of amphibians have been documented. About 20 species are thought to have gone extinct over this period. Of most concern to ecologists is that these declines and extinctions are not restricted to disturbed areas; many of them have occurred in national parks and protected lands in the United States, Costa Rica, Puerto Rico, Honduras, Panama, Venezuela and Australia. Most of the declines have been seen in higher altitude temperate and tropical zones, and species that have aquatic phases seem to be more susceptible than species with strictly terrestrial life histories.

Amphibians are one of the longest lived groups of animals, having been on Earth for 350 million years. They are extremely adaptable, occurring in tropical forests, savannas, temperate forests, prairies, deserts, lowlands and mountains. In some wetland and forest ecosystems, amphibians can make up the dominant vertebrate group in terms of biomass or numbers. They have the most diverse life history within the terrestrial vertebrates, many undergoing an aquatic larval (tadpole) stage that metamorphoses into an adult terrestrial phase. Yet, the current rate of extinction and decline appears to be higher than at any other period of their history.

Approximately 4800 species of amphibian have been described. They are divided into three main groups, the frogs (including toads), salamanders (including newts) and the caecilians (worm-like tropical animals). Amphibians are dependent upon water due to their permeable skin. Many amphibians use their skin as the major means of transport for water and air, with one group of salamanders lacking lungs entirely. Amphibians are ectothermic, so they respond strongly to surrounding temperatures and microenvironment. Many amphibians display thermoregulatory behaviours similar to those described in this chapter. However, their propensity to lose water rapidly in exposed conditions makes them seek shaded, moist microhabitats to an even greater extent than most organisms.

A novel pathogen has been targeted as the cause of many declining frog populations. This pathogen, a chytrid fungus (*Batrachochytrium dendrobatidis*) had not been known previously to affect vertebrates. This fungus is normally a saprobe, feeding on dead plants and animals. It is capable of decomposing animal chitin, keratin and plant cellulose. Its infectious spores are dispersed in water. The fungus blocks gas and water exchange in infected animals and may secrete a toxin that causes neurological dysfunction.

The question is why this fungus has recently changed in nature from an occasional pathogen affecting weakened frogs to a deadly infectious pathogen of apparently healthy frogs. There are several hypotheses, including mutation of the fungus to a more virulent form, and viral transformation of the fungus that introduced more lethal traits. Other hypotheses focus on changes in the environment and frog vigour. Environmental factors may be increasing the density of fungal populations or enhancing its infectiousness and toxicity. Alternatively, frogs may be more susceptible to the fungus due to a suppression of their immune system or increased stress. However, if the immune system was suppressed, frogs should have been more vulnerable to any number of different pathogens in addition to this specific fungus. Scientists are therefore examining other factors that may have led to local and regional changes in frog resistance to this fungal pathogen. These include climate change, pollutants, airborne agrochemicals and increased levels of UV-B radiation. It is likely that the interaction of several of these environmental factors act to stress frogs to the point that they are more susceptible to infection. These factors, along with other potential causes for the worldwide decline in amphibian populations, are listed in Table 1.

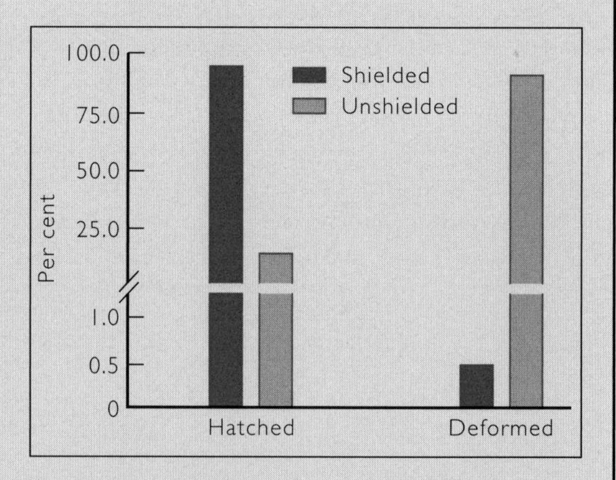

Fig. 1 Exposure of the eggs of the long-toed salamander (*Ambystoma macrodactylum*) to UV-B radiation affects both hatching success and the number of deformities displayed by hatched salamanders. (Data from Blaustein *et al.* 1994, 1997.)

(cont'd)

Box 5.2 (cont'd)

Table 1 Causative factors in worldwide declines of amphibians.

Habitat loss	**Wetlands** have experienced major declines in the last 200 years. The continental United States alone lost over 54% of its original wetlands between 1780 and 1980. Small wetlands and ephemeral (temporary) wetlands used for breeding by many species are particularly vulnerable to road grading, mosquito control and development. The common forestry practice of **tree and deadfall removal** from forests alters microenvironments that many amphibians rely on by increasing ground temperature, reducing moisture and eliminating refuge sites. **Deforestation** of tropical rainforests is destroying this most diverse of amphibian habitats. The **loss of large bison herds** in North America, and the wallows that they created, reduced breeding sites and habitat for prairie amphibians. Likewise, **reductions in beaver populations** reduced the formation of ponds and wetlands
Habitat fragmentation	Many amphibians travel long distances to and from breeding sites. Mortality of both breeding adults and dispersing juveniles is greatly increased due to automotive traffic on **roads**. **Drains and culverts** in urban environments trap amphibians, subjecting them to drowning, starvation or movement into inappropriate habitats. Roadside curbs can act as dispersal barriers. **Patches of agricultural or suburban land** containing pollutants, herbicides or pesticides can provide barriers to amphibian dispersal
Introduced species	Amphibians are sensitive to **introductions** of gamefish, baitfish, fish used to control mosquitoes, crayfish and bullfrogs because these species feed upon amphibian eggs and larvae. This is especially true in aquatic systems that were historically without fish, because native amphibians may have no defensive mechanisms against predators. Introduced fish may also carry **pathogens** that infect amphibians
Climate change	Scientists believe that the golden toad (*Bufo periglenes*) of the Monteverde Cloud Forest Reserve in Costa Rica disappeared because of a shift to **warmer, drier conditions**. Steady **increases in winter and spring temperatures** over the last two to three decades in the United Kingdom are thought to have led to earlier breeding times in newt and frog species. It has also changed the order of breeding, with newts that used to breed after frogs, now breeding before frogs. This has exposed the frog eggs and tadpoles to increased predation by the newts. **Prolonged droughts** and/or severe **floods** can have disastrous effects on local amphibian populations. Habitat fragmentation and loss combine with local disasters to reduce the re-establishment of locally extinct populations, increasing the risk of species extinction
Ultraviolet radiation	Amphibious species with low levels of DNA repair enzymes are particularly susceptible to UV-B **damage and mortality**. Habitats such as clear lakes, at higher elevations, and at higher latitudes appear to be more prone to UV-B damage (Fig. 1). UV-B radiation can **interact with other environmental factors** such as pH and pathogens, leading to further declines in amphibian populations. Amphibians may be especially susceptible to UV-B radiation, since many lay their eggs in exposed, open, shallow water in the spring before deciduous trees have expanded their leaves. This time of the year is when UV-B levels are highest due to stratospheric ozone destruction. Additionally, amphibians often have clear eggs with little pigmentation to block UV-B radiation, and some have low levels of photolyase, which is the principal enzyme for repairing damage to DNA caused by UV-B radiation
Contaminants	**Pesticides, herbicides, heavy metals, petroleum products**, and other pollutants that commonly contaminate aquatic systems can all damage, deform or kill amphibians. Many **aquatic insects** that serve as vital food for amphibians are susceptible to pesticides and herbicides. Low pH, UV-B radiation and pathogens can **interact synergistically** with contaminants to increase amphibian sensitivity to one or more of these factors
Acidic precipitation and soil	Acidic pH below about 4–5 can be **lethal** to amphibians, especially tadpoles and embryos. Low pH may also increase mortality in the **invertebrate food resources** of amphibians. UV-B radiation, pathogens, contaminants and low pH **interact synergistically** to increase amphibian susceptibility
Disease	Bacterial, fungal, viral and algal pathogens are often the **ultimate cause** of amphibian mortality. However, the **increased incidence and spread** of these diseases implies an increased susceptibility of amphibians to disease. The **stress factors** causing this increased susceptibility are often unknown. Introduced fish, **human transport** of aquatic debris and organisms, and livestock use of wetlands all increase the spread of pathogens
Trade	The use of amphibians as **food**, as a source of **medicines** and **aphrodisiacs**, as **pets** and for biological **research** puts pressure on wild populations throughout the world

microclimate of five different streams of 2–4 m width in western Washington, United States. The study was conducted in the foothills of the western slope of the Cascade Mountains, in a forest dominated by Douglas fir (*Pseudotsuga menziesii*) and western hemlock (*Tsuga heterophylla*). Microclimate before and after logging around the streams was measured by placing weather stations at different distances from the stream within and outside the remaining vegetation. After logging, surface temperatures were 4–5°C higher in the remaining vegetation zones around the streams. Immediately outside the vegetation, daytime surface temperatures were 10–12°C higher than pre-logging values. Solar radiation levels at the surface and wind speeds also increased after harvesting. In this area, rainfall percolates through the soil until it eventually flows into the streams. Thus, soil and surface temperatures are highly correlated with stream water temperatures. Stream water temperatures rose to much higher levels after logging.

Thermal refugia may protect biotic communities from some of these human-derived thermal disturbances. Fish are capable of responding to thermal heterogeneity in their environment, although many require specific ranges of temperature to survive and reproduce. Several species of salmon and trout thermoregulate behaviourally by moving to cooler areas, such as seeps, confluences with cold streams, or deep, shaded pools. Torgersen *et al.* (1999) investigated salmon habitat preference in two forks of the John Day River in Oregon, United States. The middle fork of the John Day had a greater degree of human habitation and was warmer by about 3°C than the north fork. Salmon in the middle fork preferentially aggregated in deep pools and in areas of river that were cooler than average (Fig. 5.17). Fish in the north fork also aggregated in pools, but did not show as marked a distribution in cooler than average temperatures.

In these northwestern North American streams, salmon show a two-stage habitat selection. In spring, when salmon migrate from oceans to streams, they choose areas chemically similar to where they were spawned. But, they also choose areas of appropriate substrate. They must spawn in a gravel substrate with enough porosity to allow movement of oxygenated water through it, since their eggs need high oxygen concentrations to support embryo metabolism. In summer, after migrating, salmon do not feed and must conserve energy. By living in pools with slow currents, and

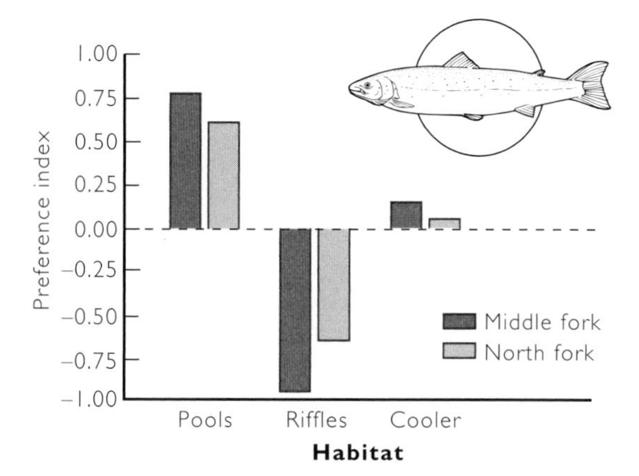

Fig. 5.17 Habitat selection in two forks of the John Day River, Oregon. Salmon in the warmer middle fork preferentially aggregated in deep pools and cooler than average stretches of the river. Salmon avoided riffle habitats in both forks, but had a stronger avoidance response in the middle fork. (Data from Torgersen *et al.* 1999.)

in cooler areas, they can reduce total energy needs. For example, a reduction of 2.5°C in temperature lowers their basal metabolic rate by 12–20%. Therefore, within chemically suitable areas with the appropriate substrate, salmon aggregate in cooler thermal refuges. Future and current land-use changes are reducing the extent of these refuges and raising average stream temperatures beyond tolerance levels, leading to large reductions in native salmon stocks.

5.8 Chapter summary

The physical environment (sunlight, wind, temperature, humidity, salinity, oxygenation and substrate) immediately surrounding an organism is its microenvironment. The microenvironment will have a major impact on an organism's temperature, metabolism, growth and reproduction. Organisms have a variety of means of altering the immediate energy exchange environment around them to create conditions that are more favourable. For example, leaf traits such as angle from the horizontal, colour, thickness and surface curvature change along environmental gradients of sunlight and water availability. Mobile animals behaviourally thermoregulate by restricting activity patterns, choosing different microsites and altering body posture. Often, these thermoregulatory behaviours are modified by other ecological factors such as food availability, predators, competitors or social interactions.

Caddisflies provide an ideal system in which to examine the relationship between an organism and its microenvironment. Caddisfly larvae of different species are found on different substrates (rock, leaf litter, mud, sand or gravel), in different flow regimes (upper surfaces of rocks, lower surfaces of rocks, pools, riffles), and in different geochemical regimes (oxygen concentration, pH). They are capable of modifying their microenvironment through the construction of cases that aid in respiration and predator avoidance. They further alter their microenvironment by building silk nets for food collection and protection. These nets will affect water flow immediately around and downstream of them, further altering microenvironments on a small scale. All of these traits have combined to make caddisflies one of the most successful and diverse of insect orders inhabiting aquatic habitats.

Human activities are altering microenvironments for an increasing number of species, especially those in riparian, forest and/or wetland habitats (Box 5.2). Microenvironmental refuges provide buffers for species against these changes, but these refuges become fewer and more limited in extent as land-use changes increase. A more complete understanding of the interrelationships between organisms and their microenvironment will allow ecologists to understand mechanisms behind organismal distribution and phenomena such as worldwide declines in amphibian populations. Increased understanding will provide a basis from which to predict and manage further changes in biodiversity caused by human activity.

Recommended reading

Chen, J., Saunders, S.C., Crow, T.R. *et al.* (1999) Microclimate in forest ecosystem and landscape ecology. *BioScience* **49**: 288–297.

Huey, R.B. (1974) Behavioral thermoregulation in lizards: importance of associated costs. *Science* **184**: 1001–1003.

Murcia, C. (1995) Edge effects in fragmented forests: implications for conservation. *Trends in Ecology and Evolution* **10**: 58–62.

Packard, G.C. & Packard, M.J. (2001) The overwintering strategy of hatchling painted turtles, or how to survive in the cold without freezing. *BioScience* **51**: 199–208.

Parsons, P.A. (1990) The metabolic cost of multiple environmental stresses: implications for climatic change and conservation. *Trends in Ecology and Evolution* **5**: 315–317.

Pechmann, J.H.K., Scott, D.E., Semlitsch, R.D. *et al.* (1991) Declining amphibian populations: the problems of separating human impacts from natural fluctuations. *Science* **253**: 892–895.

Chapter 6

Introducing biotic interactions and population models

6.1 Trying to simplify the real world

The previous chapters investigated the interactions between organisms and their physical, or abiotic, environments. Building on that basis, Chapters 6–10 will begin to examine interactions with the biotic, or living, environment. These interactions fall under the headings of predation, herbivory, competition, mutualism and parasitism; each of these will be discussed in turn. However, before proceeding to an understanding of these different kinds of interaction, it is necessary to acknowledge that ecology tends to be a very complex subject and that any population of organisms is likely to interact with many other populations in different ways.

Let us suppose that ecologists are interested in explaining the abundances of species in a small garden in the northern temperate zone of Europe or North America. They want to know why some kinds of animals and plants are less common than others at this very local scale. There will be hundreds of species of insects, as well as many other invertebrates, such as spiders, snails, woodlice and worms. There may be a few different sorts of mammal (although they will be rarely observed because they are probably nocturnal) and two or three breeding bird species, as well as many others that occur irregularly or only in winter. So far, only the animals have been considered, with no thought for the plants, fungi, bacteria and so on.

It would simplify things to begin by concentrating on the robins, two pairs of which nest each year in the garden (in Europe the name robin applies to *Erithacus rubecula*, in north America the same name is given to *Turdus migratorius*, but the point is the same, whichever one we imagine we are dealing with). To understand why the robins have an abundance of two pairs in the garden, it is necessary to know something about their food, which consists of many different sorts of invertebrates (say, for simplicity, a dozen different species). It is also desirable to know what chance the adult robins have of being eaten by other animals, so there is a need for some sort of study of the local hawk population. It is also essential to understand the level of predation that the eggs or nestlings will suffer, so an investigation must be carried out into the abundance and behaviour of squirrels and crows. Nestlings can also suffer mortality due to parasitic lice and mites, so in this imaginary example, it would be necessary to study these. There may be also another type of insectivorous bird with which the robins compete for food or nesting sites.

At every stage of this imaginary study, we have deliberately chosen to keep things simple—a decision was made to concentrate only on robins, to pretend that our robins eat nothing but 12 types of invertebrate, and to assume that the adults have only one species of predator. Moreover, all organisms other than animals were ignored. This last limitation means that it has been impossible to consider deadly infections caused by bacteria and protozoa, or the different availability of nest sites in different kinds of shrubs or trees. Even with these unrealistic simplifications, information is required about at least 18 other animal species in order to begin to understand the abundance of the robin population.

One way in which ecologists try to simplify their thoughts so that they can understand why different species have the abundances they do is by developing mathematical descriptions that give a relatively simple way of counting up the number of individuals in a particular population. This is one of the more well-known uses of mathematical models in biology.

A model aeroplane or train is a simplified, smaller version of the original, which shares some essential properties with the real thing but also differs from it in many ways that make it easier to handle. If a visitor from another planet wanted to know what a train was, and did not speak enough of our language to understand a description, we could give the visitor a model train. Although it would be simplified, it would convey a lot of information about trains. In the same way, mathematical models convey some essential information about populations of organisms, without the need for all the precise details.

Many students who dislike mathematics choose to study a biological subject, only to find that modern biology involves what appears to be complicated algebra and numerical analysis. In fact, although the mathematics that ecologists encounter can appear to be difficult, generally it turns out to be conceptually simple. In reality, it is ecology itself that is inherently complex, and the algebra is an attempt to distil out comprehensible, manageable parts of the whole subject in the belief that by understanding individual pieces, ecologists will eventually come closer to explaining the complete story. It is important to remember this during the development of mathematical models of ecological interactions. The equations are not especially interesting in themselves but they might help to explain what happens when a population becomes too large for available resources, when two different species compete for the same resources, or when predators affect a population of prey.

6.2 Models of a single population: how populations grow

Cladosporium is a kind of fungus that lives on the leaves of rye grass (*Secale cereale*). Figure 6.1 shows how the number of individual tiny colonies of the fungus increases through the summer, starting from the beginning of June. Initially the rate of growth of the population is very slow—after 2 weeks there are still fewer than 50 colonies per square centimetre of the rye grass leaves. Then the rate of growth increases very rapidly; in days 14–28, the number of colonies per square centimetre grows from around 50 to over 10 000. Following this period of rapid population growth, the rate of increase slows down. After about 45 days, it appears to stop altogether, when the population has reached a density of about 50 000 colonies per square centimetre, which appears to be the maximum number that can exist on the leaves.

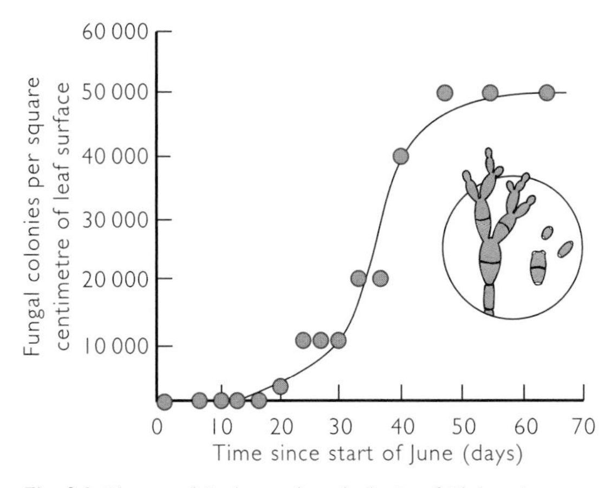

Fig. 6.1 The growth in the number of colonies of *Cladosporium* living on the leaves of rye grass throughout the northern hemisphere summer. (Data from Hudson 1986.)

The curve shows the change in abundance of the species as time progresses and this is one of the commonest types of graph that is encountered during the study of population ecology. As it happens, the shape of the curve looks approximately like a stretched letter 'S', and this kind of curve is known as a sigmoid curve, after the Greek equivalent—the letter sigma. For reasons that will be discussed in later chapters, this shape of graph is commonly a good description of increasing populations of organisms.

The sigmoid curve shown in Fig. 6.1 is concerned with just one species, and it assumes that the population starts out with enough resources for it to expand to some limit. This is more simple than the majority of situations in the real world, and one of the things that ecologists really want to understand is what is happening when populations of one species are interacting with populations of other species. To do this, they start with a mathematical model that captures the essential features of a sigmoid, S-shaped, curve. Later in this chapter and in Chapters 8–10, different models will be developed to describe what happens if a population encounters predators, or prey or parasites. By comparing the two models, ecologists might be in a position to come up with some general principles about how predation or parasitism affect the abundance of populations.

Having established that ecologists do not need to be exceptional mathematicians in order to use models in understanding populations, it is now possible to develop equations that describe the ways in which many populations change in the absence of any significant ecological competition or predation. This might appear to be so unrealistically simplified as to be meaningless but after the models have been worked through, data will be introduced to show how wild, free-living organisms do sometimes follow the predictions of such models. Once this very simple population model has been built, it will be adapted in later chapters to take account of competition, predation or indeed anything we choose.

6.3 Modelling as a tool

6.3.1 Building blocks for models

Mathematical models use letters and symbols to stand in for quantities. The sorts of models that interest ecologists usually start by using the letter N to stand for the number of individuals in a population. Thus, if a model is concerned with the human population of England in the year 2000, N is about 50 million. In mathematical terms, this can be written as:

$$N = 50\ 000\ 000 \quad \text{(eqn 6.1)}$$

Of course, N can change; there have not always been 50 million people in England and 20 years from now, there will probably be many more. Because N changes, there is a need for some method by which mathematical equations can denote this. A common way of doing this sort of thing is to mark the letter N with an extra label. In this case, ecologists could simply use the date, like this:

$$N_{2000} = 50\ 000\ 000 \quad \text{(eqn 6.2)}$$

This distinguishes the value from the number of people living in England in the past, which might be written like this:

$$N_{1900} = 28\ 500\ 000 \quad \text{(eqn 6.3)}$$
$$N_{1850} = 16\ 000\ 000 \quad \text{(eqn 6.4)}$$
$$N_{1800} = 8\ 700\ 000 \quad \text{(eqn 6.5)}$$

The equations indicate that there were 28.5 million people in England in the year 1900, about 16 million in 1850 and 8.7 million in 1800.

In order to make the equation more general, modellers might want to give the date a symbol. The most commonly used symbol for measures of time is the letter 't'. So N_t now means the number of people living in England at time 't' and the value of N depends on the value of t.

In this example, 'real' dates have been used, although these are in fact arbitrarily calculated for the calendar that happens to be commonly used over much of the world. But for people who use different calendars, the dates would not be the same. So, in order to make sure that everyone understands the equations, it is necessary to give some reference point by which to calibrate the values of t. In many biological examples this is easy because nobody cares what the 'real' value of t is. In the case of a model of an ecological population, it is simplest to say that t is zero at some starting point, which might be the first time a species occurs in a particular place or might simply be when ecologists begin counting. For example, in Fig. 6.1, 1 June was called 'day 1' because it happens to be the time of the year when *Cladosporium* populations start to grow.

Let us suppose that we are only interested in the population of England since the time when Queen Victoria came to the throne in 1837. This means that $t = 0$ in the year 1837 and $t = 162$ in the year 2000, so that our new notation is:

$$N_{162} = 50\ 000\ 000 \quad \text{(eqn 6.6)}$$

Mathematical models are built for the convenience of the ecologists, so they can do whatever they choose with them. The model has been set so that t is zero in the year 1837, but any date could have been chosen. American students may not be interested in Queen Victoria (even if they are interested in the population of England) and may choose to start their counting from 1776 when the Declaration of Independence was signed. The American equations will therefore set $t = 0$ in 1776, which means that t will be 224 in the year 2000, and if an ecologist was interested in the population in that year, the equation would be:

$$N_{224} = 50\ 000\ 000 \quad \text{(eqn 6.7)}$$

Of course, all that equations like eqn 6.7 have done is to create a way of writing down how many people lived in England at any particular point in time. It is still necessary to look in a reference book to find out what the answer is. One point of models in population biology is that ecologists can use them to work out what the answer is likely to be when they do not already know, and when there is no reference book to tell them. Thus, someone might want to know how many people there will be in England in 20 years' time, or how many there were in 1958, when there was no census. To do this, it is necessary to develop a model, which will start with census information that already exists, and to look for patterns in the data to invent a method

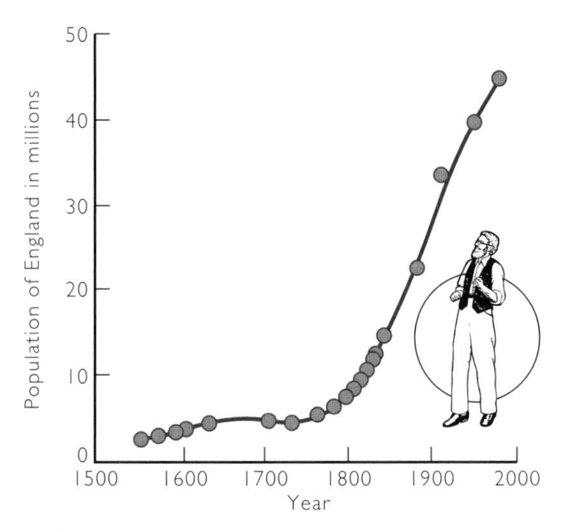

Fig. 6.2 The growth of the human population of England for the past 500 years.

of estimating how many people lived in England at times when no count was or has yet been taken.

One such pattern has already been mentioned. As time has gone by, there has generally been an increase in the population of England, so it is easy to guess that there will be more than 50 million people in 20 years' time. This guess might be wrong, the population does not always go up (it went down during the Black Death outbreak of bubonic plague in the fourteenth century) and things may well change in the future. But most probably it will be correct to assume that the number of English people will rise over the next 20 years. This assumption is, in essence, a model. It is a very simple model, and it is not very precise; 'more than 50 million' could mean 50 000 001 or it could equally well mean 1 billion. Nevertheless, nothing that will be encountered in this chapter will be conceptually much more complicated than this model that has already been developed.

Figure 6.2 shows a graph of the population of England through time. Each point on the graph represents a time in history for which there is sufficient information so that demographers can make an estimate of the population. This means that we know how many English people there were in 1601, 1841, 1911 and so on. But if somebody wanted to know what the population was in 1958, some sort of model would be required to give a good estimate. It is very simple to draw a line in Fig. 6.2 that roughly joins all the points on the graph, and this allows us to read off the estimated population in 1958. In this case, the line was not difficult to draw because there is a very clear pattern in the data. It is not normally so easy when dealing with populations of wild organisms. Luckily, there are statistical techniques that do the job, and most ecologists do not need to understand the methods in detail. In this sort of situation, they would generally use a statistical method known as regression to decide exactly where to put the line, and they can simply trust their computers to perform the regression analysis for them. The only difficulty is likely to be choosing which of several different types of regression to use, and if an ecologist is in doubt, it is always possible to consult an expert statistician.

6.3.2 Density and number

Before proceeding to build models, it is important to be clear about precisely what quantity is being denoted by the letter N. In theory, N stands in for the total number of individuals in a population, as in the example of the English human population, where N was used simply as the number of people in England.

However, ecologists will often actually imply **population density** by N. This is the number of individuals divided by the area occupied, so in the human example, it would be the number of people divided by the land area of England. Clearly, the total population and the average population density are different measures of the same thing, as long as the area of habitat stays the same in each measurement. Over the 500 years shown on Fig. 6.2, England has not changed in size, so its total population size and its population density have always gone up by the same factor. The total population doubled between 1700 and 1810, from 5 million to 10 million. The population density also doubled in the same time period.

Sometimes, it will be clearer to talk about population size and at other times it will be easiest to think in terms of population density. The following discussion may switch between the two different measures but this is merely for convenience, and everything that is said about one is true of the other, because all of the populations that are dealt with will occupy defined habitats that do not change in size. Thus, the population density is always directly proportional to the number of individuals in the population.

6.3.3 Starting to build a simple model

Sections 6.2–6.3.2 have provided all the basic theory that is needed to develop mathematical models of the way that populations change over time and also a way of writing down thoughts in mathematical form. One of the simplest models that ecologists use attempts to count the individuals at some defined point in each breeding season. The most obvious thing to do would be to count the number of individuals either at the start of each breeding season or at the end.

This section will put together a model that describes a population of frogs in a lake and will start with N_t. N_t means the number of frogs in the population 't' breeding seasons after ecologists start counting. So N_8 would be the number of frogs in the lake 8 years after the starting point, assuming that frogs only breed once each year. To begin with, the model will pretend that there is no particular limit to the number of frogs that can live in the lake. This is not really true and later on it will be essential to alter the model to correct for this error, but that is a little way off yet. Making the assumption that the population can grow indefinitely, then the model will be very simple.

To start with, the number of frogs in the lake next year will be based on two things—the number that are already present this year and the number of young that each female can produce. The mathematical model will assume that every female can produce 10 eggs, and for the time being, it will also assume that no frog ever dies or leaves the lake. There are 10 frogs this year, of which five will be females producing 10 eggs each. So they will produce 50 eggs in all, and next year there will be a total of 60 frogs—the original 10 plus the new 50.

When it comes to the following year, 30 of the 60 frogs will be females, and they will lay a total of 300 eggs, leading to 300 new frogs in the third year to add to the 60 that were there in the second year. In other words, each year the population becomes six times as great as it was the year before. A mathematical way of expressing this would be:

$$N_{t+1} = N_t \times 6 \qquad \text{(eqn 6.8)}$$

It does not matter what value of t is put into this equation; if the value of N_t is known, then the value of N_{t+1} can be calculated. If ecologists know how many frogs there are in one year, then they can work out how many there were in any year in the past, or how many there will be in any year in the future. For example, suppose that the frogs were counted this year and there were 60. Now suppose that ecologists want to know how many frogs there will be in 2 years' time. They know that next year there will be 6 × 60 frogs and this is equal to 360. Then in 2 years' time there will be 6 × 360 frogs, a total of 2160.

Note that the model already appears to be making all sorts of assumptions: half of all frogs are female, all females lay 10 eggs, all eggs are successfully fertilized, and so on. In fact, those assumptions could be dispensed with, simply by counting the average number of young successfully produced per female frog each year. It may be that, on average, each female actually lays 11 eggs but that only 10 of them are fertilized. It may be that they each lay 100 eggs, 90 of which are fertilized, and of the fertile ones, only 10 produce tadpoles that eventually turn into froglets. Any such considerations can be ignored—all it is necessary to know is that the average number of froglets produced per female frog each year is 10.

The model also assumes that frogs do not die of old age, although that does not really affect the overall form of the model. Imagine, for example, that frogs only ever breed once and then die immediately. Now, if the population starts with 10 frogs, five of which are females, they will produce 50 young. But next year, the original 10 will be dead and there will just be the new 50 or:

$$N_{t+1} = N_t \times 5 \qquad \text{(eqn 6.9)}$$

All that has happened is an alteration in the value of the constant by which the population size is multiplied each year. In this sort of model, ecologists normally call this multiplier 'R' (standing for **reproductive rate**). Thus, a general equation for the kind of population growth that is seen in the imaginary frog population is:

$$N_{t+1} = N_t \times R \qquad \text{(eqn 6.10)}$$

Figure 6.3 shows how a population would increase if all the assumptions that were used to create eqn 6.10 were true. In this case, the value of R has been set to be equal to 2. In other words, the population doubles during each time period. Notice that the curve in Fig. 6.3 is rather similar in shape to that in Fig. 6.2, although the curve is not smooth—the population increases in discrete jumps. That is because models such as eqn 6.10

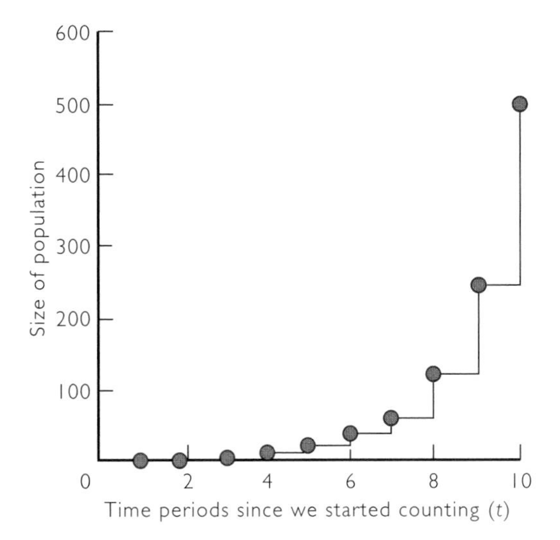

Fig. 6.3 Growth of a population following exactly the pattern described by eqn 6.10, where $R = 2$.

are suitable for populations with discrete generations—frogs only breed once each year in a clearly defined season. In many situations this may be inaccurate and it is often simpler to assume that populations breed continuously throughout the year. In such populations, there is no defined breeding season but mature individuals can breed at any time.

The effect of continuous breeding is that ecologists must think of the size of the population itself changing continuously. In other words, there is no need to wait for the next breeding season for a population to increase—a new individual could be born, or germinate or hatch at any time. When thinking about frogs, it was sensible to use equations that described the population once each year, after each breeding season. With a continuously breeding population, equations are needed that give information about what the population is like every hour, or every minute, or every second, or every millisecond, or at whatever instant ecologists choose to measure the size of the population.

Theoretically, ecologists need equations that describe changes occurring over unimaginably small timescales. Fortunately, there is a mathematical technique for doing this and, luckily, ecologists do not need to understand it fully.

The important factor that ecologists need to know about is dN/dt, which is normally called 'dN by dt'. It means the rate of change in the number of individuals (N) in the population. So when dN/dt is large the population is increasing rapidly and when dN/dt is small the population is increasing slowly. If dN/dt is a negative number, then the population is decreasing, and if dN/dt is zero, then the population is not changing in size.

To help understand competition, predation and other interactions between a population of one species and a population of another, ecologists need a set of equations that tell them about the rate of change of populations. The sort of question that might be asked is: What is the rate of change of a population of a plant species in the presence of a herbivore population and how does this compare with the rate of change when the herbivore is absent? This helps ecologists understand the process of herbivory and its affect on populations of plants.

The starting point is to build a model, like eqn 6.10, that describes a population growing with a constant reproductive rate. As with the model for organisms with discrete breeding seasons, the rate of increase of the population depends on two things, the current size of the population (N) and the reproductive rate. By convention, the reproductive rate in models for continuously breeding populations is normally given as a lower case 'r', but, for most purposes, it can be thought of as being the same as the upper case 'R' in the previous equations. Thus, the model for continuous breeders is:

$$\frac{dN}{dt} = rN \qquad \text{(eqn 6.11)}$$

In other words, the rate at which the population increases depends on how many individuals there already are (N), multiplied by how quickly they reproduce (r). It is like the money in a bank account—the amount of interest that a depositor earns depends on how much money he or she already has and what rate of interest the bank is offering. With a bank account, a customer can withdraw money at any time, and he or she will be paid their interest up to the time when the money is withdrawn. Likewise, with eqn 6.11, it is possible to count the number of individuals in the population at any time; there is no need to wait until some predetermined point, as was the case with eqn 6.10, which only gave information about the size of the population at the end of each breeding season.

6.3.4 Assumptions and limitations

One of the things ecologists must realize about their models is that they make certain assumptions about the things that organisms do. Strictly, the models will only work for organisms where the assumptions turn out to be true, but in practice ecologists can often get away with some deviations from the assumptions. The principal assumption made by the first model (for the frog population) was that the population had clearly defined, discrete bouts of breeding, interspersed with periods in which the young develop before any more breeding takes place. This is true for many types of frog living in temperate parts of the world; there is a spawning period in the spring, followed by a long gap through the autumn and winter before any more breeding. When the new breeding season starts, last year's young may spawn (in fact they may wait until they are 2 or 3 years old), as will those surviving adults that were born in previous years.

Some types of organism do not have discrete generations, and the first model that was developed (eqn 6.10) is not really suitable for describing their populations. A population of bacteria growing in infected food, for example, will not have discrete breeding periods. Each bacterium could reproduce at any time. For these species, ecologists would need to make different assumptions, like those in the second model (eqn 6.11).

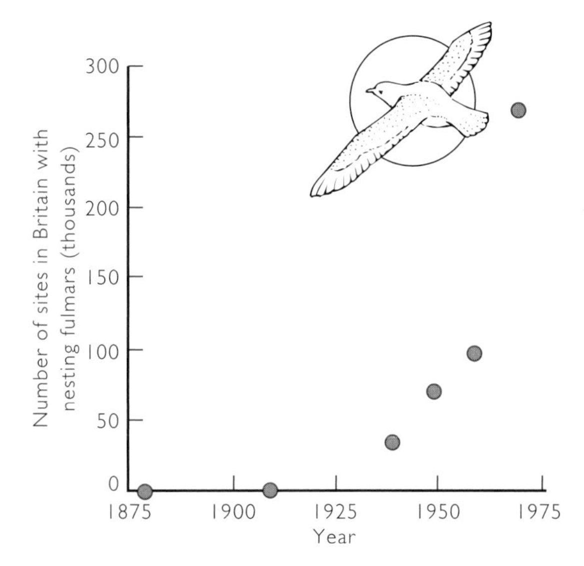

Fig. 6.4 The growth of population of fulmars in Britain. (Data from Cramp & Simmons 1977.)

6.3.5 Making the models more realistic

The main problem with models such as eqn 6.11 is that they describe populations that can continue to increase without limit. Obviously, this situation cannot persist forever, although some populations do grow in this way for a time, as long as there continues to be enough resources for all of the individual organisms. For example, until 1878, the only British nesting place for the fulmar (*Fulmaris glacialis*, a seabird belonging to the petrel family) was St Kilda, a small and relatively remote island off the coast of northern Scotland. In 1879, the fulmar occupied 24 sites elsewhere in Scotland and by the turn of the century, it was nesting in 2000 different places in Britain. Its expansion increased throughout the decades so that by 1970 it was using almost 270 000 sites on suitable rocky cliffs in the British Isles. Nobody really knows why the expansion occurred when it did, although it may have been associated with altered fishing practices (fulmars eat offal discarded by trawlers) or perhaps with climatic change. Whatever the cause, the population grew for a century exactly as the model would have predicted and in this case R was approximately 1.11. In other words, the population grew by about 11% each year (Fig. 6.4).

But resources are not unending and it is clearly important for the equation to be modified to take account of the fact that populations eventually stop growing, just as the population of fungi in Fig. 6.1 stopped growing. In the case of the imaginary frog population, for example, there will simply not be enough space in the lake to go on accommodating more and more frogs. The fulmar population will eventually run out of nest sites, if it does not run out of food first.

More generally, population growth slows down when some resource starts to be in short supply. It does not matter which resource runs out, nor is it important how much of a surplus of other resources exists; as soon as one resource becomes limited, it will limit the population. For instance, a grass plant needs water, sunlight, nitrogen and phosphorus, as well as other trace elements. If the existing population of grass in a field is utilizing all the available nitrogen, then the population cannot grow. It does not matter that the field has been covered with a phosphorous fertilizer and that it rains every morning and is sunny every afternoon. If there is not enough nitrogen, the population cannot increase. The maximum number of individuals of a species that can be accommodated in a particular patch of habitat is known as the **carrying capacity** and is determined principally by the amount of the limiting resource. In most population equations, carrying capacity is normally denoted by the letter 'K'.

For the moment, it is sensible to leave aside the model of populations with discrete generations and move on by incorporating the idea of the carrying capacity into the model of continuously breeding populations. This is because, in developing equations, many people find that these are slightly easier to understand than the equations for populations with discrete generations.

In order to develop the appropriate equations, it is essential to begin with an equation that expresses dN/dt in terms of the current population size, the reproductive rate and the carrying capacity. The equation that ecologists generally use is called the logistic equation as it is usually written as:

$$\frac{dN}{dt} = rN\left(1 - \frac{N}{K}\right) \qquad \text{(eqn 6.12)}$$

Investigating this equation, it is clear that when the population size, N, reaches it maximum carrying capacity, K, then:

$$N = K \qquad \text{(eqn 6.13)}$$

Because N and K are now the same, ecologists can substitute one for the other, as they choose, so that:

$$\frac{N}{K} = \frac{N}{N} \qquad \text{(eqn 6.14)}$$

Any number divided by itself is equal to 1, so:

$$\frac{N}{K} = \frac{N}{N} = 1 \qquad \text{(eqn 6.15)}$$

In other words:

$$\left(1 - \frac{N}{K}\right) = (1 - 1) = 0 \qquad \text{(eqn 6.16)}$$

If this value is put into eqn 6.12, then:

$$\frac{dN}{dt} = r \times N \times 0 \qquad \text{(eqn 6.17)}$$

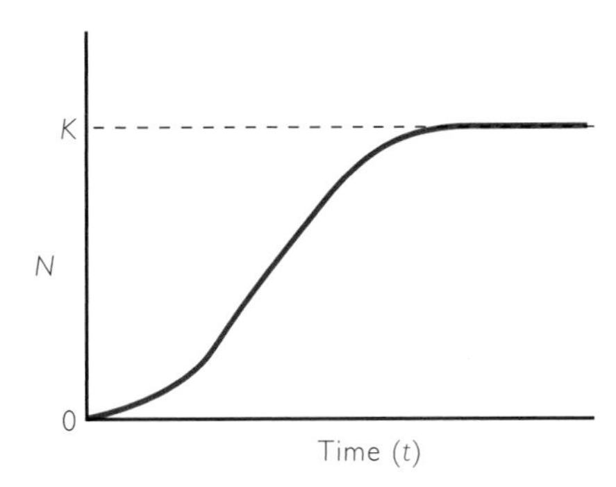

Fig. 6.5 Sigmoidal population growth for a population with continuous breeding. After a period of rapid growth, the rate of increase slows down as the population size approaches the carrying capacity, K.

Any number multiplied by zero is equal to zero, so it no longer matters what the values of r and N are in the equation:

$$\frac{dN}{dt} = r \times N \times 0 = 0 \qquad \text{(eqn 6.18)}$$

Under these conditions the rate of change is zero—the population has stopped growing, as it must do when it reaches the carrying capacity. Note that this is exactly what happened in the population of *Cladosporium* described in Fig. 6.1, which stopped growing once it reached a density of 50 000 fungal colonies per square centimetre.

For the sake of completeness, it would be appropriate to look at how the concept of the carrying capacity is incorporated into eqn 6.10, which describes a population with discrete generations. In fact, all that it is really essential to note is that the model must be altered in such a way that when a population reaches its carrying capacity:

$$N_{t+1} = N_t \qquad \text{(eqn 6.19)}$$

In other words, there are the same number of individuals at time $t + 1$ as there were at time t. The population has stopped growing, as it must do when it reaches its carrying capacity.

Figure 6.5 shows the growth of a population that is following eqn 6.12 exactly. It is almost exactly like the sigmoidal curves in Fig. 6.1. Thus, the very simple equations encapsulate something about real populations, even though they have been developed by imagining a population of frogs or bacteria.

As Section 6.1 noted, mathematical equations are not especially interesting in themselves, but they help ecologists to understand what might be the most important factors that they need to know about. Thus, the most important lesson from the development of eqn 6.12 is that it defines what factors need to be put into the equation to produce something that replicates real, wild populations of organisms. The two values that were most important were the **reproductive rate** and the **carrying capacity**. There are many occasions in ecology when investigators will be interested in these two features of populations.

6.4 What the models can teach us

Equation 6.12 is a model that describes a single population. There would be no point in such models if they did not tell ecologists anything about real populations of organisms. As Section 6.3.5 has already noted, the models describe populations that stop growing when they reach their carrying capacity, *K*. In fact, it is possible to tell from the models that the situation is slightly more complicated because the population does not simply increase at a constant rate and then stop when it reaches it limit. To demonstrate this, the term N/K in eqn 6.12 is one of the most interesting. Equation 6.15 has already noted how:

$$\frac{N}{K} = 1 \qquad \text{(eqn 6.20)}$$

when the population is at its carrying capacity, and that this means a population growth of zero, or no change.

It is also possible to observe that when the population is very small, relative to the carrying capacity, the value of N/K will be small. Imagine a population of 10 individuals in a patch of habitat where there are actually enough resources for a carrying capacity of 1000. In this case:

$$\frac{N}{K} = \frac{10}{1000} = 0.01 \qquad \text{(eqn 6.21)}$$

This means that in eqn 6.12:

$$\left(1 - \frac{N}{K}\right) = (1 - 0.01) = 0.99 \qquad \text{(eqn 6.22)}$$

which implies that the population increases at 99% of the rate it would do if it were growing at the speed implied by its reproductive rate.

Now imagine that after some time, the population has increased to 500 individuals. In this situation:

$$\frac{N}{K} = \frac{500}{1000} = 0.5 \qquad \text{(eqn 6.23)}$$

which means that:

$$\left(1 - \frac{N}{K}\right) = (1 - 0.5) = 0.5 \qquad \text{(eqn 6.24)}$$

In other words, the population is now increasing at 50% of the potential implied by its reproductive rate.

Real populations often behave roughly like this. The number of individuals increases relatively rapidly when the population is small but once the number of individuals becomes much larger, then the rate of increase slows down. If the population becomes 'too large' (i.e. it exceeds the carrying capacity), then the rate of increase becomes negative, and the size of the population decreases. This is called **density dependence**—the rate of change in the size of the population depends on the current population density.

6.5 What the models have not taught us

Because mathematical models help clarify thoughts, there is a danger that ecologists may fail to treat them with the scepticism they deserve. There are two important lessons that ecologists must be careful to learn about what models may have failed to tell them. These two lessons will be dealt with in turn in Sections 6.5.1 and 6.5.2.

6.5.1 Describing, not discovering, the real world

The first reason for caution is that models have not given any information about how real populations of organisms behave. The model above was built in a way that accorded well with the prior knowledge that real populations often grow in the way described by sigmoid curves. Other populations do not grow in this way, and for these populations, models such as eqn 6.12 are useless in their current form. To understand populations that do not grow like those in Fig. 6.1, ecologists might need different models. These may involve starting afresh or, preferably, modifying the existing models.

6.5.1.1 Other kinds of populations

Section 6.3.3 noted that in theory, dN/dt could be a negative number. In other words, the population could decrease in size, whereas in the sigmoidal population growth that has been considered so far, populations always increase or remain constant. Also, as noted above, not all populations can necessarily be described by sigmoidal curves. However, for now, it is sensible to continue with the assumption that most populations, when they are small relative to the available resources, increase just as the left-hand side of the sigmoidal curve describes. Even if this is true, it does not imply that all populations approach the carrying capacity in the same way.

6.5.1.2 Time lags and population cycles

In some populations, the number of individuals does not remain relatively constant, as it has in the models that have so far been developed. Instead, the size of the population might show large regular increases followed by equally regular decreases. The cause of such fluctuations is often a delay in the time that it takes for a density-dependent factor to operate.

Imagine a population of rodents, say voles, living in an area of grassland that is also occupied by a population of foxes. The fox is the only predator species that preys on the voles. To begin with, there are few foxes, so the voles suffer little predation and can increase dramatically in number. Now there are so many prey around that the foxes have bountiful food, and the fox population also increases. Thus, the voles begin to experience much greater levels of predation and their numbers decrease rapidly. The outcome of these events is that there is a large fox population but too few voles to feed them, so the fox population crashes, bringing the cycle back to where it began.

The reason such a cycle could occur is that the fox population could not initially keep pace with the increase in vole numbers. Most foxes have only one litter each year and the young may take a year or more to mature, whereas rodents can breed several times in the course of one spring and summer. Moreover, young voles born in the spring could themselves be reproducing long before the end of the summer. Thus, the size of the vole population can for a while grow unchecked, and can reach a level that will prove unsustainable once the fox population catches up.

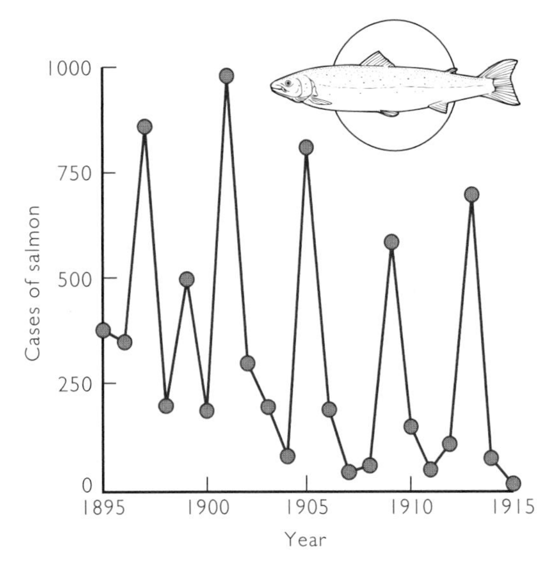

Fig. 6.6 An example of a population that shows fairly regular population cycles: the number of cases of sockeye salmon caught on the Fraser River in each year from 1895 to 1915.

This kind of explanation may not be the only reason why populations show regular cycles, but it is one of the best known, and ecologists have built this kind of time-lagged density dependence into their models, and have created modified models that describe cyclical populations.

In describing such cyclic populations, ecologists often choose to say that the population is fluctuating around some average value of the carrying capacity, K. Such an approach can prove useful but it is sometimes more helpful to recognize that the real value of K is constantly changing.

Figure 6.6 shows an example of a population that shows cycles. The number of crates of sockeye salmon (*Onchorhynchus nerka*) caught between 1895 and 1915 on the Fraser River in British Columbia shows extreme peaks once every 4 years (Smith 1994).

6.5.1.3 Other kinds of fluctuation

In the populations described in Section 6.5.1.2, the number of animals was constantly changing but it did so in a highly predictable manner. Other populations fluctuate more irregularly, like the population of flour beetles shown in Fig. 6.7.

This kind of irregular fluctuation has become known as chaos, although in this context, the word is used

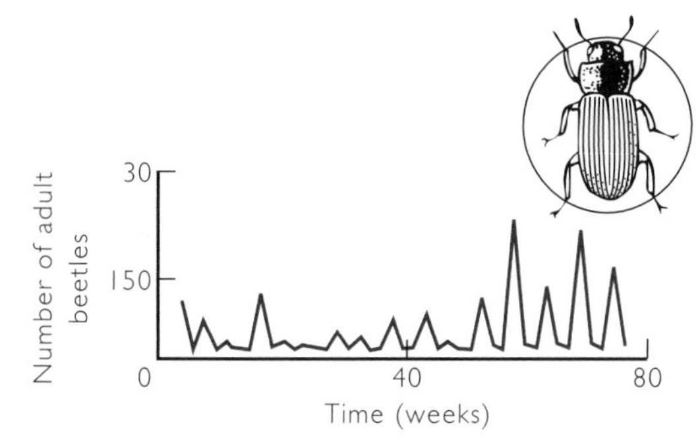

Fig. 6.7 Population dynamics of a population of the flour beetle *Tribolium castaneum*. The increase or decrease in the population is not related to the existing size of the population, nor are there regular cycles. Statistical analysis shows the fluctuations to be 'chaotic'. (Reprinted with permission from Costantino *et al.* (1997) Chaotic dynamics in an insect population. *Science* **275**: 389–391. Copyright 1997, American Association for the Advancement of Science.)

slightly differently from its everyday meaning of 'disorder'. In fact, the evidence suggests that much of the fluctuation of chaotic populations may be governed by strict rules. The problem is that a small number of slightly complicated rules can produce the appearance of mayhem.

The classic example used to illustrate this feature of chaos is that of predicting the weather. Meteorologists understand the most important rules that dictate whether or not the sun will shine tomorrow morning or whether it will rain. Such an understanding means that they are rather good at forecasting what the weather will be like for the next few days. But when they look more than a week into the future, the accuracy of the meteorologists' predictions breaks down and their attempts are little better than guesses based on the experience of what normally happens in a particular place at a particular time of year.

The reason for this is that the laws that govern the weather are not straightforward, which makes forecasting very sensitive to the accuracy with which the meteorologists can measure what is happening in the present. It is extremely difficult to measure precisely the speed at which an area of high pressure is moving. If meteorologists get the measurement slightly wrong, it may have little effect on their forecast for tomorrow. But because of the complexity of the rules governing the weather, the slightest error or imprecision can rapidly become enhanced as they try to predict further and further into the future.

The same is true of chaotic populations of organisms. The slightest gap or error in our knowledge or understanding will prevent us from predicting accurately what will happen to the size of the population in the future.

As with cyclic populations, it is again possible to define some average value of the carrying capacity, K, and to view the changes in population size simply as variation around this handy summary statistic. But the problem with this approach is that whatever the pattern of fluctuations, it is always possible to calculate an average value whether it has any real biological meaning or not. Ecologists might easily fool themselves into thinking that if the average value of K was very large, then they had no reason to worry about the long-term survival of the population. But if the size of the fluctuations was also very large, then the population could be in serious danger.

A single fluctuation large enough to reduce the population size, N, to zero would wipe it out forever and it would be of no comfort to say that, statistically, the population was just as likely to have doubled as it was to have become extinct. Thus, one important feature of chaotic populations is the average size of the fluctuations relative to the average value of population size, N.

6.5.2 Individual organisms

The second reason ecologists must be careful not to read too much into their models is that they do not provide any information at all about individual organisms. Recall that Chapter 1 discussed two different approaches to defining ecology. One of them was a broad approach dealing with abundance and distribution. That is the approach adopted by many models, which are interested merely in N, the total number of individuals in a population of organisms, and in dN/dt, the speed at which N changes. The other approach to defining ecology centres around the individual organisms themselves. After all, the reason why a population stops growing when it reaches carrying capacity is that a new individual entering the population does not have access to enough of some limiting resource. To be certain that ecologists have understood what their models have told them, they need to think about individual organisms.

What does it mean to say that a population has stopped growing? In terms of the model for organisms with discrete breeding seasons, the change in the size of the population between two generations is defined as:

$$N_{t+1} = N_t \quad \text{(eqn 6.25)}$$

This means that the population size at time '$t + 1$' is the same as that at time 't'. Note, however, that the fact that the size of the population is the same does not imply that the population is made up of the same individuals. All the individuals that were alive at time t could have died and been replaced by an identical number of new individuals by time $t + 1$.

Likewise, for a population of continuously breeding organisms, the equation:

$$\frac{dN}{dt} = 0 \quad \text{(eqn 6.26)}$$

means that the rate of change of the population size is zero, i.e. the population size is constant. But again, this does not mean that the make-up of the population is necessarily constant. It merely implies that the number of individuals entering the population (by being born or immigrating from a different population) is balanced by the number leaving (by dying or emigrating). It is one thing to write a good mathematical description of the number of individuals in a population, but it is quite another to understand what is actually happening to the individual organisms that make up the population.

6.6 Investigating individual organisms

6.6.1 How density dependence might work

Section 6.4 illustrated that populations may be regulated by density dependence. In other words the rate of increase (or decrease) of the population depends on the current size of the population. Section 6.5.2 also observed that ecologists must anchor their models in reality by thinking about how the models relate to the lives of individual organisms. It is now possible to look at what density dependence actually means for real organisms.

Think of a population that has been increasing, perhaps like those in Figs 6.2 or 6.3. Recall that Section 6.4 described how the rate of increase in the population might fall off gradually until the increase stops altogether when the population reaches its carrying capacity. How might this come about? The rate of change of the size of the population is a balance between the recruitment of individuals (by birth and immigration) on the one hand and the departure of others (by death or emigration) on the other. Thus, there are logically two ways in which the population increase could slow down. Either the rate of recruitment of new individuals slows down or the rate of departure speeds up. In some cases, both the rate of recruitment and the rate of departure could change at the same time.

6.6.2 Density dependence in sandflies

The case of the sandfly *Lutzomyia longipalpis*, which lives in northern Brazil, is a clear example for observing the effects of density dependence on real individual organisms (Kelly *et al.* 1996). Female sandflies (but not the males) suck the blood of vertebrates, and one of their most successful habitats is chicken sheds. The female sandflies need to feed on the blood of the poultry before they can reproduce successfully, because laying eggs uses a great deal of energy and protein.

Each shed can be thought of as containing a separate population of sandflies. A standard method of trapping was used in each shed and the first question that was asked was: Is the density of sandflies the same in each of the poultry sheds? The answer was clearly 'no'. In the dry season, the fewest female sandflies caught in a shed in a single night was 30, with most sheds producing a few hundred sandflies. However, a small number of sheds produced thousands of sandflies, with the greatest number of females trapped in a single shed in a single night being 3117.

The next issue to investigate was whether or not individual sandflies that share their sheds with just a few others behave in ways that are different from sandflies that live in more densely populated sheds. In particular, ecologists are interested in whether they are more successful at feeding. Feeding success among the female sandflies does indeed depend on the number of other sandflies with which they share their shed. The size of the bloodmeal they obtain is smaller for sandflies in densely populated sheds than it is for sandflies in more sparsely populated ones (Fig. 6.8). There is, therefore, a density-dependent effect on the feeding success of individual organisms. The most likely explanation for this pattern is that when poultry are subjected to a greater number of bites, they are more likely to indulge in bouts of preening, which will force sandflies to fly off part way through their meal.

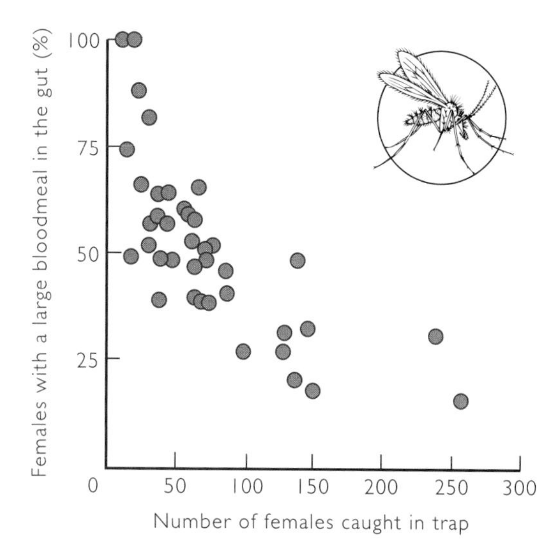

Fig. 6.8 Female sandflies that share a poultry shed with just a few others are more likely to obtain a large bloodmeal than those in densely populated sheds.

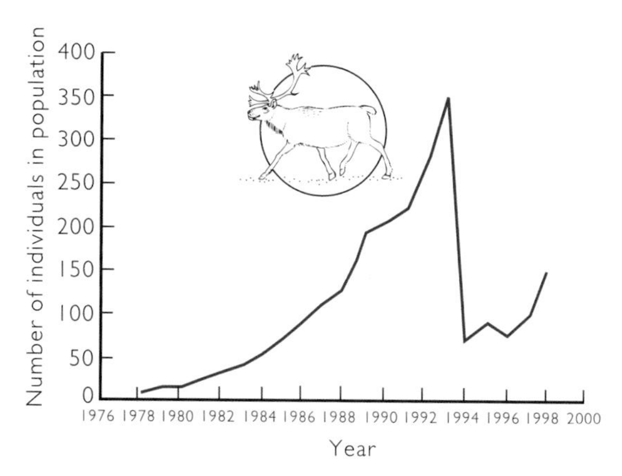

Fig. 6.9 Density-independent population fluctuations of a population of reindeer (*Rangifer tarandus*) in Svalbard. The growth or fall in the size of the population is not determined by the existing size of the population. In 1990, a large population grew quickly and in 1996, a small population grew just as quickly. The best predictor of changes in the population size is the weather in winter. Falls in the size of the population, such as those in 1993 and 1995, are associated with the wettest winters. (From Aaenes *et al.* 2000.)

Of course, it is necessary to examine how this differential feeding success affects the rate of change of the population, which is the main purpose of the study. Section 6.6.1 described how variations in the rate of change in the size of the population could come from a change in the rate of births and immigrations or from a change in the rate of deaths and emigrations. Little is known about the rates of immigration and emigration in these populations of sandflies, so they are discounted as possibilities, and instead, the study concentrated on births and deaths in the populations.

Does the birth rate of sandflies depend on the feeding success of female sandflies? In a separate experiment, different groups of female sandflies were allowed to swallow different amounts of blood (Carneiro *et al.* 1993). Of the sandflies that were fully fed, 100% produced eggs; the average number of eggs per female was 75. However, of the females that were only partially fed, only 6% produced any eggs at all and among those that did lay eggs, the average clutch size was only 20. Thus, the feeding success of a female sandfly does indeed affect whether or not she lays any eggs and also how many she lays.

So, when populations of sandflies are large, the individuals feed less well and this affects their breeding success. In terms of the original model described by eqn 6.12, this means that when N is large, the birth rate (a measure of the reproductive rate, r) is low, which slows down the rate of increase of the population, dN/dt. Thus, it is clear how the terms in the models of populations are relevant to these populations of sandflies.

6.6.3 Density independence

Not all fluctuations in the size of a population are necessarily governed by density-dependent factors. Some causes of change in the size of a population may act equally on large and small populations. For instance, bad weather, such as a cold winter, kills many invertebrates, plants and small vertebrates, and the rate of death is not necessarily correlated with the density of the population.

This can be seen in the tick *Rhipicephalus appendiculatus* in South Africa. At least one of the stages in its life cycle is governed by density-independent factors. Adult ticks are subject to density-independent mortality correlated with climatic factors that vary geographically. In some places the death rate of ticks is associated with the minimum temperature—the colder the weather, the higher the death rate. In other places, it is the minimum relative humidity that correlates with the death rate of adult ticks (Randolph 1997).

Figure 6.9 shows an example of another population that is strongly affected by density-independent effects.

It is not possible to build general models that describe density-independent effects on populations, unless of course ecologists can model the causes of these effects accurately. As this chapter has described, these effects are often caused by the physical environment, such as changes in the weather. Since it is not yet possible to make accurate predictions about the weather for more than a few days into the future, it is not feasible to incorporate information about its effects into general models of populations of organisms. This is another limitation of the models and another reason why ecologists must be careful when using them.

6.7 Chapter summary

Ecologists can build models that contain very few different terms, and these models have many limitations and must always be used carefully. But relatively simple models can point an ecologist's attention towards the features of populations that might be most important in understanding them. Of these, the reproductive rate and carrying capacity are important quantities of populations, which will be investigated further in later chapters.

Density dependence is a crucial concept in understanding populations of organisms—populations tend to grow in size when the number of individuals is below the potential carrying capacity, and to decrease in size when the number of individuals is higher than the carrying capacity. However, the ways in which populations fluctuate around the carrying capacity varies; some populations show stable cycles while others appear chaotic, even if they are in fact governed by strict rules.

Significantly, ecologists must also anchor their abstract models and concepts in reality. Populations are made up of individual organisms, and ecologists must be able to explain their models in terms of what they mean for those individuals. This approach, of thinking about individuals, will be important in the next few chapters, as models are developed as a way of understanding real populations in the real world. In the real world, populations of organisms do not exist in isolation. Instead, they experience predators, prey, competitors and parasites, all of which make up the biological environment of a population. Chapters 7–10 will elaborate on the basic model of populations, and incorporate the important features of these biological interactions.

Recommended reading

Björnstad, O.N., Ims, R.A. & Lambin, X. (1999) Spatial population dynamics: analyzing patterns and processes. *Trends in Ecology and Evolution* **14**: 427–432.

Gotelli, N.J. (2001) *A Primer of Ecology*. Palgrave Press, Basingstoke.

Lundberg, P., Ranta, E., Ripa, J. & Kaitala, V. (2000) Population variability in space and time. *Trends in Ecology and Evolution* **15**: 460–463.

Chapter 7

Population demography and life history patterns

7.1 The age, sex and survival of individual organisms

Imagine an ecologist in Zimbabwe charged with the task of reversing a decline in the local population of hippopotamuses (*Hippopotamus amphibius*). In order for the ecologist to achieve this goal, it would be necessary to know something about the current population. For example, how many adults and young are currently in the population? How long does an average hippopotamus live? How many young does a typical female hippopotamus have? What are the chances of surviving the next year for a young hippopotamus compared to the chances for an adult hippopotamus? These types of vital statistics about populations are addressed in the study of **demography**.

In demographic studies, ecologists record the distribution of individuals among different age categories, map the fate of individuals of different sexes or ages, record the mating patterns of individuals, and project future population growth. The age categories that are used depend on the species that is being studied; ecologists may simply divide individual organisms into two categories (young and old) or they may use a more elaborate system with more precise information about the age of individuals.

Demographic studies are invaluable for ecologists interested in the dynamics of natural populations, preserving endangered species, managing game species, investigating population response to the environment, or comparing populations of different species.

7.1.1 Life tables

By recording the survival, reproduction and deaths of individual organisms of different ages and sexes, the demographer can predict how long an average individual is likely to live, at what age it will breed and how many offspring it will have, and how the population size will change over time. These vital statistics are usually arranged in **life tables**.

To construct life tables, the ecologist must first determine whether age, size or some other life stage (such as the pupal, larval or adult stages of insects) is the best measure of an organism's demography. For example, age is a very good measure for most mammals, since reproduction, size and survival are all highly related to age. However, for many long-lived plants, size is more closely related to performance than is age. This is because many plants can gain or lose size in any particular year, and plants located in unfavourable habitats may take many more years to reach reproductive size than plants in favourable habitats.

Once a demographer decides whether to use size or age, then the population can be divided into a number of different classes (categories). This division is usually dependent upon the characteristics of each class. For example, a tree population may be divided into seed, seedling, sapling and adult stages, since each of these size categories may differ in their survival and reproductive characteristics. After determining the categories to be used, the demographer will survey the population for the number of individuals in each age or size class.

For clarity, we will look at examples of age-structured populations. Age in life tables is designated by the symbol x, and population statistics associated with that age are indicated by the subscript x. A summary of some important variables used in life tables is presented in Table 7.1.

The data collected from individuals that are placed in life tables can then be used to determine important properties of the population. For example, survivorship and fecundity can be used to calculate the **net reproductive rate** of the population:

$$R_0 = \sum(l_x \times b_x) \qquad \text{(eqn 7.1)}$$

The symbol $\sum$ means to add up, over the whole range of possibilities, so in this case, we would calculate

Table 7.1 Life table variables.

N_x	The number of individuals of age x
l_x	Survivorship—probability of an individual surviving from birth (age 0) to age x
b_x	Fecundity, or birth rate for individuals of age x
s_x	Survival, or the probability of an individual of age x surviving to age $x + 1$
q_x	Mortality, or the probability of an individual of age x dying before reaching the age of $x + 1$

$(l_1 \times b_1)$ and add it to $(l_2 \times b_2)$, then add $(l_3 \times b_3)$, then $(l_4 \times b_4)$, and so on, up to whatever the oldest or largest age class happens to be.

So, net reproductive rate (R_0) is calculated by multiplying survivorship (l_x) to each age by fecundity (b_x) for each age. Then the result for each age is added together with all other ages to calculate the growth rate for the population as a whole. If the net reproductive rate is less than 1, the population is decreasing in size. If R_0 is equal to 1, then the population is neither increasing nor decreasing in number. Finally, if R_0 is greater than 1, the population is growing in size.

The **reproductive value** of an individual to the population can also be derived from life table data. The reproductive value of an individual at age x is given by the symbol V_x. This term represents the relative contribution of an individual that is currently of age x to the growth of the population over the rest of the individual's life. It is used by ecologists to evaluate those age classes that are most important to the growth of the population. For example, for hippopotamuses, a young female just entering reproductive age might be more important to population growth than an older female that is past reproductive age. Reproductive value is calculated using:

$$V_x = b_x + \sum \left(\frac{l_{x+i}}{l_x} \right) b_{x+i} \qquad \text{(eqn 7.2)}$$

Or, in words, the reproductive value of an individual of age x is equal to the number of young it will produce now (b_x) plus the number of young that it will produce in the future (the sum of all future reproductive events). The number of young produced in the future will be a function of the individual's survivorship to older ages (the longer it lives, the more young it is likely to produce) and its fecundity in each succeeding age class. Because there is usually a low probability of young organisms surviving to reproductive age, reproductive value is usually low for young age classes. In other words, a young organism is, on average, unlikely to have too many offspring, because most organisms die young. Reproductive value then peaks at maximal values for those individuals who do survive to reach reproductive age. It then stays high over the reproductive years, and declines in the older, post-reproductive years.

7.1.1.1 Cohort and static life tables

There are two ways that ecologists construct life tables. One way is to follow an entire **cohort** of individuals, from birth to death, by direct observation. A cohort includes all of the individuals born at one time in the population. A life table constructed by observing this cohort throughout their lives is called a **cohort life table**. This approach is only feasible for short-lived organisms that do not disperse great distances, such as mice. Another method is to survey populations at an instant in time, and count the number of newborns and the number of individuals of known age. The number of individuals of age x is assumed to be a measure of the probability that a newborn organism will survive to that age. This method is called the time-specific or **static life table** method. It is best applied to long-lived organisms where direct observation of the entire life of a cohort is impractical.

7.1.1.2 Differences between the sexes in life table variables

Many animal populations show large differences in the behaviour of males and females. Often these differences are associated with mating behaviour. Differences in behaviour can lead to large differences in the survivorship and reproductive success of males and females. One study that focused on such differences between males and females was that of Sherman and Morton (1984). This demographic investigation spanned 11 consecutive field seasons, and studied a population of Belding's ground squirrels (*Spermophilus beldingi*) at Tioga Pass in the central Sierra Nevada Mountains of California.

Belding's ground squirrels are diurnal, social rodents that inhabit meadows in mountainous areas. Squirrels

Table 7.2 Population life table results from a study of Belding's ground squirrels (Sherman & Morton 1984).

	Females				Males	
Age (years)	Number of age x (n_x)	Survivorship to age x (l_x)	Fecundity of age x (b_x)	Survivorship × fecundity (l_xb_x)	Number of age x (n_x)	Survivorship to age x (l_x)
0–1	337	1.000	0.00	0.00	349	1.000
1–2	252	0.386	1.04	0.401	248	0.350
2–3	127	0.197	2.171	0.428	108	0.152
3–4	67	0.106	2.45	0.255	34	0.048
4–5	35	0.054	3.113	0.168	11	0.015
5–6	19	0.029	1.875	0.054	2	0.003
6–7	9	0.014	1.575	0.022	0	
7–8	5	0.008	2.00	0.016		
8–9	4	0.006	1.44	0.009		
9–10	1	0.002	1.632	0.003		
				R_0 = 1.356		

emerge from their burrows and stay active above ground from mid May until early October. They hibernate for the rest of the year. Each spring, sexually mature males that are over 2 years old tunnel through more than 2 m of snow to emerge 1–2 weeks before the females.

Males gather on snow-free ridge tops near female burrows and defend small mating territories. In the presence of females, males threaten, chase and fight other males that invade their territory. Fights are violent and involve grappling, kicking, scratching and biting. Females run through the groups of males, promoting male aggression, and usually mate with the winners of fights. Because of the preference of females for fight winners, each season a few males who are good fighters have many matings, but most males seldom or never mate. Typically, the heaviest, oldest males win the most fights and mate with the most females.

Besides fighting, another difference in behaviour between the sexes is the movement of squirrels. Males typically leave the population that they are born into either before or at the time of sexual maturity, while females stay in the birth population. Additionally, many males will leave the population after the mating season.

The life table generated from Morton and Sherman's observations of Belding's ground squirrels is shown in Table 7.2. The study follows a single cohort of 337 female squirrels and 349 male squirrels, all born in the same year, throughout their lives. From Table 7.2, it is immediately obvious that the numbers of males and females at birth are similar (the n_x columns), highlighted in light purple for females and dark purple for males. By the second year of life, the original number of females has decreased by 85, to 252, while males have decreased by 101, to 248. After this point, differences between the number of males and females become larger and larger. By the 6–7-year age class, 328 of the original 337 females have disappeared, but nine are left, and a small number of females survive each year until the age of 9 or 10. In contrast, no males live to be older than 5 or 6 years.

The net reproductive rate (R_0) for this population can be obtained by summing the product of female survivorship times fecundity ($\sum l_x b_x$) for each age. This value was 1.356. Because this figure is larger than 1.0, it is possible to conclude that the females produced more than enough daughters to replace themselves, and the population was actively growing over the study period.

The difference in survivorship between the sexes from Table 7.2 is illustrated graphically in Fig. 7.1. Along with the steeper decline in survivorship for males is a shorter **life expectancy** (Fig. 7.2). Life expectancy is the number of additional years an individual of age x is expected to live. Males usually die before they reach the age of 6 years, while females can live past 10 years.

What were the factors that contributed to differences between the sexes in mortality? One possibility would

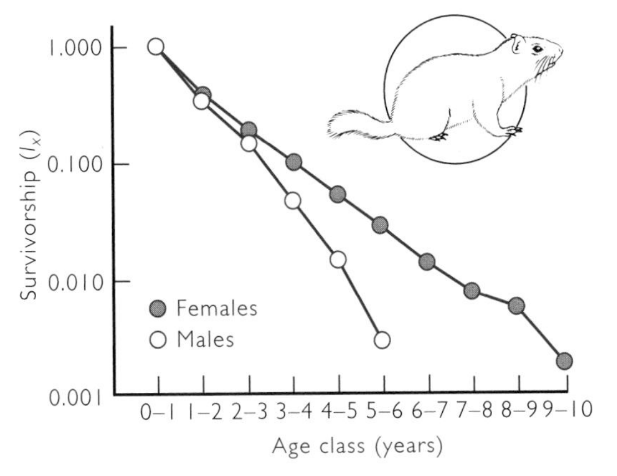

Fig. 7.1 Survivorship over 11 years for a population of Belding's ground squirrels in the Sierra Nevada of California.

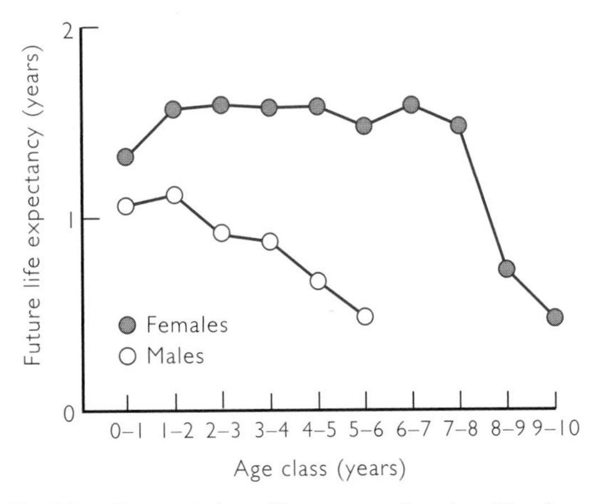

Fig. 7.2 Differences in future life expectancy for male and female Belding's ground squirrels in the Sierra Nevada of California.

be that males and females differed in survival over winter due to differential costs such as fighting by males or bearing of young by females. However, Morton and Sherman found no differences between the sexes in winter survival. Rather, the difference in survival for males and females was generated during the active, summer season. During this season, predators such as badgers (*Taxidea taxus*), coyotes (*Canis latrans*), pine martens (*Martes americana*) and raptors (hawks and eagles) preyed upon ground squirrels. In addition, vehicular traffic on nearby roads killed many squirrels.

Behaviour patterns of the males, such as their higher mobility, increased their exposure and loss due to predation and vehicles. In addition, because males emerged earlier than females, they were exposed to predators and traffic for longer. Another male behaviour that increased exposure to predators was fighting, which took place on the exposed tops of open ridges, where predators found it easy to spot them. Combat among males also directly and indirectly increased male mortality. Males could injure each other severely, sometimes killing each other by severing major blood vessels in the throat. Wounds from fights also increased mortality by encouraging infections or generating injuries that weakened the male squirrels and increased their susceptibility to predators. In other words, in an attempt to secure opportunities to mate successfully, male squirrels engaged in behaviour that reduced their survival.

7.1.2 Survivorship patterns

Differences in survival can also be seen between different species. Sensitive stages in the lives of individuals can be distinguished by examining the pattern of survival for different ages. This is achieved by plotting the logarithmically transformed number of individuals surviving to age x against current age (Fig. 7.3). These curves are usually divided into one of three main types, type I, type II and type III, depending upon the shape of the curve. Type I indicates a population with high survivorship among the young, but lower survivorship at older ages. This is characteristic of humans. Type II survivorship curves indicate a relatively constant survivorship throughout the lifetime of an individual. Some annual plants, lizards and rodents, such as the

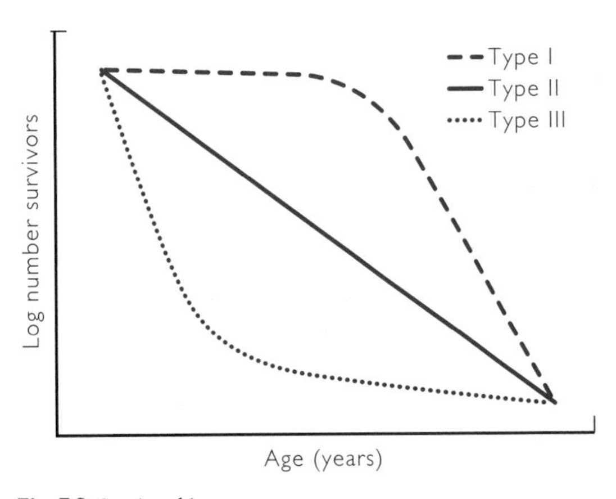

Fig. 7.3 Survivorship curves.

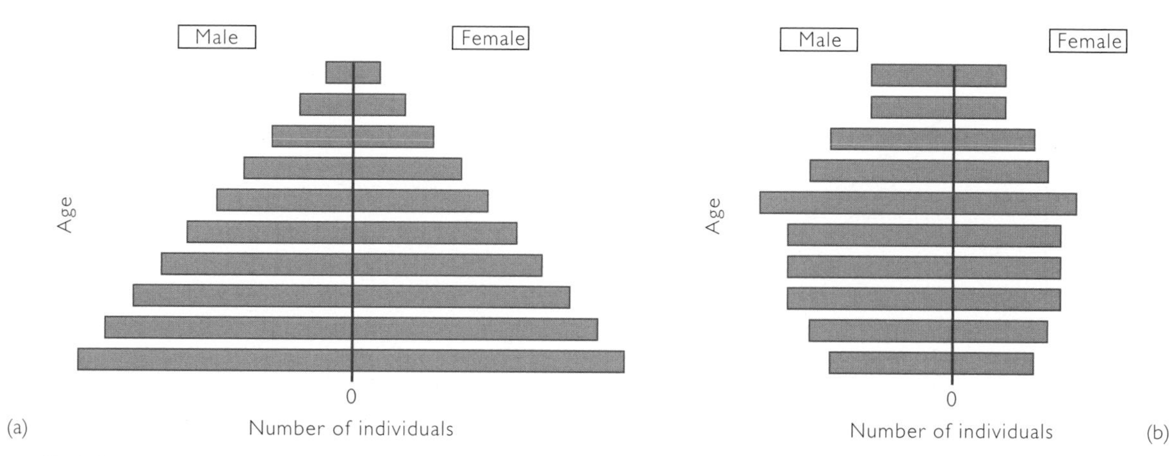

Fig. 7.4 Hypothetical population pyramids. (a) An example of many young individuals. The population will probably experience rapid growth in the future. (b) A population with fewer young individuals, and more intermediate-aged individuals. This population may not increase as fast as the population in (a). Additionally, the population in (b) has half as many females as males. This may result in even slower growth, with some males failing to reproduce.

grey squirrel (*Sciurus carolinensis*), show this type of survivorship. Type III survivorship curves are indicative of low survivorship in young age classes, and higher survivorship for older age classes. These types of curves are characteristic of long-lived forest trees, many fish and some marine invertebrates like oysters (e.g. *Crassostrea* spp.).

In nature, many populations of organisms show a combination of one or more of these survivorship curves. Therefore, it is difficult to place most organisms in one type only. However, by observing the age at which mortality is most severe, ecologists can identify critical time periods in the life of an organism. For endangered species, management plans may be designed to minimize mortality during these critical periods.

7.1.3 Age structure and sex ratio

Another characteristic of a population that can be obtained from life table data is the age structure of a population. Demographers often use what are termed 'population pyramids' to show age structure. These are graphs showing the number of individuals in each age class, starting with the youngest age class at the bottom and proceeding to the oldest age class at the top of the graph. They are called pyramids because many populations have more young individuals than old ones, hence the bottom of the graph is wider than the top (Fig. 7.4). If one side of the graph is used for males and the other for females, population pyramids can also be used to look at the ratio of sexes in a population.

A population age structure is important in determining the rate at which a population grows, and the future direction of growth in the population (Box 7.1). A population with a large percentage of its individuals either in the prime reproductive ages or in slightly younger age classes will grow faster than populations with many older individuals.

The sex ratio, or proportion of individuals in each sex, is another important demographic feature. The number of females is usually related directly to the number of births expected. The number of males can be less important, depending upon the **breeding system** of the population. For example, if a single male mates with many females, as happens in the white-tailed deer (*Odocoileus virginianus*), then the number of males is less important to total births in the population. However, in breeding systems where one male mates with one female, the number of males would be equally as important as the number of females to population growth. Wildlife management biologists often consider this in developing plans to manage populations. For example, hunting of white-tailed deer is often restricted to males, since harvesting males will have less of an effect on the population than harvesting females. However, if populations of white-tailed deer are

Box 7.1 Matrix models of population growth

A matrix is a rectangular array of numbers divided into columns and rows. For population projections of an age-structured population, a matrix containing a single column with the number of individuals in each age group at time t is multiplied by a **transition matrix** containing information on age-specific survival and birth rates. The result is a new **column matrix** representing the number of individuals in each age group for the following time period, $t + 1$.

To project future population growth for the hypothetical population with three age classes from Section 7.2, assume that the population has 100 individuals in age class 1, 50 in age class 2, and 20 in age class 3. The column matrix would be:

$n_{1(t)} = 100$
$n_{2(t)} = 50$
$n_{3(t)} = 20$

Next, this column matrix would be multiplied by a transition matrix made up of the probability that an individual in one age class at time t would change to a different age class at time $t + 1$. Because there are three age classes, there would be three rows and three columns in the transition matrix. The first row would represent the contribution of each age class at time t to age class 1 at time $t + 1$. Since it is impossible for organisms to get younger, this transition has to be made through reproduction. Thus, the first row of the transition matrix is b_1, b_2, b_3. There are no survival terms in this first row, because this example assumes the simplest case where reproduction occurs before any mortality.

Since reproduction only adds individuals to the youngest age class, no reproduction terms are needed in the second and third rows of the transition matrix. The three entries for row 2 represent the contributions of individuals of age 1, age 2 and age 3 at time t to age 2 at time $t + 1$. Only individuals in age class 1 can make this transition, since age 2 individuals cannot stay the same age, and age 3 individuals cannot get younger. The only entry is from the survival of individuals of age 1 from time t to time $t + 1$. So, the three entries would be s_1, 0, 0.

The three entries for row 3 represent the contribution of each age class at time t to age class 3 at time $t + 1$. Only individuals in age class 2 make this transition, so the three entries would be 0, s_2, 0. The entire matrix would be:

Age class next year	Age class this year		
	1	2	3
1	b_1	b_2	b_3
2	s_1	0	0
3	0	s_2	0

In this example, assume that survival of the youngest age class, s_1, equals 0.2. In other words, 20% of age 1 individuals survive to age 2. Assume that survival of the next age class, s_2, equals 0.3. Finally, assume that survival of the last age class, s_3, is equal to 0.

For fecundity, b, in each age class assume $b_1 = 0$, $b_2 = 2$ and $b_3 = 3$. In other words, the youngest individuals do not reproduce, intermediate-aged individuals have on average two young and the oldest individuals have three young.

The calculation of the population size at time $t + 1$ is shown in eqn 1.

Population growth over 10 years for this population, along with the proportion of the population in each age category, is presented in Fig. 1. Notice that the population is decreasing in size, down to approximately 20 individuals after 10 years.

To illustrate the types of comparisons ecologists can make using matrix models, assume that there is a second population that inhabits an environment more favourable to survival of age 1 individuals. Here, s_1 equals 0.4, rather than 0.2. Population growth in this population, along with the proportion of the population in each age class, over 10 years is also presented in Fig. 1.

Notice the huge difference in population growth in these two examples. This occurred with a simple change in survival of only one age class! This change converted a population that was clearly going extinct, with only 20 individuals after 10 years, to one that was growing steadily, with 356 individuals after 10 years. This is nearly an 18-fold difference in 10 short years.

Column matrix × Transition matrix = New column matrix

$$\begin{pmatrix} n_{1(t)} \\ n_{2(t)} \\ n_{3(t)} \end{pmatrix} \times \begin{pmatrix} b_1 & b_2 & b_3 \\ s_1 & 0 & 0 \\ 0 & s_2 & 0 \end{pmatrix} = \begin{matrix} n_{1(t)}b_1 + n_{2(t)}b_2 + n_{3(t)}b_3 \\ n_{1(t)}s_1 + n_{2(t)} \times 0 + n_{3(t)} \times 0 \\ n_{1(t)} \times 0 + n_{2(t)}s_2 + n_{3(t)} \times 0 \end{matrix} \begin{matrix} = n_{1(t+1)} \\ = n_{2(t+1)} \\ = n_{3(t+1)} \end{matrix} \quad \text{(eqn 1)}$$

$$\begin{pmatrix} 100 \\ 50 \\ 20 \end{pmatrix} \times \begin{pmatrix} 0 & 2 & 3 \\ 0.2 & 0 & 0 \\ 0 & 0.3 & 0 \end{pmatrix} = \begin{matrix} 100 \times 0 + 50 \times 2 + 20 \times 3 \\ 100 \times 0.2 + 50 \times 0 + 20 \times 0 \\ 100 \times 0 + 50 \times 0.3 + 20 \times 0 \end{matrix} \begin{matrix} = 160 \\ = 20 \\ = 15 \end{matrix}$$

Total = 195

(cont'd on p. 114)

Box 7.1 (cont'd)

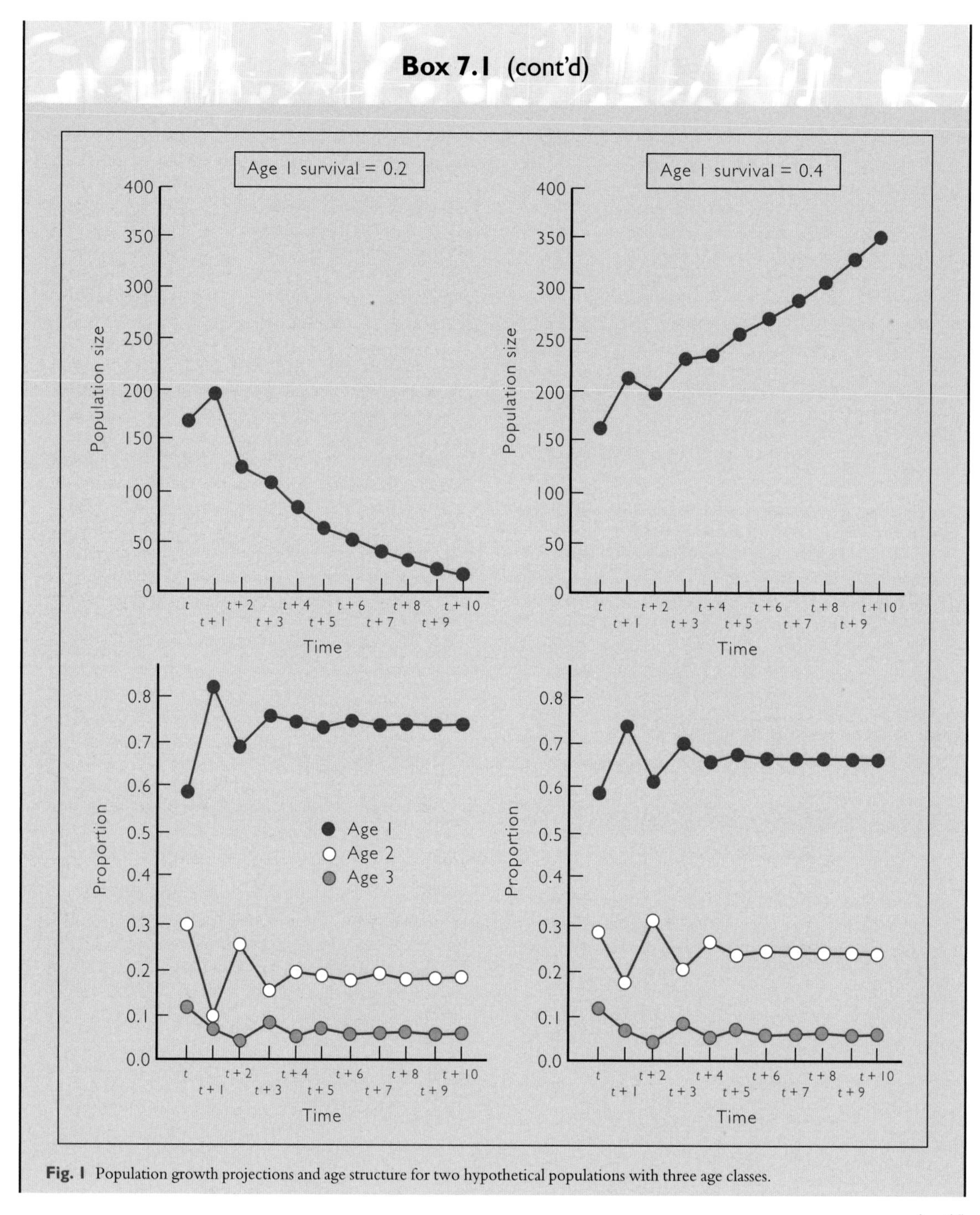

Fig. 1 Population growth projections and age structure for two hypothetical populations with three age classes.

(cont'd)

Box 7.1 (cont'd)

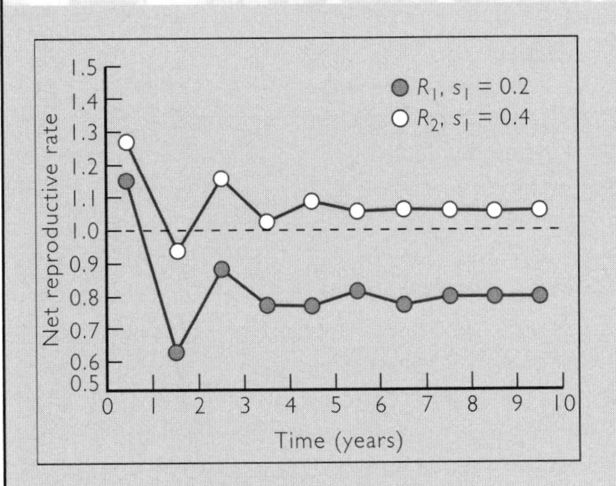

Fig. 2 Net reproductive rate over time for two populations differing in the survival of age class 1.

There are a couple of other concepts illustrated by these examples. Notice how the proportion of individuals in each population changed up and down in the first couple of years. Obviously, the original numbers chosen were not in a stable proportion to each other. However, within a few years, the **age structure** (proportion of individuals in each age category) stabilized. This is a common feature in age-structured models: given constant survival and birth rates, they will approach a **stable age distribution**.

The **net reproductive rate** of these populations, R_0, can be calculated by dividing population size at time $t + 1$ by population size at time t (N_{t+1}/N_t). Net reproductive rate varied in the first few years, but then converged into steady values (Fig. 2). This occurred when the populations reached stable age distributions. As long as the age-specific survival and birth rates remain the same, the age distribution and growth rate will stay the same. For the first population, net reproductive rate stabilized at 0.79. For the second population, R_0 stabilized at 1.07. Thus, comparing the two populations, it is evident that survival of pre-reproductive individuals was very important to the magnitude and direction of population growth.

growing rapidly, then harvesting of female deer may be initiated to reduce large herds.

7.2 Age and size-structured population growth

In Chapter 6, elementary mathematical models of population growth were introduced. These models were based on simplifying assumptions, such as the suggestion that each individual makes an equal contribution to the growth of the population. However, as we have seen in Section 7.1, individuals are not equal in most populations. Older individuals may be past the age of reproduction, younger individuals may not have begun to reproduce, and larger individuals may produce more offspring than smaller individuals. These differences were evident in the example of the Belding's ground squirrel (Section 7.1.1.2). Notice from Table 7.2 how the fecundity of 4–5-year-old female squirrels was much higher than that of 0–1 or 1–2-year-old squirrels.

The age structure and sex ratio of a population are two important factors in determining future population growth (Section 7.1.3). How do ecologists incorporate such effects into mathematical models? One way to do this is by dividing individuals into different age or size classes, and then adding the reproduction in each class together to get reproduction for the entire population.

For example, assume that a population has three different age classes, 1-year-old individuals, 2-year-old individuals and 3-year-old individuals. All 3-year-old individuals in this hypothetical population die before they reach the age of 4 years, so there are no older individuals. The number of 1-year-old individuals at time t, could be represented by $n_{1(t)}$. Total population size at time t (N_t) would be represented as:

$$N_t = n_{1(t)} + n_{2(t)} + n_{3(t)} \qquad \text{(eqn 7.3)}$$

In the interval between one year and the next, some individuals will die, some will reproduce and some will mature into the next age class. Therefore, more variables must be added to the model to predict population growth into the future. The variable 'b' is often used to represent the birth rate, or fecundity (Table 7.1). The addition of new individuals by birth would be represented by the number of individuals in each age class multiplied by the birth rate for that age class:

births from individuals of age 1 = $n_{1(t)}b_1$
births from individuals of age 2 = $n_{2(t)}b_2$ (eqn 7.4)
births from individuals of age 3 = $n_{3(t)}b_3$

In many organisms, the youngest age classes do not reproduce, so b_1 may equal 0. Each birth would result in individuals joining only the youngest age class (n_1) at the next succeeding time interval, $t + 1$. Therefore, total births would be added to the first age class, n_1, not to other age classes. However, other age classes would change in number of individuals because some individuals would die, and others would make the transition into the next older age class.

Thus, maturation (age transitions) and mortality have to be accounted for in the population. The variable 's', is used to represent survival, the proportion of individuals surviving from one year to the next (Table 7.1). The number of individuals in each age class at time $t + 1$ would be represented as:

$$n_{1(t+1)} = n_{1(t)}b_1 + n_{2(t)}b_2 + n_{3(t)}b_3$$
$$n_{2(t+1)} = n_{1(t)}s_1 \quad \text{(eqn 7.5)}$$
$$n_{3(t+1)} = n_{2(t)}s_2$$

There is no term for a fourth age class because the population was defined as having only three age classes. Thus, survival of the third age class must be $s_3 = 0$, and $n_{4(t+1)} = n_{3(t)} \times 0 = 0$.

Accounting for maturation, birth rate and mortality in different age or size classes can become very complex. Some ecologists have dealt with this complexity by using matrix algebra to make future projections about the growth and structure of a population (Box 7.1).

7.3 Life history traits

At the beginning of this chapter, an imaginary example was introduced, involving an ecologist interested in populations of hippopotamuses. To develop a more effective plan to increase population numbers, this ecologist might ask three more questions about this organism: (i) How often do hippopotamuses reproduce? (ii) At what age do hippopotamuses begin to reproduce? (iii) How many offspring do hippopotamuses produce each time they have young? The patterns of survival and reproduction documented in life tables by this ecologist can then be combined with breeding patterns to form a **life history**. Life histories include the survival and fecundity information of life tables, as well as information on the size and number of young, age at first reproduction, number of reproductive episodes during a lifetime, dispersal and resting stages, and the amount and kind of parental care of the young.

Life history traits will have a large effect on the rate of growth of populations. For example, a life history where individuals begin reproducing at an early age, produce many offspring each time they reproduce, and reproduce many times in a lifetime, would result in a very high rate of population growth. However, because organisms only have a limited amount of resources available to apportion to different functions, we rarely see this combination of traits in one organism. Thus, the life histories found in nature appear to represent a compromise between the conflicting demands placed on an organism's resources by growth, reproduction and survival (for example defence against predators).

Ecologists are interested in the patterns found in life histories because life history traits respond to natural selection. Thus, they represent adaptations that organisms have made to their physical and biological environments. Life history patterns often parallel environmental factors; for example, birds in temperate habitats have larger clutch sizes than tropical birds. David Lack (1968), a British ornithologist, hypothesized that this was because higher latitudes have longer summer daylengths than tropical latitudes, thus allowing temperate species to gather more food in a day. More food can support more young. Tropical days are roughly the same length (close to 12 h) all year round.

There are also broad patterns of **trade-offs** in life history traits. For example, delayed reproductive maturity and high levels of parental investment in young tend to be correlated with low fecundity and low adult mortality. To understand the bases for trade-offs found in life histories, ecologists assume that the time, energy and nutrients used for one purpose cannot be used for another purpose. For example, current reproduction may deplete resources, which are then not available for body maintenance, and this may result in lowered survival, or in lower future reproduction. Greater current growth can result in lower or delayed reproduction.

Of course, one of the most fascinating aspects of ecology is that there are mixtures of different life histories within the same environment, and even within the same type of organism. For example, in a study of nine

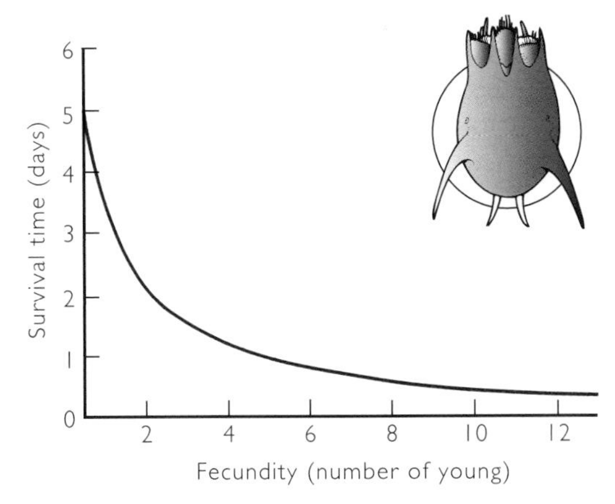

Fig. 7.5 The relationship between survival time and fecundity found for nine species of planktonic rotifers in response to starvation.

different species of planktonic rotifers (small, floating zooplankton that graze on algae), Kirk (1997) found a variety of responses to starvation. The availability of the phytoplankton on which these zooplankton feed is quite variable. There are frequent periods when food is scarce and organisms have a reduced intake of energy. Kirk hypothesized that the response of zooplankton to periods without food might be important to their success.

He found that there was a trade-off between survival and reproduction during starvation. Thus, if resources were allocated to reproduction, survival decreased (Fig. 7.5). However, species differed in the way they handled this trade-off. Some species ceased all reproduction and could survive up to 5 days without food. Other species maintained or actually increased reproduction, but did not survive very long (less than half a day).

7.3.1 Number of reproductive episodes

Some plants and animals devote energy and resources only to growth and development for an extended period, and then expend huge amounts in a single reproductive effort. Reproduction only once during a lifetime is termed **semelparity**. Many insects, salmon and annual plants are semelparous. There are also extreme examples of semelparity, such as that found in bamboo and agave plants. These plants spend many years growing vegetatively before they have one massive reproductive episode, and then die. This life history type also occurs in some animals, such as periodical cicadas (*Magicada* spp.), and has generated a lot of curiosity; there are several hypotheses about the factors that led to its evolution. For example, in bamboo plants and cicadas, ecologists think that by delaying reproduction for many years, few predators can specialize on these species for their food. Then, by synchronizing reproduction throughout the entire population, there are so many seeds or young insects produced at one time that the few predators around become quickly satiated, and many of the young escape predation.

Other organisms produce fewer offspring at a time, but reproduce several times. This pattern of reproduction is called **iteroparity**. The relative advantages of semelparity versus iteroparity are thought to depend on the probability of survival of the adults and immature young. Semelparity is favoured when the cost to parents of staying alive between reproductive events is great, or if there is a large trade-off between fecundity and survival. Iteroparity is favoured if individuals survive well once they are established, but immature individuals are unlikely to survive. At any time in the life history, the general pattern is that organisms with a low probability of surviving to another year may maximize their reproductive success by investing more in current reproduction. For example, environments characterized by high predation or high overwintering mortality are often associated with organisms that invest more in current reproduction than in future survival.

Some organisms may respond to failures in reproduction, or increased chances of survival to the next reproductive bout, by changing from semelparity to iteroparity. For example, scarlet gilia (*Ipomopsis aggregata*), a plant found in high elevations in the western United States, is normally semelparous, with each individual reproducing just once in its lifetime. However, some individuals will produce an ancillary rosette of leaves at the base of a flowering stalk that will survive to flower in a subsequent year. The production of this rosette, and iteroparity, is associated with the success a plant has experienced in its first attempt at reproduction.

Scarlet gilia plants are pollinated by hummingbirds and hawkmoths. These pollinators are not always numerous in the mountain habitats of scarlet gilia, especially at higher altitudes, and their presence varies with the time of season. Pollinators are more numerous early in the season and most scarlet gilia plants that flower early are semelparous. As the season progresses,

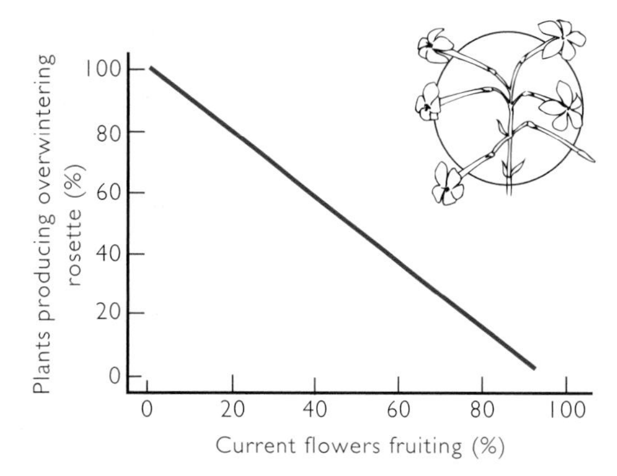

Fig. 7.6 The trade-off between the amount of fruit set in the current year and the chances of producing an overwintering rosette that will flower in future years for scarlet gilia (*Ipomopsis aggregata*).

pollinators become scarcer, and late-flowering individuals have reduced fruit and seed production, and increased probability of forming ancillary rosettes. When Paige and Whitham (1987) excluded pollinators from flowers, they found that plants were more than five times as likely to produce an ancillary rosette as plants with natural pollination. When flowers were removed to prevent fruit production, plants were almost eight times as likely to produce an ancillary rosette. (Fig. 7.6).

Paige and Whitham hypothesized that the flexible life history traits of scarlet gilia represent an adaptive response to changes in pollinator abundance. In response to low numbers of fruits setting, plants were able to reallocate resources and shift from current reproduction to current maintenance and growth, i.e. from semelparity to iteroparity. This flexibility appears to be an adaptation to the unpredictability of pollinators at higher altitudes, since scarlet gilia rarely shows iteroparity at lower elevations where pollinator densities are more reliable.

7.3.2 Number and size of offspring

A pattern often seen in life histories is a trade-off between the number and the quality (or size) of offspring produced during a single reproductive event. Generally, high numbers of offspring are associated with smaller sized offspring and less parental investment in each. This is seen in barnacles and other marine invertebrates that cast many larvae to the sea to be dispersed to new anchoring sites. Many, small offspring are often associated with environments characterized by high mortality in younger age groups. For example, rabbits and mice, which suffer high levels of predation, produce relatively large numbers of relatively small young.

In contrast, a low number of offspring is usually associated with larger individual size and more parental investment per offspring. Where increased investment will lead to better establishment of young, such as in the forest understorey, large offspring are produced. The forest understorey has low light, and plants grow slowly. To become established in a mature forest, a seedling needs a large store of energy, represented by the endosperm in seeds. Plants such as oak, walnut and chestnut all have large seeds with a store of energy that seedlings can use to establish themselves. Primates and large mammals, such as elephants, have young that need several years to mature to adulthood. These animals also have extended periods of parental care that are important to the survival of their offspring.

7.3.3 Age at first reproduction

The age at which an individual first reproduces can have a large effect on the rate of population growth and an individual's lifetime reproductive output. Organisms such as mice, cockroaches and zooplankton reach reproductive age very early, and their populations are characterized by rapid growth. However, if reproduction is costly in terms of resources, then early reproduction may not be possible. Organisms that delay reproduction can use for growth resources that would otherwise have been allocated to gamete production, nest building or production of nectar and fruit. If size is positively related to fecundity (as it is in many trees and fish), and if there is a good chance of surviving to older age categories, then lifetime reproduction may be maximized by delaying the age of first reproduction.

7.3.4 The response of life histories to the environment

An example of how environment can influence life history traits is seen in the work of David Reznick, John Endler and their various colleagues. For years, these researchers studied the life history characteristics of guppies, *Poecilia reticulata*, small freshwater fish that are common in the rivers and streams of Trinidad.

In the Aripo river system of Trinidad, populations of guppies occur in isolated pools. Considerable variation in the life histories of the different populations is associated with other fish found in particular pools. There are two major predators of guppies, killifish (*Rivulus hartii*) and pike-cichlids (*Crenicichla alta*). Killifish are smaller than pike-cichlids and their mouth size limits them to preying upon smaller guppies. The pike-cichlids feed on larger, adult guppies. In pools with killifish, guppies are larger, have delayed the age of first reproduction, and have a smaller number of young per reproductive event. In pools with pike-cichlids, adult guppies are smaller, reproduce at a younger age, and have a larger number of young per reproductive event.

Reznick and Endler (1982) hypothesized that differential predation caused the life history differences. To test this, they moved guppies from pools containing only pike-cichlids to pools with only killifish. What they found was that over time, the populations of transplanted guppies evolved so that individuals resembled those that were found naturally in the pool type to which they had been moved. In pike-cichlid pools, there was a low probability of adult guppies surviving to reproduce several times. Therefore, the guppies with the greatest reproductive success would be those that matured at a young age and produced at least one brood before growing into the size range preferred by the predator. In contrast, reproductive success was increased in killifish pools by allocating more resources to growth. By delaying reproduction, guppies could grow more rapidly through the size range preferred by killifish. Once they were too large to be eaten, they could then allocate resources to reproduction.

Guppies are not the only organisms that show flexible life histories. It is common among organisms that inhabit temporary, disturbed or unpredictable environments to show plasticity in the timing and size at which reproduction and other life transitions occur. For example, annual plants that grow in deserts after rainfall show a great deal of flexibility in the size at which they flower. If conditions are dry, and no further rain occurs, they flower at very small sizes. If several rains occur in succession, and soils remain moist, these plants may reach larger sizes and maintain flowering for longer periods.

Many terrestrial amphibians that rely on water during the first stages of life show flexibility in size and age at metamorphosis. For example, the tadpoles of Couch's spadefoot toad (*Scaphiopus couchii*) are capable of changing their size and time before metamorphosis in response to food availability and the rate of pond drying. These toads lay eggs in temporary ponds that vary in depth, and hence in the time it takes them to dry. After the eggs hatch, the tadpoles must feed and grow before metamorphosing into terrestrial adult toads. Larger size at metamorphosis is advantageous, so longer developmental times are preferable. However, if the pond dries out too quickly, developmental time can be shortened, and metamorphosis can occur at smaller sizes. Plasticity in development may be adaptive in this variable environment, because pond depth and duration vary unpredictably.

Life histories may also respond to rare events in the lifetime of individuals. Severe climatic conditions can increase mortality and reduce population numbers. The timing and frequency of large changes in population size, especially decreases, can influence the levels of genetic variation in populations. Conversely, periods of favourable climatic conditions may lead to large population increases and increased levels of competition. They may also result in increased reproductive success for individuals that can exploit the improved conditions. This differential reproductive success among organisms is the basis for evolution by natural selection, as described in Section 1.4.

El Niño events are combined oceanographic and meteorological phenomena characterized by high sea temperatures in the central Pacific. They occur irregularly at intervals of between 2 and 11 years, with an average periodicity of about 7 years. El Niño years have major effects on marine organisms such as fish, plankton and seabirds. The El Niño in 1982–1983 was one of the strongest of the twentieth century.

The islands in the Galapagos Archipelago are subject to the strong effects of El Niño. These islands receive very little rain in normal years, but during the El Niño of 1982–1983, rainfall was 1359 mm, 10 times the previously recorded high point. Most plants on the islands responded to this high rainfall by increasing seed production, some by over 10-fold. The increased plant production led to increased numbers of insects, especially caterpillars. Caterpillar numbers were almost six times higher in 1983 than in 1981.

Isla Daphne Major is a small island in the middle of the Galapagos Archipelago. Populations of Darwin's ground finches have been studied on this island since

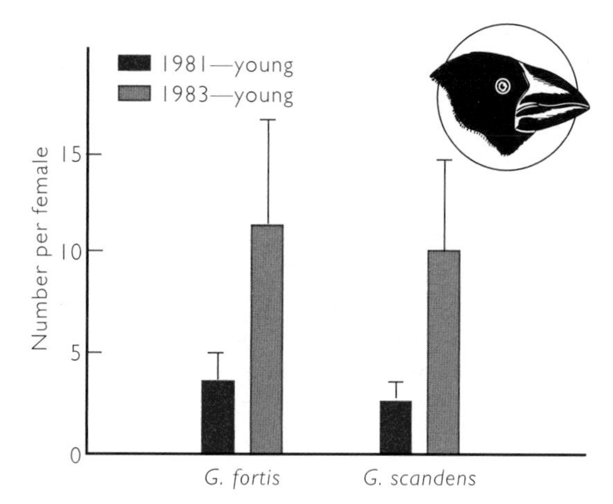

Fig. 7.7 Comparison of the reproductive success of two finch species on Isla Daphne Major in the Galapagos during an El Niño year (1983) and during a year with normal precipitation (1981). The increased number of fledglings was not due to larger clutch sizes, but rather repeated breeding and egg laying.

1976. The most abundant species are the medium ground finch (*Geospiza fortis*) and the cactus ground finch (*G. scandens*). Caterpillars are an important food that finches feed to their young. In addition, the primary food for adult finches is seeds. The high seed and insect food levels in response to increased rainfall resulted in finches breeding continuously for 9 months in 1983. The previous maximum observed breeding period was only 4 months. There were six times as many clutches produced by the medium ground finch in 1983 as in 1981, and three and a half times as many young. The cactus finch produced eight times as many clutches and four times as many young (Fig. 7.7) (Grant & Grant 1992).

The high levels of reproduction during 1983 made up a large part of the lifetime reproduction of the female birds. For females of both species born in 1978, over half of their lifetime production of young occurred in 1983. The increased reproduction in 1983 led to large increases in the populations of both species. *G. fortis* numbers increased from 206 in January 1983 to 733 in June 1983. Numbers of *G. scandens* rose over the same period from 172 to 603.

Finches are capable of living for 12 years or more, so an individual will normally experience only one El Niño. But, these once-in-a-lifetime events can result in a substantial proportion of an individual bird's lifetime reproductive success. The plasticity in life history traits that allowed these finch species to extend breeding beyond the normal end of the season represents an important adaptation to this variable, rain-limited environment.

7.3.5 Classification of life histories

The changes in life histories that occur with changes in the environment and the repeated occurrence of particular combinations of traits have led ecologists to attempt to classify life histories into different types. One classification system arrays life histories along a gradient from '*r*-selected' species to '*K*-selected' species (Fig. 7.8). The *r* and *K* come from the logistic model of population growth (see Chapter 6). The term *r* is the intrinsic rate of growth, and *K* represents the carrying capacity of the environment. Accordingly, *r*-selected species are those that combine life history traits that maximize reproduction, leading to rapid population growth. *K*-selected species are those that combine life history traits that maximize the growth rate of individual organisms, leading to success in competitive or crowded conditions.

Traits found in *r*-selected species include early maturity, the production of many young during a single reproductive bout, low levels of parental investment in the young, and short generation times. These species tend to occur in unstable, temporary habitats. Examples include small, weedy plants, aphids, mice and zooplankton.

In contrast, traits associated with *K*-selected species include delayed maturity, few young per reproductive episode, high parental investment in the young, and long generation times. These species are associated with stable, predictable habitats. Examples include coconut

r-selected species
Unstable, temporary habitats

K-selected species
Stable, predictable habitats

Fig. 7.8 *r*–*K* gradient in life history traits. Traits may vary from those of an extreme *r*-selected species to those of an extreme *K*-selected species across environmental gradients of habitat stability and disturbance.

palms, whooping cranes, whales and elephants. Many of the most endangered species worldwide are *K*-selected species. Their slow reproductive rates and long generation times make them especially susceptible to habitat loss and reductions in population numbers.

Most species possess life history traits somewhere intermediate between the two extremes of *r*- and *K*-selected species. As seen in Section 7.3.4, many species also show some plasticity in life history traits that allow them to alter their position along the *r–K* gradient (Fig. 7.8) in response to environmental conditions.

7.3.6 Life history studies as tools to conserve biodiversity

There are some final steps needed before our imaginary Zimbabwean ecologist from Section 7.1 can develop a comprehensive management plan for local hippopotamus populations. In order to apply life history and demographic information to this problem, the ecologist would need information on habitat requirements (landscape ecology is discussed in Chapter 11) and sources of mortality. However, even if habitat is set aside for hippopotamuses and sources of mortality such as hunting, disease and competition with domestic animals are reduced or controlled, there may still be limits to what can be done to help a population recover. Populations, especially for an animal such as a hippopotamus, may take a long time to recover. This is because hippopotamuses take years to become sexually mature, have only one or two young at a time, and invest several years in caring for their young, preventing them from having more young in succeeding years. Information on hippopotamus life history traits and population demography will, however, allow the ecologist to project how rapidly the population can increase, and what life history stages are the most crucial to protect.

At least 97 species of bird have become extinct since 1600, and 90% of these were restricted to islands. Most of these extinctions can be attributed to human introductions of diseases and alien species, and habitat alteration or destruction. Because of reductions in population size, many more bird species are currently endangered, and 39% of these are restricted to islands. Peter and Rosemary Grant have been observing Darwin's finch species in the Galapagos Archipelago for over 30 years (Section 7.3.4). They have shown how Darwin's finches are ideal for understanding how small populations respond to events that could lead to extinction. Because the birds have small population sizes and because they do not disperse readily to other islands, the Grants have been able to conduct complete counts of births and deaths in these birds over long periods. These studies have become classic examples of population demography and the study of life histories in birds.

Four cohorts (1975–1978) of two species of finch *Geospiza fortis* (medium ground finch) and *G. scandens* (cactus ground finch) were followed by the Grants every year up to 1991 on Isla Daphne Major. Life tables were then constructed to follow reproduction and outline the life history of these birds (Table 7.3). These life tables can also be used to identify especially sensitive stages in the life history. As described in Section 7.3.4, Isla Daphne Major experiences extreme variation in rainfall, from 0 mm per year in drought years, up to 1359 mm per year during an El Niño. The effects of El Niño years have already been described, but individual reproduction and population size also respond strongly to dry years, especially when there are two or more consecutive years of drought.

From Table 7.3, it can be seen that the finches suffered high mortality in the first year (almost a half of all birds died), and lower mortality afterwards. Survival of young birds was especially affected by variation in rainfall and food supply. These birds eat seeds and feed caterpillars to their young, so with high rain there is high seed supply, high caterpillar numbers and a high survival of finch juveniles. With drought, there is reduced seed supply, increased competition for seeds, fewer caterpillars, increased mortality and lower survival, especially of younger birds. There are also differences in life history between the sexes, with males living longer than females, taking longer to produce young and having generally lower reproductive values (Fig. 7.9). This is because some males fail to mate each year, while every female is able to mate.

The higher reproductive values for females compared with males stemmed from their earlier maturity and an unequal sex ratio, with more males than females. This pattern of fluctuation in reproductive value is somewhat unusual for bird species. A more typical pattern is a low initial reproductive value, rising to a peak at the beginning of reproductive age, and then staying near the peak until the birds reach post-reproductive stages.

Table 7.3 Life table of the 1975 cohort of *Geospiza scandens*. (Adapted from Grant & Grant 1992.)

Age (x)	Number of individuals of age x (N_x)	Survival from birth to age x (l_x)	Mortality of individuals from age x to $x + 1$ (q_x)
0	82	1.00	0.49
1	42	0.51	0.21
2	25	0.31	0.10
3	17	0.21	0.04
4	14	0.17	0.00
5	14	0.17	0.01
6	13	0.16	0.00
7	13	0.16	0.00
8	13	0.16	0.07
9	7	0.09	0.04
10	4	0.05	0.00
11	4	0.05	0.00
12	4	0.05	0.02
13	2	0.02	0.00
14	2	0.02	0.01
15	1	0.01	0.01
16	0	0.0	0.0

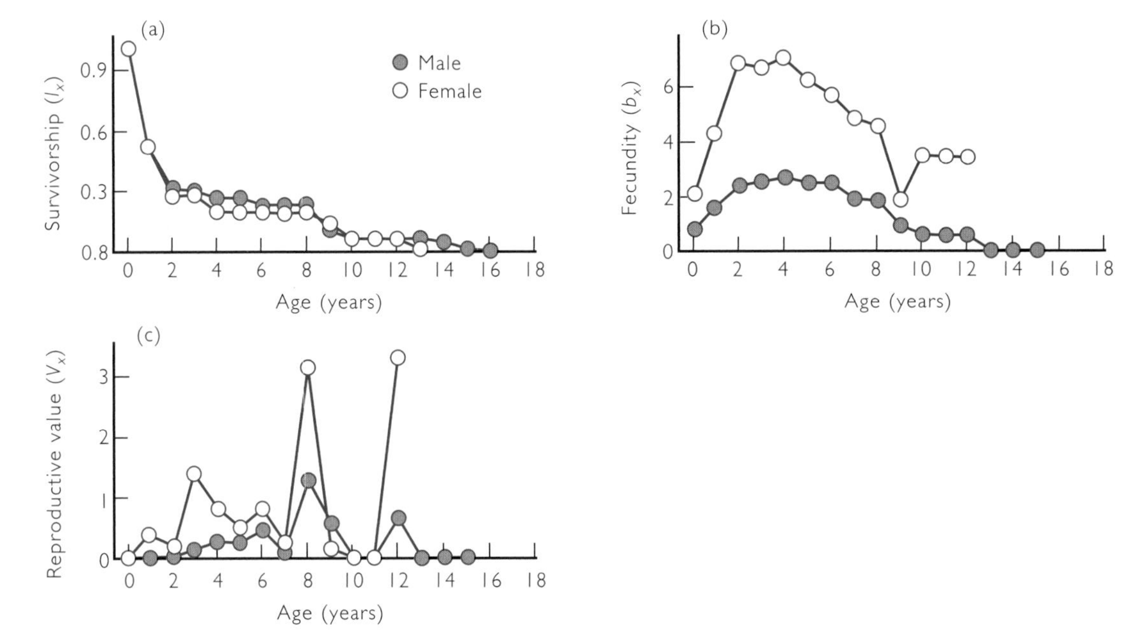

Fig. 7.9 (a) Survival, (b) fecundity and (c) reproductive value for males and females of the medium ground finch cohort born in 1975 on Isla Daphne Major. Peaks in reproductive value for females are associated with years of high precipitation.

Thus, the demographic structure of Darwin's finches is dominated by the fluctuating climate in their habitat. Adult numbers can change by a factor of five or more, and mortality of young can vary from 0% to 50% or more. The features that are most important for the persistence of these finch populations include a long lifespan, a flexible period of maturity, a high survival of older adults during drought, and a high reproductive rate under favourable conditions. From a conservation biology perspective, maintaining healthy populations of these birds entails sustaining high enough numbers to allow for the effects of dry years. Habitat destruction, introduced diseases or introduced competitors reduce native populations of these birds. Thus, Darwin's finches are especially vulnerable to a combination of extended droughts with habitat disturbance, introduction of diseases or the introduction of competing exotic species.

7.4 Chapter summary

Demography is the study of a population's vital statistics. These statistics are used to construct life tables that contain information on the age structure, sex ratio, survival and fecundity of the population. Stage-specific reproduction and survival data from life tables can be used in matrix models to project future population growth.

Suites of traits such as longevity, breeding patterns, number of lifetime reproductive episodes and stages of growth are grouped together by ecologists to form a life history. Patterns in life history traits can be used to examine trade-offs in the allocation of resources to the competing functions of growth, survival and reproduction. Life history traits have also been useful in investigating species' adaptations to environmental gradients and variation.

Demographic and life history studies have been used extensively by ecologists in the past to examine population dynamics. Studies have identified crucial life stages for organisms, differences among the sexes in behaviour, the importance of plasticity in life history responses, and the relative contribution of individuals in different stages to population growth. These studies have and will play an important role in preserving and understanding local and global biodiversity.

Recommended reading

Benton, T.G. & Grant, A. (1999) Elasticity analysis as an important tool in evolutionary and population ecology. *Trends in Ecology and Evolution* **14**: 467–471.

Higgins, S.I., Pickett, S.T.A. & Bond, W.J. (2000) Predicting extinction risks for plants: environmental stochasticity can save declining populations. *Trends in Ecology and Evolution* **15**: 516–520.

McHugh, D. & Rouse, G.W. (1998) Life history evolution of marine invertebrates: new views from phylogenetic systematics. *Trends in Ecology and Evolution* **13**: 182–185.

Partridge, L. & Harvey, P.H. (1988) The ecological context of life history evolution. *Science* **241**: 1449–1455.

Chapter 8

Interspecific competition

8.1 Competition between different species

When different members of the same species are living in the same place, they tend to require identical resources. As the population grows, these resources will inevitably end up in short supply, so the main effect of competition between individuals of the same species (known as **intraspecific competition**) is to control the total number of individuals in a population. This is reflected in the concept of the carrying capacity (see Section 6.3.5), which is the number of individuals of a particular species that can exist on the available resources in a particular patch of habitat. If intraspecific competition is considered in terms of individual organisms, it is clear that its main effect is to determine which particular individuals persist within the population.

Competition between different species (known as **interspecific competition**) is entirely different. Individuals of the different species require similar resources, but they do not necessarily have identical requirements. Their niches may be extremely similar, in which case competition may be intense, or the niches may be almost completely different, with just a small amount of overlap. In this case, competition may not be especially important in the dynamics of either population. The variation in the overlap of the requirements of the different populations, and the ways in which they use resources, mean that interspecific competition can create a huge diversity of outcomes for the populations involved.

Thus, as this chapter develops a perception of interspecific competition, it will be necessary to look closely at how the mathematical models built in previous chapters can be altered to help understand the effects of competition on populations of organisms and on the individuals of which they consist.

Once the chapter has investigated the details of how competition operates between different populations, it will be possible to treat interspecific competition as part of the biological environment, just as rainfall is part of the physical environment. In doing so, it will be desirable to examine the ways in which the effects of competition on individual organisms translate into consequences for the distribution and abundance of populations.

8.2 Interference and exploitation

Two individual organisms can compete with one another in two different ways. The first is a very direct form of interaction and, in the case of animals, this may even result in physical fighting. In the other kind of competition, the two organisms may not even be in the same place at the same time, but if one of them uses a resource, at a particular point in time, then that resource will not be available for the other organism to use in the future. These two kinds of interactions are sometimes referred to as **direct** and **indirect** competition, but these terms can be confusing because the words can have different meanings, so some ecologists prefer to call them **interference** and **exploitation** competition, which is the practice followed here.

In fact, ecological competition comes in so many different forms that it is impossible to categorize every example into one of these two types, and they are probably best seen as two ends of a spectrum. However, the terms make a useful shorthand to summarize the variety of competitive interactions that occur in the ecological world, and examining these two extremes will be a reasonable start in developing an understanding of competition between different species.

8.2.1 Interference competition

One of the most dramatic forms of competition between species only occurs in the animal kingdom, but it is so severe as to serve as a useful extreme case. Some animals have evolved to make life easy for themselves by allowing members of other species to find and even collect their food for them. Then, they simply steal the food from the organisms that did all the work. In the most extreme cases, the losers even partially digest the food before the winners steal and eat it.

This kind of theft occurs in some seabirds, such as the skuas and jaegers, which will chase another bird that has recently caught some fish, until the victim regurgitates its meal for the jaeger to eat. The parasitic jaeger (*Stercorarius parasiticus*), also known as the arctic skua, provides one of the most dramatic examples of this kind of behaviour. Outside the breeding season, these birds obtain most of their food by behaving as pirates and taking the food of a variety of gulls, terns, auks and petrels. The jaegers force their victims to drop their meals as they are flying back to land, and the pirates then catch the food before it hits the water. Puffins (*Fratercula arctica*) may lose as much as 5% of the food they catch in this way.

This kind of competition is important because it forces the victim to spend more time and use more energy looking for food than it otherwise would, which reduces the time it has for other important activities (such as guarding its young or removing parasites from its body), and also reduces the energy available to its body to repair wounds, lay down a layer of fat or produce eggs and sperm. This lack of energy may have a variety of effects, such as making the organism's immune system less able to fight off an infection.

This disruption of the energy balance is especially dramatic in the African wild dog (*Lycaon pictus*), which often has its food stolen by spotted hyenas (*Crocuta crocuta*). Wild dogs hunt in packs and eat small antelopes, and they normally hunt for between 3 h and 4 h each day, expending about 15 megajoules (MJ) of energy per dog per day. But if they lose 25% of their food to hyenas (which is not unknown), then there is a major alteration to the balance between the energy they obtain from eating and the energy they use in hunting; this means that they must hunt for 12 h a day (Gorman *et al.* 1998). However, they are already operating close to their physiological limit, so this increased amount of activity cannot be sustained over long periods. This may be one reason why wild dogs are rare, and why they seem to be especially susceptible to this form of competition from hyenas.

A less spectacular form of interference competition in animals comes from the African elephant (*Loxodonta africana*) and the black rhinoceros (*Diceros bicornis*) in sub-Saharan Africa. When female elephants encounter rhinos, they stand their ground and refuse to be moved away from whatever resource it is they are using (water, food, etc.), and the rhinos are eventually forced to move on. For some reason, male elephants are less likely to win in this kind of competition and they are often displaced by rhinos (Berger & Cunningham 1998). Although this can be classed as interference competition, no physical contact need take place.

In fact, although interference competition is sometimes thought of as 'behavioural', and hence perhaps limited to motile organisms such as animals, other types of organisms can compete in ways that are just as direct or interfering. For example, many plants are suspected of producing chemicals that inhibit the germination or growth of their neighbours. It is extremely difficult to demonstrate this effect in wild populations of plants, but laboratory experiments leave little room for doubt that the process occurs, and that it can be a powerful feature of interactions between species.

The annual wormwood (*Artemisia annua*) is a native of China that produces a chemical known as artemisinin, which may be important in competition between the shrub and other plant species. Artemisinin has a bitter taste and is used in the preparation of drinks such as vermouth, and also in the production of some medicines. Lydon *et al.* (1997) demonstrated that artemisinin and other chemicals produced by the wormwood bush inhibit the germination of a variety of other plant species. When they mixed leaves of the wormwood into the soil in which they grew their seeds, they found that the presence of the leaves inhibited more than 80% of the germination of the redroot pigweed (*Amaranthus retroflexus*), even when the wormwood leaves constituted less than 1% of the weight of the soil mixture.

The closely related species *Artemisia campestris* also shows some inhibitory effects on the germination of other species of plants, as well as exhibiting interference competition with fungi. When the important chemicals are extracted from the plant and applied to the soil, they appear to prevent some fungi from growing properly (Yun & Maun 1997). This affects the local grass plants, which normally have a favourable interaction with the fungi. The fungi link up with the grass roots, and it is believed that they provide the grass with minerals in return for carbohydrates given to the fungus by the grass. In other words, the wormwood plant not only affects the grass directly (by reducing the success of its seeds in germinating) but also indirectly, by interfering with its relationship with the fungus.

8.2.1.1 Territories defended against individuals from other species

Many species of animals defend territories against other individuals of their own species. This is true of many birds, mammals, insects and a large variety of other kinds of animals. Defending a territory uses energy and can be dangerous, but evolution has favoured territoriality in some species because of the benefits of exclusive ownership of resources.

In a few cases, animals also expend energy excluding individuals from other species from their territories. Individual white-tailed damselfish (*Pomacentrus flavicauda*), which live on the Great Barrier Reef in Australia, are known to evict individuals of more than 30 species from their territories. Whenever people study the ecological interaction between the damselfish and these other species, it generally turns out that there is competition for food (e.g. Roberts 1987).

8.2.2 Exploitation competition

Not all competition is as obvious as the violence and poisoning involved in some examples of interference competition. In many cases, it may happen that the competing organisms do not even encounter one another. This kind of competition is often called exploitation competition, because it is concerned with one organism exploiting resources that would otherwise be used by another organism.

Exploitation competition is reasonably easy to demonstrate because, essentially, all that is needed to show its effects is that the presence of one population negatively affects a second population. It is not necessary to prove which resource is being competed for, but it is possible to be certain that one of the requirements of the competing organisms is indeed in short supply.

The karoo is a dry shrubby habitat in South Africa, where the rust fungus *Ravenelia macowaniana* induces the formation of galls on *Acacia* bushes. The tissue that forms the galls is used by a variety of species of moths as a food resource for their developing caterpillars. McGeoch and Chown (1997) looked at 12 different species of moths from seven different families, including the tortricids, the noctuids and the pyralids. Each of the 12 species had caterpillars that live inside the galls. In a typical living gall, there was an average of 11 species, so that the potential for interspecific competition was considerable.

The researchers asked whether the different species of caterpillars competed with one another for the resources inside the galls, and two lines of evidence led them to suggest that competition is indeed important in this assemblage of species.

First, they found that, for each gall, there was a limit to the number of individual caterpillars that could eventually emerge as adult moths. In galls with fewer than 13 individual caterpillars there was a simple relationship that the greater the number of caterpillars in the gall, the greater the number of adult moths that eventually emerged. But for galls with more than 13 caterpillars, this relationship broke down. Whatever number of extra caterpillars were living in the gall, they could never gain enough resources to develop fully. All of the available food in each gall was used by the initial 13 caterpillars.

The second line of evidence noted by McGeoch and Chown was that the moths that emerged first from the galls tended to be larger than those that emerged later. The earliest moths appeared to be using a larger proportion of the resources, and each successive emergent moth left fewer and fewer resources for the moths that followed.

In other words, at the natural densities at which they occur in the karoo, the 12 species of caterpillars competed for food, with some individuals being denied access to essential resources that had been used by individuals of other species.

Some of the best examples of exploitation competition come from studies of competition among plant species. Here, contact is often very important, and competition is usually a local phenomenon, occurring between immediate neighbours. For example, there is a prairie habitat in Minnesota that is low in nitrogen, and the key to competitive success there is the depletion of nitrogen in the patch of soil immediately surrounding an individual grass plant. The most successful competitors are those plants that can develop the most extensive root systems, enabling them to exploit the soil's resources fully. However, nitrogen reduction by these successful competitors is restricted to a region within 1 m of the plant's centre (Tilman 1989). By exploiting the nitrogen in its own vicinity, an individual plant is using resources that are not then available to other competitors.

8.3 Describing interspecific competition with a mathematical model

Chapter 6 developed a mathematical model of how populations behave, and Section 6.2 noted that it would be necessary to alter the model to take account of interactions with the biological world. This section will begin to incorporate interspecific competition into the model.

The model described by eqn 6.12 uses the reproductive rate, r, the population size, N, and the carrying capacity, K, to describe a single population:

$$\frac{dN}{dt} = rN\left(1 - \frac{N}{K}\right) \qquad \text{(eqn 6.12)}$$

To take account of interspecific competition, it is necessary to write down two equations, one for each of the two competing species. For example, it is possible to write down the equations for two of the moth species that compete for food in the galls of *Acacia* bushes in the karoo of South Africa. This is achieved by applying a marker to each of the elements in the equation, so that it is clear which of the two populations is being referred to. So r_1, N_1 and K_1 are the reproductive rate, population size and carrying capacity of one population (perhaps the population of one of the species of noctuid moths) and r_2, N_2 and K_2 are the values for the second population (the population of one of the pyralid moths). The model will be developed by concentrating on species 1.

Section 6.3.3 dealt with an imaginary population of frogs, and it was clear that the reproductive rate of the population depended on inherent biological characteristics such as the number of eggs that were laid by each female every year. Likewise in plant species, the number of seeds or spores produced by each plant every year will be the greatest determinant of the basic reproductive rate. In altering the model to take account of competition, the role of the reproductive rate has not been changed. It remains part of the term 'r_1N_1', which describes the total number of potential new recruits to the population. It does this very simply by multiplying the number of organisms present by the average number of young or seeds produced by each organism.

Note, however, that it is not necessary to assume that the value of the reproductive rate is unchanged. One of the effects of competition with other species may be that organisms are unable to produce the same number of offspring as they would in the absence of competitors. For example, if a species of toad were to be introduced to the pond in which the imaginary frog population is living, the toads may utilize some of the same foods as the frogs. Although the frogs may be able to persist, it is entirely possible that there is a reduced amount of energy available to each female, so that instead of having enough to maintain her own body and to produce 10 eggs, she has only enough for her own maintenance and to produce three eggs.

Recall that the carrying capacity, K_1, represents the total number of individuals of a species that can occupy a particular patch of habitat, and that this limit is set by the total amount of whichever resource is in shortest supply; once all of this resource is being used, the population simply cannot expand further. The amount of this resource already being utilized is set by N_1, the total number of individuals currently in the population. In eqn 6.12, **N_1/K_1 is a measure of what proportion of available resources is currently being used.** When there are few individuals in the population, N_1 is small and N_1/K_1 is a small fraction. But when the population approaches its carrying capacity, N_1 is almost as large as K_1, so N_1/K_1 approaches 1.

8.3.1 Competition coefficients

Where the population of one species (called species 1) is competing with a population of another species (species 2), the second species is using some of the resources. In other words, N_1 is no longer a good measure of the resources that are already being used. It has become essential to include the resources being used by the second species. The amount of resources used by the population of species 2 will depend on two things. First, it depends on the number of individuals of species 2 that are present, N_2; the more individuals of species 2 that are present, the more resources they will use. Second, the amount of resources used by the population of species 2 will depend on how much resource each individual uses.

For current purposes, it is desirable to measure this relative to how much resource would be used by a member of population 1. To do this, ecologists use the value $\alpha_{1,2}$ which represents the effect of adding one individual of species 2 onto an existing population of species 1. This is the **competition coefficient**, and the larger the value of α, the greater the intensity of the competition.

If $\alpha_{1,2} = 3$ then each individual of species 2 uses three times the amount of critical resources that would be used by a single individual of species 1. In effect, this means that, for every individual of species 2 present, the carrying capacity for species 1 is decreased by three individuals. And if there are six individuals of species 2 present (i.e. $N_2 = 6$), the carrying capacity for species 1 must be reduced by $3 \times 6 = 18$.

More generally, the overall effect of species 2 on the population of species 1 is

$$\alpha_{1,2} \times N_2$$

Thus, eqn 6.12 can be modified by changing the term N_1/K_1, which gives information about how full the available carrying capacity currently is, for the first species. This is achieved by adding the effect of population 2 ($\alpha_{1,2} \times N_2$) to N_1.

Thus the overall effect of all the individuals of species 1 and species 2 is

$$N_1 + (\alpha_{1,2} \times N_2)$$

so that overall, the equation now looks like this:

$$\frac{dN_1}{dt} = r_1 N_1 \left(1 - \frac{N_1 + (\alpha_{1,2} N_2)}{K_1}\right) \qquad \text{(eqn 8.1)}$$

It is now possible to see that when the population of species 2 is large (i.e. N_2 is large), or if the competition coefficient ($\alpha_{1,2}$) is large, then the effect on species 1 will be considerable. In other words, if two species are competing, the greatest effect on one species will occur when the second species is common or when individuals of the second species use large quantities of the resources required by the first species.

Making direct measures of the value of competition coefficients is difficult. The only real way to ascertain the effect of individuals of one species on the population of a second species is to manipulate populations so that they contain more or fewer individuals than they otherwise would, and to observe the effects on populations of other species. Such experiments are difficult, and are rarely performed.

8.4 Variations on the basic, simple concept of competition

The models that have been developed, and the examples that have been investigated, provide a basic and relatively simple concept of competition, in which two populations of reasonably similar species share a single easily identifiable resource such as food, water or nutrients. In fact, the study of ecological competition has revealed a more complex picture. Nevertheless, it is generally not difficult to see how unusual examples of competition can fit into the simple theoretical framework described by the mathematical model captured in eqn 8.1. Sections 8.4.1–8.4.4 will examine some of the peculiarities revealed by ecologists' studies of competition between populations of different species.

8.4.1 Asymmetrical competition

The two populations in a competitive interaction may not be affected equally. One of the two may suffer much more than the other. For example, a small herb beneath a tree may experience a significant decrease in growth due to shading, but light availability for the tree is not affected by the presence of plants below it. In some circumstances, it is even possible that a population of one species suffers competition from a population of a second species, but that the second species benefits from the presence of the first.

The seaweed flies *Coelopa frigida* and *C. pilipes* live on the kinds of brown seaweeds that are common on rocky shores in northern Europe. Similar seaweeds live on rocky shores in other temperate parts of the world, and are the habitat for various kinds of flies. In the laboratory, the flies will thrive in plastic pots, where they are fed on chopped pieces of the alga known as saw-wrack (*Fucus serratus*). Competition can be demonstrated by comparing the success of *C. pilipes* when a population is kept in a pot of its own and when it shares its pot with a population of *C. frigida* (Fig. 8.1). Over 24 days, about 45 adults emerged from populations of *C. pilipes* kept on their own. In the presence of *C. frigida*, however, *C. pilipes* fared very badly, and over the course of 24 days only a handful of *C. pilipes* adults emerged (Hodge & Arthur 1997).

In other words, the effect of adding a number of *C. frigida* into the 'habitat' is likely to be a reduction

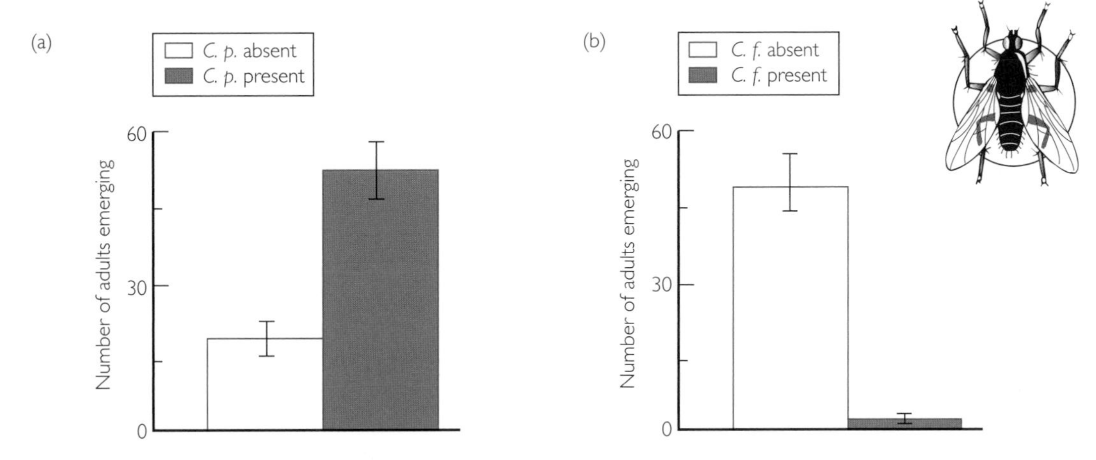

Fig. 8.1 The number of adult seaweed flies emerging from standardized experimental tubes when given chopped seaweed to feed on. (a) The number of *Coelopa frigida* emerging in the presence and absence of *C. pilipes*. (b) The number of *C. pilipes* emerging in the presence and absence of *C. frigida*. The presence of *C. frigida* dramatically reduces the success of the *C. pilipes* population, but the presence of *C. pilipes* increases the success of the *C. frigida* population.

in the eventual size of the population of *C. pilipes*. In terms of competition coefficients, this effect is described by a strongly negative coefficient (which is expressed mathematically by giving the coefficient a minus sign).

In studying the flies, Hodge and Arthur (1997) asked whether this competitive interaction treats the two species equally. In fact, the effect of the experiment on *C. frigida* is exactly the opposite. In single-species populations, about 15 adults were produced in the 24-day period, but when *C. frigida* were kept with *C. pilipes*, the productivity of its populations actually increased, to an average of around 50 emerging adults.

The interaction is clearly very delicately balanced, because the experimental results are different if the seaweed is finely minced before it is presented to the flies. For reasons not understood, under these circumstances, the presence or absence of *C. frigida* has no significant effect on the number of adult *C. pilipes* emerging over the 24-day period.

8.4.2 Resources can include ecological interactions

Traditionally, ecologists think of competition in terms of using resources such as food, space, minerals, water and light. This is reflected in the examples that were considered in Sections 8.2.1 and 8.2.2, involving food for jaegers and African wild dogs, or nitrogen for herbs and grasses.

But competition does not only occur for such obvious resources. If a population of one species benefits from an ecological interaction with organisms of another species, then access to the beneficial organisms can also lead to competition. This is particularly evident among the species of thorn trees of the genus *Acacia* in the East African savanna. The trees must all flower at about the same time of year because of the very seasonal nature of the climate, and they rely on insects to pollinate their flowers.

Unlike some flowering plants (such as many orchids), *Acacia* trees do not attract highly specialized pollinators, and all the species of thorn tree living in an area must rely on the same set of insects for pollination; much of the pollination is performed by solitary bees and greenbottle flies. Because of this, *Acacia* trees of different species are in competition for visits by pollinating insects (Fig. 8.2).

This causes potential problems for each species of thorn tree. If any one of the species were for some reason more attractive to the insects, perhaps because it produced large quantities of nectar, trees of the other species would fail to achieve pollination with sufficient certainty, unless they too evolved some particularly

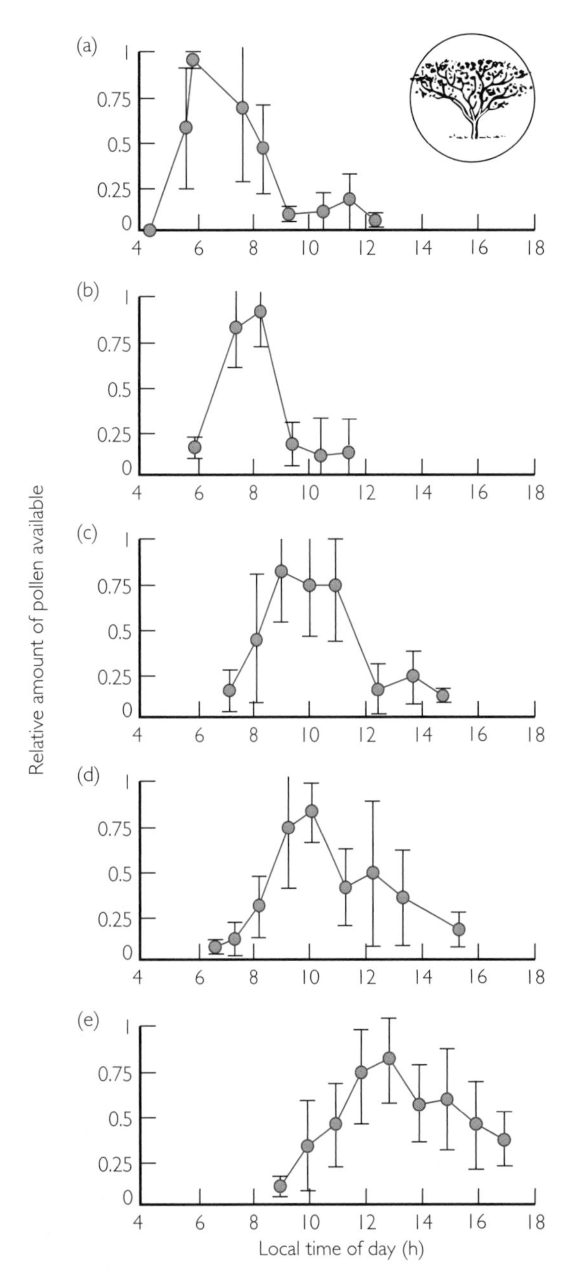

Fig. 8.2 Patterns of pollen availability through the day for trees of different species: (a) *Acacia nilotica*, (b) *A. tortilis*, (c) *A. zanzibarica*, (d) *A. drepanolobium*, and (e) *A. senegal*. Each species releases its pollen at a different time of day, which reduces competition for pollinating insects. (Adapted from Stone *et al.* 1996.)

attractive feature. Evolution would force all the species to offer equally attractive rewards in order to survive.

But assuming that the flowers of the different species are all roughly equally attractive, a further problem arises. Suppose a bee or fly visits a cluster of flowers of *Acacia senegal*, and collects pollen, and that its next visit is to a flowerhead of *A. zanzibarica*. It will deposit the pollen from the first tree onto the receptive surface of the flower of the second, where it will be incompatible. The first tree has wasted its pollen, and the second has had one of its flowers blocked with alien pollen, so that the flower cannot now be pollinated, even if some suitable pollen were later deposited by another insect. Both individual plants have suffered.

Stone *et al.* (1996) studied the *Acacia* species of northern Tanzania, and asked whether the different species can solve this problem by releasing their pollen at different times of day. This is indeed the case: *A. tortilis*, for example, only releases pollen between about 7 am and 10 am, while the flowers of *A. senegal* release their pollen between 10 am and 4 pm (Fig. 8.2).

Competition for insect pollinators has forced the different species to evolve subtle differences in the timing of pollen release, in order to avoid extinction.

8.4.3 Apparent competition

The thorn bushes described in the previous section compete for access to beneficial pollinators. But not all ecological interactions are beneficial, and when the organisms in a population experience a disadvantageous interaction, it is possible to imagine that they may seem to compete to **avoid** the organisms that affect them negatively. In effect, these situations are analogous to interspecific competition, but there is no obvious resource requirement. These circumstances are known as **apparent competition** because competition appears to occur, even though there is no 'real' resource for which to compete.

The instance of apparent competition that is most commonly cited is competition for 'enemy-free space', which is also known as 'predator-mediated competition'. This occurs when two populations are both subject to predation by the same population of predators. It is easiest to understand in terms of enemy-free space, because it is possible to view such enemy-free space as a resource, just as food, sunlight and water are resources.

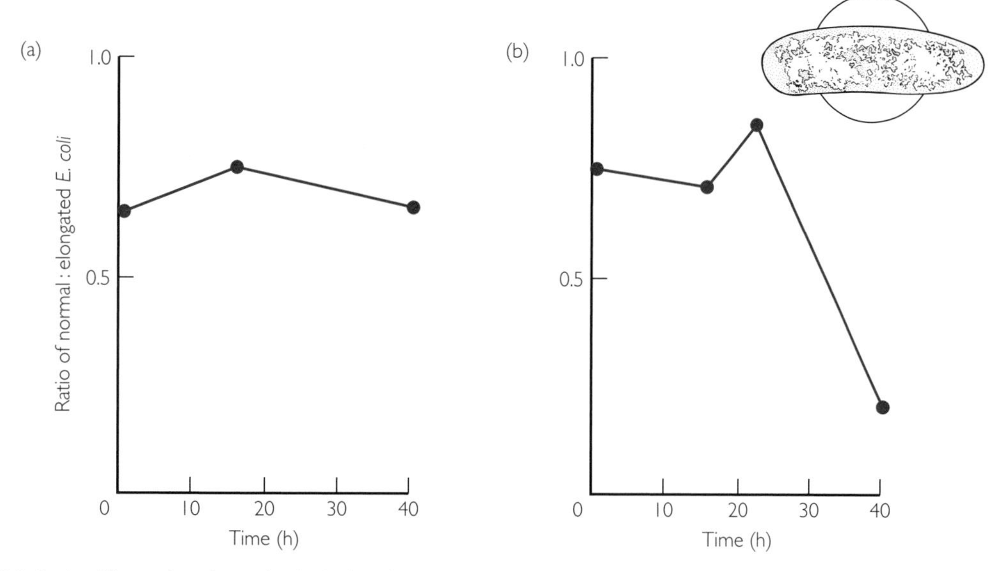

Fig. 8.3 Ratio of the number of normal individuals to elongated individuals in populations of the bacterium *Escherichia coli:* (a) in the absence of the predatory protozoan *Tetrahymena thermophila* and (b) in the presence of *T. thermophila.* The ratio of normal to elongated bacteria falls in the presence of the predator because the predator prefers the normal type. (Adapted from Nakajima & Kurihara 1994.)

In the Mediterranean, two species of hermit crab compete for real enemy-resistant locations. The two species, *Clibanarius erythropus* and *Calcinus tubularis*, both require some kind of protective home because they have soft bodies, and unlike true crabs, do not have a shell of their own. The ideal home is the shell of a dead whelk or similar gastropod, and the number of such shells limits the number of ideal homes. *Calcinus* can find alternative shelter in the empty tubes constructed by marine worms, but laboratory experiments have shown that there is indeed interspecific competition for the shells, which can be regarded as enemy-free spaces (Busato *et al.* 1998). Looked at in this way, there is no reason to think of this competition for empty shells as being qualitatively different from competition for food, and there is in fact no logical difference between apparent competition and the other examples of competition that were investigated in Section 8.2, and which were described with a mathematical model in Section 8.3.

Although enemy-free space is a helpful way of understanding predator-mediated competition, the interaction does not rely on the prey having to seek physical shelter from the predator. *Escherichia coli* is a common bacterium that lives almost everywhere, including inside the intestines of humans. It exists in several forms, two of which can be described as 'normal' and 'elongated', and it suffers from predation by the protozoan *Tetrahymena thermophila.* The predator prefers normal bacteria to elongated ones.

Nakajima and Kurihara (1994) asked whether the presence of the predator acts as a source of competition between the two different strains of bacteria. In the presence of the protozoan predators, the elongated bacteria become much more common than they otherwise would (Fig. 8.3). The elongated bacteria grow faster than the normal ones, so that as their density increases, the amount of food available for the predators also increases. But even though the predators benefit from the increased growth rate of the elongated bacteria, they maintain a preference for the normal type. Thus, as the population of protozoan predators increases, there is greater predation pressure on the normal bacteria.

In other words, the dynamics of the populations of the two bacterial types are not only affected by competition for resources. The competitive interaction between the two types of *E. coli* is strongly affected by the presence of a third population—that of the predators—and the bacteria compete for enemy-free space. At any given time, the space in parts of the tanks that are not occupied by the predatory protozoa is essentially a resource, for which the two types of bacteria compete.

Chapter 10 will return to the concept of apparent competition because it is in fact a good example of what are often called 'indirect' interactions. Here the indirect nature of the competition implies simply that an interaction between two populations is mediated by their interactions with a third population. The word 'indirect' is being used in a different sense from the definition in Section 8.2, where the term 'indirect competition' was used as alternative to 'exploitation competition'.

8.4.4 Different resources may become limiting

One of the effects of competition with another species may be to change which one of many resources is responsible for limiting the size of a population. For example, it is possible to imagine a population of plants that is limited by the availability of phosphorus; there is only enough phosphorus in a field to support 100 individuals of the species, and there is plenty of water, nitrogen and potassium for those 100 individuals. Now imagine that a second species is added to the field, and that this species has different physiological needs. Perhaps it has a life history that involves a massive production of seeds very quickly. To do this, the plants of this second species must produce a mass of flowers, for which they require large amounts of potassium. Thus, plants of the second species appropriate most of the available potassium and lock it into their own biological systems. Consequently, there is only enough potassium left available to support 50 individuals of the original population, and potassium becomes the limiting factor for this population. Phosphorus is no longer the limiting factor for the first population.

This situation is obvious in the plants of the hot deserts of Africa. In Namaqualand in Namibia, it is obvious that water is a strongly limiting resource for much of the time. Droughts are severe, and for long periods, plants exist only as seeds lying dormant in the sand. However, when the rains come, there is an excess of water, and the plants must flower and produce seeds within a short space of time. Space is now at a premium and is a limiting resource for some of the species, such as various *Mesembryanthemum* species and the Namaqua daisy (*Dimorphotheca sinuata*). The colourful flowering plants cover huge areas very densely, and they can only produce as many flowers (and hence seeds) as the available space will allow. A Namaqua daisy cannot use the space that is already occupied by a *Mesembryanthemum*. As the heat of the sun dries the sandy soil, water again starts to be scarce, and in the final period of the flowering, it starts once again to become the limiting resource for which the plants must compete.

8.5 Competition is part of the biological environment

A working definition of ecology was developed in Section 1.5, stressing the fact that the distribution and abundance of populations are determined by the interactions between individual organisms and their environments. Chapters 4 and 5 looked in detail at how physical factors, such as temperature, have a strong effect on the evolutionary fitness of individual organisms, and consequently on their distribution and abundance in populations. The ecology of an organism or population is changed by the presence or absence of some physical resource, such as water or phosphorus.

Because ecological competition changes the availability of resources, it has an equally powerful potential to govern the distribution and abundance of organisms.

Now that the mechanisms of competition between different species have been examined in detail, it is possible to start to appreciate the effects of competition as a part of the biological environment of a population. In doing so, it will be essential to recall that evolutionary changes can occur because of changes in the biological and physical aspects of the environment. Interspecific competition provides examples of the ways in which the biological environment can drive such evolutionary changes in populations.

8.6 The role of competition in structuring ecological communities

Competition between two species could potentially be a strong factor governing the nature of ecological communities. Chapters 11–14 will deal in greater detail with the nature of such communities. In order for that detail to make sense, it will be necessary to have some understanding of the ways in which interspecific competition might play an important role. To begin to appreciate the potential importance of competition in structuring communities, it is essential to consider the possibility of one population entirely excluding another through competition, and also to consider the potential for the coexistence of competing populations.

8.6.1 Competitive exclusion

It is clear that competition between populations of different species is all about the needs of individual organisms. Competition occurs when resources are limited, and in any one interaction between individuals only one of the individuals can end up using the finite resources in question. This leads to the logical conclusion that the individuals of one population might consistently outcompete those of a second population, and potentially drive the second population to extinction.

This situation, known as **competitive exclusion**, is probably quite common in natural ecosystems, although in most cases, it is difficult for ecologists to be certain that it is occurring. A species could be absent from a habitat because individuals cannot cope with the intensity of competition from another species, or because the conditions in the habitat for some reason fall outside the physical conditions in which individuals of the species can survive, or for one of many other different reasons.

Parasitic organisms that cause diseases in people or farm animals make especially good subjects in which to study the total exclusion of populations because there are strong social and economic incentives for eradicating the populations. Massa *et al.* (1998) form just one among many groups of food scientists who have concerned themselves with eliminating harmful bacteria from stocks of poultry. Partly because they are kept in close confinement, commercially farmed birds such as chickens (*Gallus gallus*) and turkeys (*Meleagris gallopavo*) are particularly prone to infection by bacteria such as *Salmonella*, which can cause serious food poisoning in people who eat the poultry. Excluding these bacteria from the habitat formed by the birds' intestines is thus considered extremely important. One of the ways of attempting to achieve this is to infect the birds with other, less harmful, bacteria, which might competitively exclude the *Salmonella*.

Massa and colleagues infected a group of 1-day-old chicks with a variety of bacteria species, not including *Salmonella kedougou*. They did this by feeding the chicks with a suspension of faecal material which was rich in bacteria, but which did not apparently contain any *Salmonella*. Later, these chicks were deliberately infected with the harmful bacteria. The faecal material had conferred some protection on the chicks, and they were less likely to become infected with *Salmonella* than chicks that had not been treated with the suspension of old faeces. The micro-organisms in the faecal material had become established in the intestines of the chicks, and the populations they had formed were able to outcompete the *S. kedougou*.

The food scientists made some attempt to identify which other species of bacteria could be responsible for outcompeting the harmful *Salmonella*. They prepared different suspensions of faecal material that were particularly rich in *Lactobacillus*, *Bacteroides* or *Enterococcus* species, but these did not confer protection from *Salmonella* infection, suggesting that some other bacterium was the important competitor, or perhaps that the ability to combat *Salmonella* infection was conferred by a combination of more than one of the other kinds of bacteria.

8.6.1.1 Demonstrating the effects of competitive exclusion

It can be very difficult to demonstrate whether competition is occurring in natural ecosystems, and if so, whether or not it has a strong effect on the structure of the ecological community that is present. Very few studies are really able to demonstrate experimentally that competition is leading to the exclusion of species from ecological communities, and ecologists are forced to make inferences from other evidence.

For example, someone may observe the distributions of different species and note that two similar species seem never to occur in the same place together. Ecologists might decide that they have evidence for competition, arguing that the reason the two species never co-occur is that one always manages to outcompete the other. This is easy to do by looking at the distribution maps that occur in the field guides that naturalists use to identify organisms in the wild. Kingdon (1997), for example, shows a map of the distributions of the bush pig (*Potamochoerus larvatus*) and the closely related red river hog (*P. porcus*) in Africa (Fig. 8.4). There is a sharp dividing line in central Africa, running through the Democratic Republic of Congo (Zaïre); only bush pigs are found south of the line, and only river hogs north of the line. It would be easy to conclude that in the north, river hogs are capable of

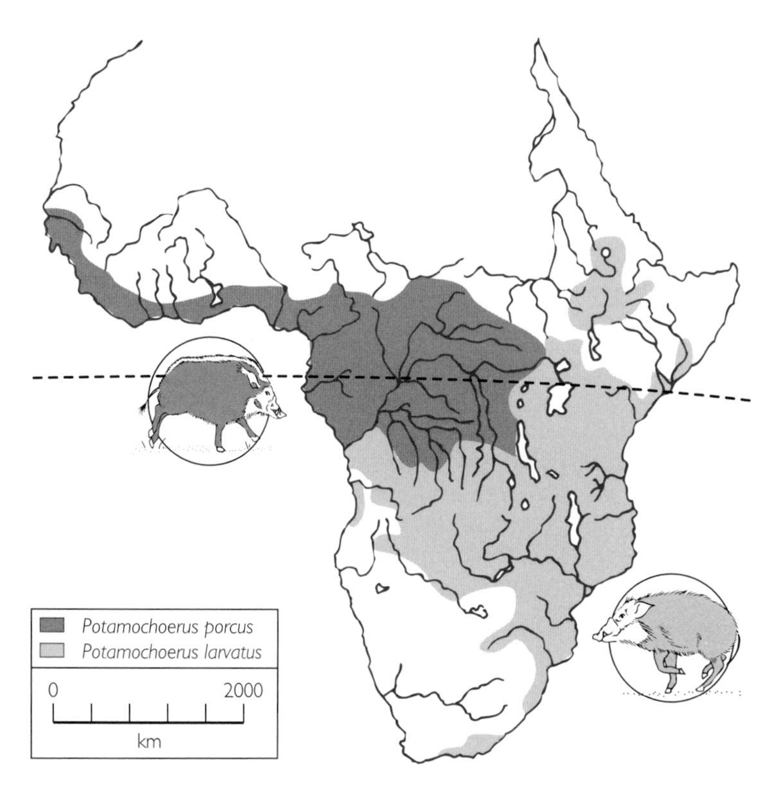

Fig. 8.4 Distribution of the red river hog (*Potamochoerus porcus*) and the bush pig (*P. larvatus*) in Africa. There is a very sharp boundary between the distributions of the two species. (From Kingdon 1997.)

outcompeting bush pigs and in the south, the reverse is true. However, the line actually marks the boundary between the rainforests of northern Zaïre and the savannas of the south. It may indeed be true that if one of the species were to be removed, competition would be eliminated and the other could successfully invade the empty habitat. But it may equally be beyond the physiological capabilities of river hogs to thrive in the arid savannas of the south, or beyond the physiological limits of individual bush pigs to live in the forests of the north. These possibilities may seem unlikely, but they were not impossible. Indeed, there were several examples in Chapter 4 of geographical distributions with sharp boundaries that are known to be determined by the extremes of a species' physiological limits.

To confirm that competition really is the force that is governing the boundaries of populations, stronger evidence is needed, and the only really conclusive proof would come from an experiment that demonstrated clearly that removing one of the species allowed the other to invade the vacated habitat.

Similar boundaries can be seen on a much smaller scale, and here it is easier to perform experiments to confirm the importance of competition. Members of the genus *Idotea* are small crustacea that live in coastal areas in many parts of the world. Where they occur together in Europe, the species *Idotea baltica* and *I. emarginata* tend to have subtly different habitat requirements. *I. baltica* lives among the seaweed that is floating on the surface of the sea, while *I. emarginata* lives among the decaying seaweed on the sea bed. This distinction between surface and sea bed is a boundary analogous to the boundary between the ranges of the bush pig and the river hog in central Africa.

It is possible to establish the crustacea in an aquarium in the laboratory, where the densities of the two species (and other conditions) can be manipulated. In such experiments, Franke and Janke (1998) questioned the extent to which competition might be responsible for creating the boundary between the two species that is observed in nature. They found that when they set up aquarium tanks, each of which contained just one of the species, the populations performed very well, although *I. emarginata* established populations that were four times as great as those of *I. baltica* (Fig. 8.5). However, when both species were allowed to live together,

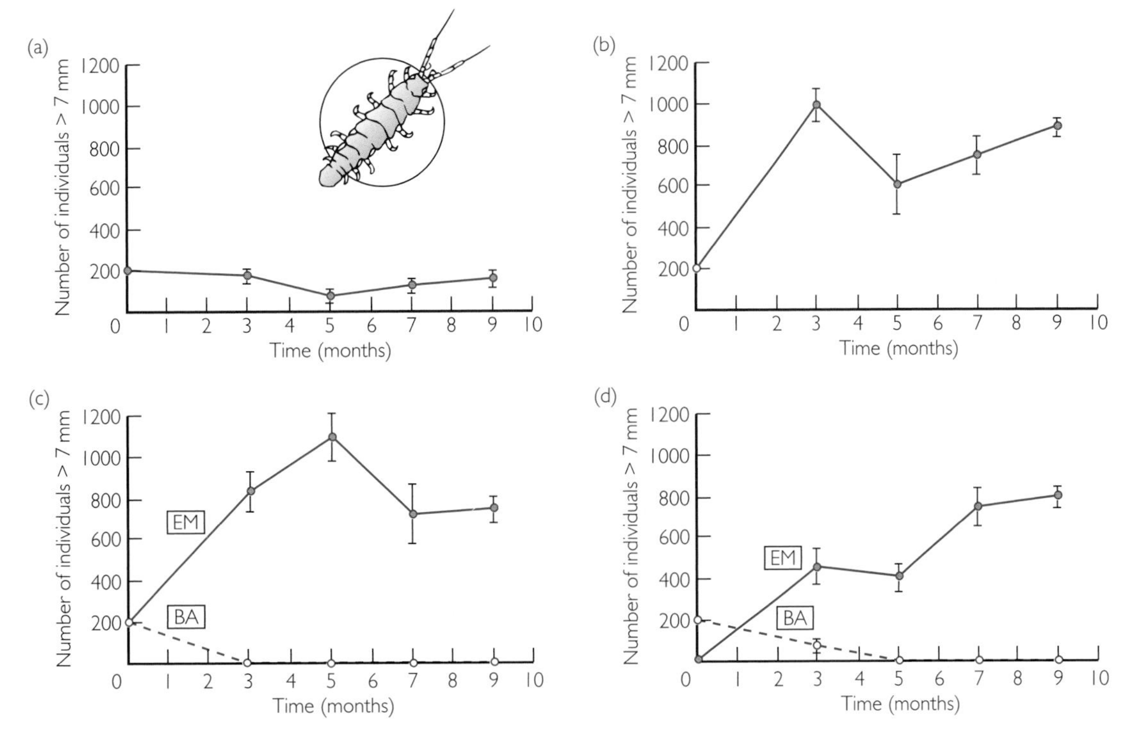

Fig. 8.5 Change in the size of laboratory populations of the two closely related crustacean species, *Idotea baltica* (BA) and *I. emarginata* (EM). (a) Single-species population of *I. baltica* starting with 200 individuals. (b) Single-species population of *I. emarginata* starting with 200 individuals. (c) Mixed population starting with 200 individuals of each species. (d) Mixed population starting with 200 individuals of *I. baltica* and 10 of *I. emarginata*. *I. baltica* cannot persist in the presence of *I. emarginata* even when it starts out with a substantial numerical advantage. (Reprinted from Franke & Janke 1998, with permission from Elsevier Science.)

I. baltica rapidly disappeared, having been outcompeted by *I. emarginata*. This was true regardless of the number of individuals of each species that the researchers used when initially setting up each aquarium. Franke and Janke concluded that *I. emarginata* competitively excluded *I. baltica* from the aquaria. It seems that in the wild, *I. baltica* is competitively excluded from the sea bed, and that this interspecific competition is a strong factor in maintaining the habitat segregation of the species.

8.6.1.2 Competitive exclusion takes time

No competitive process happens instantaneously, and this is especially clear in the case of competitive exclusion. For a variety of reasons, inferior competitors may continue to coexist with the superior ones for long periods of time. In the case of plants, which are immobile, an individual inferior competitor can continue to occupy a site as long as no seed of the superior competitor happens to germinate nearby. Likewise, a patchy habitat may permit the persistence of an inferior competitor, by allowing it to occupy a small patch of habitat that is not optimal for the superior competitor, even though the superior competitor can nevertheless thrive throughout the rest of the habitat.

Disturbance (such as fires in a grassland or storms in a woodland) may permit the continued existence of inferior competitors by creating patches of habitat that are not perfect for other species, which would otherwise dominate the ecological community of the site. Conservationists use this principle when managing nature reserves; they may cause deliberate disturbance in order to arrest the succession of increasingly strong competitor species of plants (see Section 1.10.2). Such actions may encourage the persistence of a rare or desirable species that would otherwise be outcompeted.

8.6.2 Coexistence

Previous sections have described powerful indications that ecological competition can exclude species from communities, but there are also many examples of two species experiencing strong competition with one another while continuing to coexist. There are well-studied ecological and evolutionary processes that are believed to promote the coexistence of competing species and these will be discussed in some detail in Sections 8.6.2.1–8.6.2.3. However, ecologists must bear in mind as they investigate particular instances of coexistence that in many other cases they are a long way from understanding the details of the factors that allow populations of different species to coexist in the face of strong competition.

8.6.2.1 Ecological processes that promote coexistence: niche differentiation

Exclusion is not the only way in which competition between populations of different species can affect the structure of ecological competition. There are other ways in which ecological competition between two species might have a powerful effect on the communities that ecologists observe.

When the ecological niche was defined in Section 1.6, it was noted that there is sometimes a difference between the extent of the resources that a population is theoretically capable of using and the resources that it is, in reality, able to use. The set of resources that is theoretically usable was defined as the fundamental niche, and the resources that are actually used were defined as the realized niche. However, this appeared to imply that the difference between the fundamental and realized niche was merely in terms of geographical distribution. In other words, it concentrated on the fact that there are some *places* that a population could occupy but where it is excluded by competition, predation or some other biological interaction. The previous section demonstrated that this is indeed true—competitive exclusion does restrict the distribution of populations, and keeps them out of some locations that might otherwise offer suitable habitats, and form part of their niches: *Salmonella* bacteria are excluded from the intestines of poultry by other bacteria.

Where ecological competition between two species is very intense, it is not necessarily the case that one of the species is entirely excluded from a particular place. The individuals of a species could occupy a geographical location but be excluded from part of their fundamental niche within that site, because populations of other species outcompete them for some of the resources. At its simplest, this is obvious if we imagine a small herbaceous plant growing in a forest clearing where it is able to use all of the available solar energy. When a large tree grows nearby and casts a shadow over the plant, the herb no longer has access to all of the sun's energy. Individuals of the herb species would still be able to occupy the clearing, but only those parts of it that were not overshadowed by the tree. If someone planted thousands of trees in the middle of a treeless plain, they would change the realized niche of such a herb.

When this situation occurs, the organisms in the population will only survive by focusing on using those resources that remain available to them. This alteration of the overall resource use of the population is known as **niche differentiation**, because different populations have different realized niches.

Findley (1993) reviewed the evidence of earlier ecologists in asking whether bat species use different food resources when they might be competing with other species. The long-eared myotis (*Myotis evotis*) and the southwestern myotis (*M. auriculus*) are two closely related insectivorous bats that live in New Mexico in the southern United States. In places where one of the species occurs in the absence of the other, the populations have a generalist insect diet; this is true regardless of which species is present. However, where the two species occur together, the southwestern myotis tends to specialize on catching beetles, while the long-eared myotis concentrates on eating moths. The fundamental niche of both species includes the use of both beetles and moths as food resources, but it appears that interspecific competition for food brings about niche differentiation. Populations have different realized niches according to the presence and absence of competitor populations.

8.6.2.2 Evolutionary processes that promote coexistence: character displacement

Niche differentiation could be a purely ecological phenomenon. Different populations of the same species use different subsets of the available resources because of the presence or absence of competitors. If the distribution of competitors were to change, there might be an immediate change in the resources available to a population. In the case of the bats described in Section 8.6.2.1, the complete disappearance of the southwestern myotis from New Mexico would increase the availability of beetles as a food source for the long-eared myotis.

Sometimes, however, competition with other species may lead to natural selection driving *evolutionary* change in a population. This change can be manifest as a difference in the physical appearance of the organisms in the population, in which case it is known as **character displacement**.

The whiptail lizards are insectivores, some of which are also known as racerunners; they live in central America, and are common sights on Caribbean islands such as Trinidad and Tobago. Two species of these lizards, *Cnemidophorus tigris* and *C. hyperythrus*, occur on the Baja California peninsula in Mexico. The same two species occur on oceanic islands, but they do not always occur together: some islands have only one of the species.

On the mainland, individuals of *C. tigris* are noticeably larger than those of *C. hyperythrus*. But on islands, the whiptail lizards are almost always intermediate in size, regardless of which one of the two species happens to be present (Fig. 8.6).

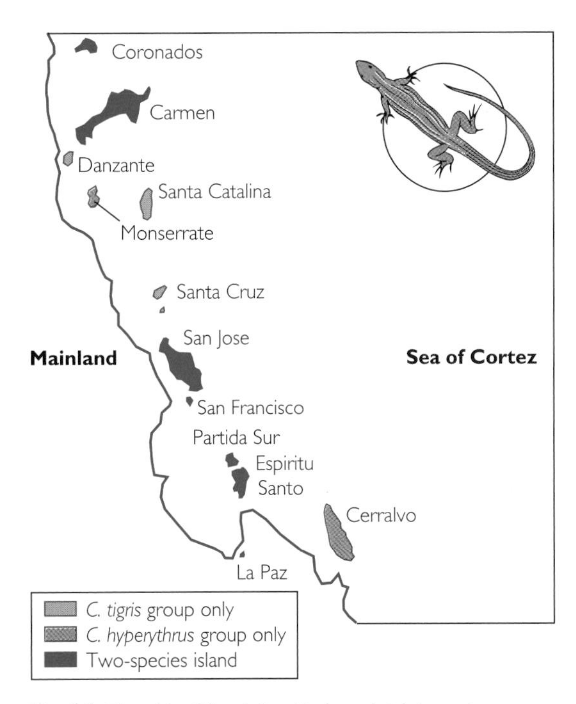

Fig. 8.6 The whiptail lizard *Cnemidophorus tigris* is larger than *C. hyperythrus* where they occur together, but where each species occurs alone on islands in Baja California, all individuals are intermediate in size. (Adapted from Radtkey *et al.* 1997.)

Radtkey and colleagues (1997) asked what could be learned about competition from the sizes and distributions of these lizards. They deduced that many island populations are completely free of competition with their close relatives, while on the mainland, the two species compete for resources, probably food. Where they compete, *C. tigris* specializes on larger prey organisms, while *C. hyperythrus* concentrates on smaller prey. Natural selection has in each case favoured those individuals that happen to be most efficient at capturing prey, so the population of *C. tigris* has evolved a larger size and *C. hyperythrus* has evolved a smaller size. But on islands, where there is no competition, lizards of either species are middle-sized and have a generalized diet. They are able to eat prey of all sizes because there are no competitors to reduce the range of available food types.

8.6.2.3 Evolutionary processes that are not evident as character displacement

Character displacement, like the evolutionary changes in size of whiptail lizards, makes a suitable topic for ecologists to study, largely because displaced characters are easy to measure. But ecologists should not fall into the trap of believing that size, or other obvious and visible characters, are the only features of organisms that can evolve in response to competition between populations of different species.

Various species of fungi in the genus *Epichloë* infect grass plants. Two such types of *Epichloë* infect the grass known as slender false-brome (*Brachypodium sylvaticum*). One of the types is known as symptomatic *Epichloë*, because it causes the set of symptoms known as choke disease; the other type of fungus, known as asymptomatic *Epichloë*, lives within the grass plants, but does not cause any apparent illness.

Bucheli and Leuchtmann (1996) investigated the two types of fungi on the slender false-brome, and although they were not studying competition, there was some evidence that competition does occur between the two types of fungus. The two forms of *Epichloë* were found to infect the same grass species in at least some of the same habitats, and since they needed the same resources, it was possible to postulate that under some conditions, they could suffer exploitation competition from each other. There was even some evidence that niche differentiation had taken place; the two different types of *Epichloë* appeared to be slightly more common in different species of grasses. Grass species that grew in clumps or tussocks appeared to be infected by asymptomatic *Epichloë*, while the symptomatic form occurred more in those grasses with a sprawling growth form. The possible ecological or evolutionary reasons for these differences remain unclear. They are not caused by firm limits to the fundamental niches of the fungi, because in some conditions, both kinds of *Epichloë* can infect both groups of grass species.

Assuming that competition was occurring (at least in some circumstances), the evolved differences between the two types of fungus seem to be equivalent to character displacement, even though no visible, measurable character is involved. The symptomatic and asymptomatic fungi are very closely related, and they had previously been thought to be merely different forms of the same species. So Bucheli and Leuchtmann asked whether they really were different, and were then able to measure sufficient genetic differences to conclude that the two populations were distinct.

Although interspecific competition was not the focus of the study, it seems likely that it caused the differences between the two types of *Epichloë* fungus.

8.6.3 Limits to the similarity of coexisting species

If populations of two different species are competing for the same resources, and individuals of one species are better adapted to obtaining those resources, it is clear that this 'fitter' population will 'win', and will exclude its competitors. But under some circumstances, one or both of the two populations may evolve, and the resultant character displacement will reduce the level of competition, and the two species may continue to coexist.

One population evolves to become a better competitor in some circumstances, and individuals of the other population are more efficient in other situations. In other words, the two populations end up concentrating on slightly different resources. Resources are always limited, and for competition to lead to such coexistence, each of the two kinds of resources must remain in sufficient quantity to support a population. This idea has led some ecologists to suggest a limit to the similarity that can occur in coexisting species. The suggestion is most often made in relation to animals, and the traditional way of expressing it is to look at closely related animal species that appear to use essentially identical resources. For example, ecologists might compare different species of seahorses (*Hippocampus* spp.) in the Indo-Pacific or Caribbean. They are all the same shape, and they all feed on minute invertebrates sucked through a tubular snout. They all use their tails to attach themselves to patches of seagrass (*Zostera* spp.) or to gorgonians (invertebrates that form fan-shaped colonies attached to the sea bed).

South American ornithologists might make similar comparisons among different kinds of Amazon parrots (*Amazona* spp.), of which there are many species, almost all of which have similar nesting and food requirements. Australian mammalogists might compare the kangaroos and wallabies (Macropodidae), which are all remarkably similar in shape, and almost all of which eat either grass or other leaves. In Africa, entomologists might make comparisons among many kinds of dung beetles (Scarabidae), which share many ecological features. In fact, wherever you go in the world, there will be suites of closely related animals that appear to share many behavioural, nutritional and ecological requirements.

The same is true of plants. The genus *Nepenthes* consists of around 70 species, all but one of which are found in similar rainforest conditions in South East Asia. There are at least 35 species of sedges in the genus *Carex* in Iceland alone, and almost all of them inhabit boggy or marshy habitats, and produce almost identical flowers in June. However, much of the ecological research that looks at the limits to similarity of coexisting species has concentrated on animals.

The original ideas about the limits to how similar two coexisting species could be suggested that, for animals, two species that were in all other ways similar could only coexist if they differed in size by a fixed ratio. Evelyn Hutchinson arrived at a specific theory by observing various kinds of closely related animals that co-occurred in nature (Hutchinson 1959). Animals of different size, it might be argued, eat foods of different sizes, so similar species that differ in size will differ in diet, and hence competition will be reduced. According to Hutchinson's ideas, the smaller of the two species would need to be roughly three-quarters the length of the larger. Any closer similarity would mean that

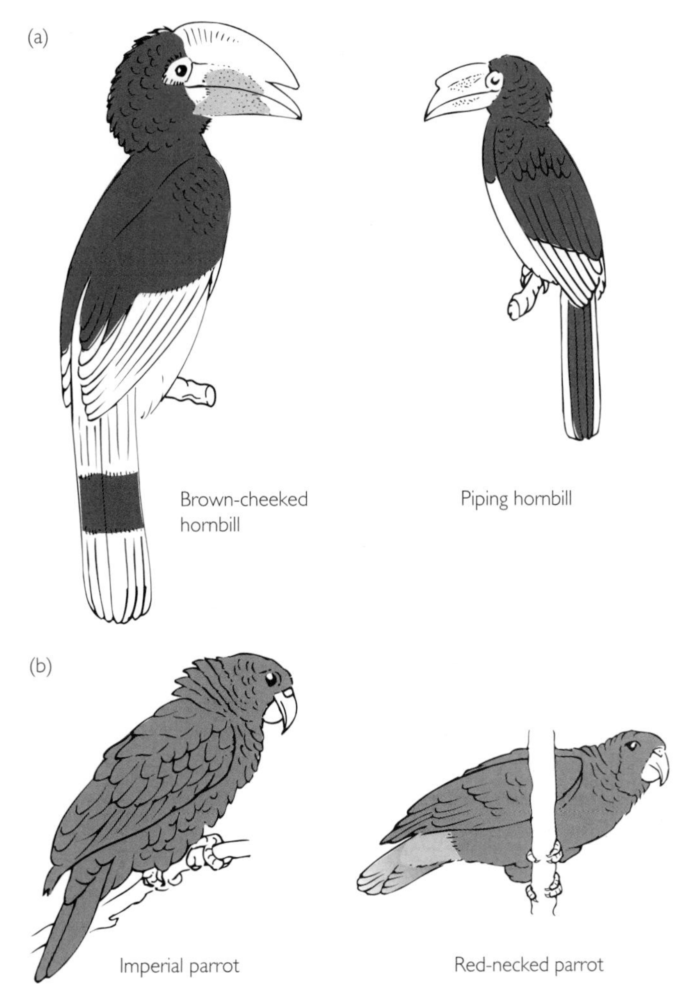

Fig. 8.7 (a) The piping hornbill (*Ceratogymna fistulator*) and brown-cheeked hornbill (*C. cylindricus*), conform to Hutchinson's suggestion that where two related species coexist, the smaller one can be no more than three-quarters of the length of the larger one. (b) The red-necked parrot (*Amazona arausiaca*) and imperial parrot (*A. imperialis*) do not conform to Hutchinson's rule. Both are restricted to the Caribbean island of Dominica, where their ranges show substantial overlap, but the red-necked parrot, at about 40 cm, is about 90% of the length of the larger imperial parrot.

competition would be too intense and one would outcompete the other. It is not difficult to find pairs of similar animals that differ by approximately this much.

The spider that in North America is known as the aggressive house spider (*Tegenaria agrestis*) is in fact native to northern Europe, where it coexists with the closely related *T. sylvestris*. Males of *T. agrestis* have a body about 8 mm long, and males of *T. sylvestris*, at about 6 mm in length, are three-quarters as long. Moreover, *T. sylvestris* does not generally coexist with a third species, *T. atrica*, which is very similar in size. *T. atrica* is restricted to the northern and western parts of the British Isles, while *T. sylvestris* lives throughout central Britain and much of mainland Europe (Roberts 1995). In other words, although the two most similar species do not exist together, the two species that differ in length by the predicted amount do indeed coexist.

Similarly, the brown-cheeked hornbill (*Ceratogymna cylindricus*), which is about 60 cm in length, has an almost identical distribution in West Africa as the piping hornbill (*C. fistulator*), which is about-three quarters of this length (Fig. 8.7) (Kemp 1995).

There are many problems with such a theory: it is not easy to apply it to organisms other than animals, it assumes that size is a good indicator of food preference, and it ignores all the other ways that organisms may differ, independently of size. It is also not difficult to think of pairs of animals that coexist but that are very much more similar in size than the theory would suggest is possible. The 13 Amazon parrot species on mainland

Fig. 8.8 The chiffchaff (*Phylloscopus collybita*) and willow warbler (*P. trochilus*) occur together in the same habitats throughout much of Europe, even though they are almost identical, demonstrating that there is no limit to the physical similarity of coexisting species. One possible explanation is that the populations of both species are kept below their carrying capacities by density-independent mortality during the winter.

South America all have average body sizes that fall in the size range 31–38 cm. Of course, not all of them exist in the same habitats at the same time, but many of their ranges overlap, and the different species manage to coexist without the theoretical minimum difference in size (Fig. 8.7).

8.6.3.1 Any limits only apply to competing species

Although there are no easy rules to define the existence of any limits to the similarity of coexisting species, one thing is clear. Any limits that may exist can only apply to species that are actually competing, not to those that merely appear to have sufficiently similar requirements to make competition possible. Take for example, two species of small songbirds that have almost identical resource requirements, the chiffchaff (*Phylloscopus collybita*) and the willow warbler (*P. trochilus*), both of which are common in northern Europe (Fig. 8.8). Both species occur in woodlands throughout Denmark, where they use almost identical food resources and occupy identical nesting sites. Chiffchaffs and willow warblers are exactly the same size and their behaviour in the breeding areas is remarkably similar. In fact, the two species are so similar that individuals are virtually indistinguishable unless they are held in the hand.

However, it is not possible to use the chiffchaff and willow warbler as an exception, in an attempt to prove that there is no limit to the similarity of competing species, because it is not known for certain that populations of one species suffer competition with the other. It is possible that neither species ever reaches its carrying capacity, perhaps because both suffer from high levels of mortality that are independent of the density of the populations (Fig. 8.8). Both of the two warblers are migratory, and suffer very high levels of mortality as they cross the harsh environment of the north African desert every spring and every autumn. If this mortality limits the size of the populations, then the populations that return to the breeding grounds will be sufficiently depleted so that they are not limited by resources in the nesting area, and hence will not compete with one another.

8.7 Competition between more than two species

Almost all of the examples that have been considered so far have involved two coexisting populations of two different species. In reality, of course, it is far more common for there to be a much larger number of species inhabiting a particular patch of habitat. Any one population might potentially experience competition with a whole variety of others. A single population may compete with a second population for one kind of food resource, with a third population for space, and with a fourth population for a different kind of food. The third population may be competing with the second for water, and with the fourth for shelter. Competition between species 1 and species 3 might be intense, while that between species 3 and species 4 may be less important.

It is easy to see that the networks of competitive interactions that can be built up in this way have the potential to be enormously complex. In fact, the number of potentially competing species in one place can be vast. In the Serengeti ecosystem in Tanzania, there are 56 different species of hawks and falcons (Sinclair & Arcese 1995). Assuming that each of the hawk species competes with only one in 10 of the other species, there would still be something like 300 different competition coefficients describing the competitive interactions among these species. There are presumably also a great many such interactions among the 21 species of antelopes that live in the Serengeti, and an uncountable number among the hundreds of species of plants, or innumerable species of bacteria and micro-organisms.

There is no way that ecologists can ever hope to comprehend the individual sets of interactions that make up such a nexus, although modern computing techniques may allow calculations about what might happen to the network as a whole if one of the interactions were to be upset.

Although ecologists may not understand the precise way in which all the interactions are interwoven, they can look for patterns in the overall structure of the networks, and that may help them to understand their ecology. In fact, these suites of coexisting populations are generally known as **communities**, although different ecologists choose a variety of different names for slightly different kinds of community. Chapters 11–14 will look in greater detail at communities and the interactions that structure them.

8.7.1 Biodiversity

The details of competition described in this chapter make it possible to begin to see how competition is intimately linked with the vast diversity of species that was encountered in Chapters 1 and 2. Section 8.6.2.2 described how competition between populations of different species can lead to evolution within those populations, and hence ultimately to the formation of new species. Sections 8.6.1 and 8.6.3 dealt with how competition can cause the exclusion of species from particular habitats, and how the evolution that competition drives can lead to the coexistence of similar species. In other words, competition between populations of different species appears to have a profound effect on the diversity of species that exist together in space and time.

In fact, ecologists have long debated the importance of competition in structuring ecological communities. Many examples that appear to demonstrate the effects of competition may in reality have been misinterpreted for a variety of reasons. When Radtkey and colleagues (1997) inferred that competition caused the patterns in the sizes of whiptail lizards (Section 8.6.2.2), they did so without any firm evidence that the two species of lizards ever actually compete for resources. It is by no means impossible that some local feature of the physical environment (such as temperature or humidity) led natural selection to favour middle-sized lizards on oceanic islands.

It could be true that populations of *Cnemidophorus tigris* and *C. hyperythrus* are subject to such intense mortality from the harsh dehydrating summer weather of Mexico that their populations never reach a high enough density for the two species to come into competition with one another.

Many unanswered questions arise from the issue of whether or not interspecific competition is an important force in structuring the ecological communities that we see around us. Chapters 11–14 will examine more closely the nature of ecological communities and their composition. In doing so, they will pay particular attention to the potential role of competition in determining whether or not those communities are in an equilibrium, and in governing the distribution and abundance of species within the communities. Understanding the ways in which ecological processes determine the biodiversity of a community can be important in conserving important wildlife habitats.

8.8 Human activities affect competitive interactions

Human activity has two different kinds of effect on competitive interactions. The most direct effects come from human interference with the distribution of species, either by causing the extinction of species in a particular community, or by introducing exotic species. The other set of effects of human activity stems from the observation in Section 8.4.4 that, at different times and in different circumstances, different resources can be the limiting factor in competitive interactions.

A clear example of the first kind of effect comes from many instances of pollution, such as when crude oil is spilled near the coastline, as happened in the case of the tankers *Exxon Valdez* in Alaska, *Amoco Cadiz* in France and *Torrey Canyon*, *Sea Empress* and *Braer*, all three of which released oil onto the coast of the United Kingdom. Between them, these tankers spilled more than half a million tonnes of oil onto the shoreline. On some rocky coastlines, the oil killed many molluscs, such as limpets, which graze some kinds of seaweed, such as *Enteromorpha* species. In some places, the populations of limpets were so reduced that there are large areas of rocky shore where they are almost extinct. This changed the nature of the competition between *Enteromorpha* and other coexisting species of algae. Removed from grazing pressure, *Enteromorpha* grew quickly and spread widely, occupying areas that would normally be occupied by other, slower growing species of algae.

Until the mollusc populations recover as the oil is broken down, the competitive interactions among the algae remain greatly altered from their original state. The appearance of these habitats can change very dramatically during the period that the ecological community returns to its original composition.

Equally significant and dramatic effects occur when humans introduce alien species into communities where they have not previously lived. Examples of introduced species include the water hyacinth (*Eichhornia crassipes*) introduced to the United States from South America, domestic cats and pigs introduced on many islands in the Pacific, and many European birds introduced into Australia and the Americas (see Section 13.4.1 and Fig. 13.5). Such introductions can strongly alter the balance of competitive interactions, and they will be considered in greater detail in Section 13.4.1.1, which will discuss their effects on the composition of ecological communities.

Examples of more diffuse human effects come from global climate changes. Chapters 1, 4 and 5, described the difference between the fundamental niche and realized niche of a population of organisms. There are some physical conditions under which a particular species can exist, even if they are not the optimal conditions for the species. That existence is not, however, possible in the presence of another species for whom the conditions are optimal. In these circumstances, the second species outcompetes the first. Connell (1961) described the interaction between two species of barnacles, one of which (*Chthalamus stellatus*) can live at any height in the intertidal zone in the absence of the second (*Semibalanus balanoides*), but only occupies half the range in places where *Semibalanus* is present. This is because individual *Semibalanus* barnacles are better at exploiting the habitat in the lower part of the beach, where the physical conditions are optimal for them.

It follows that when human activities alter the physical environment, they will alter the nature of competitive interactions because conditions will become less than perfect for some species that previously enjoyed optimal conditions. An example of this was described in Section 3.9, where probable increases in temperature due to the greenhouse effect are predicted to change the nature of ecological competition between grasses and other plants in the mountains of Ethiopia. This is likely to change the distribution of the grass species, with consequent effects on other organisms that rely on the grass for food, such as the gelada baboon (*Theropithecus gelada*).

8.9 Chapter summary

Competition between species comes in two forms. The first is interference competition, in which individuals of one species directly reduce the ability of another species to utilize resources. The second is exploitation competition where individuals of one species use resources that then cannot be used by those of another.

Competition may not be equally severe for the two species involved, and the interaction may be asymmetrical. Apparent competition is a form of indirect interaction, of which competition for 'enemy-free space' is an example.

Competition can lead to niche differentiation and to character displacement, and can therefore have a strong effect on structuring ecological communities. This phenomenon, coupled with the fact that human activities can alter a competitive balance, means that ecological competition is an important factor in the study of biodiversity.

Recommended reading

Kaye, J.P. & Hart, S.C. (1997) Competition for nitrogen between plants and soil microorganisms. *Trends in Ecology and Evolution* **12**: 139–142.

Keddy, P. (1989) *Competition*. Chapman & Hall, London.

Chapter 9

Predation, herbivory, parasitism and other interactions between populations

9.1 Introduction

The last three chapters have looked in considerable detail at competition, concentrating both on competition between individual organisms of the same species and also on competition between organisms of different species. These chapters have developed models that help ecologists to think about the ways in which individual behaviour affects the dynamics of whole populations, and they have also dealt carefully with the ways in which resources such as food, light, water and space form the basis of many interactions.

This chapter will look at a range of other types of interactions. It will not be necessary to work through each one in the same detail as Chapter 8 investigated competition, because the previous chapters have already established the basic ground rules, and much of the understanding that has already been developed in studying competition will apply equally strongly for other interactions.

The following sections will make use of concepts that have traditionally been the basis for classifying ecological interactions: predation, herbivory, parasitism and mutualism. However, it is essential to bear in mind that, to a certain extent, these concepts are artificial. The words 'herbivorous' and 'predatory' were both used to describe animals more than 300 years ago, and their meanings may not be as precise as modern scientific ecology demands. Looking more deeply into the details of the interactions, it will become clear that some herbivores affect the population of their food plant in exactly the same ways that parasites generally affect their hosts, but that two examples of interactions with the same name (say two examples of herbivory) may be completely different in their effects.

Chapter 10 will look at the similarities and differences between the various types of interactions, but this chapter will concentrate on understanding the four main varieties, which are given the names predation, herbivory, parasitism and mutualism.

9.2 How organisms obtain organic resources

All organisms need a supply of various chemical elements, with oxygen, nitrogen and carbon being the most important; they also need a supply of energy with which to manipulate and use these chemicals. Some types of organisms, notably green plants, use the sun's energy directly, but most others (animals, fungi, a few plants and many micro-organisms) cannot do this. Since the sun's radiation is the only external source of energy that the Earth has, any organisms that cannot use the sun's energy directly must use it indirectly, and they do this by consuming other organisms, or parts of them; sometimes they wait for the food to die, and sometimes they kill it themselves, or eat it alive. We normally think of 'eating' as something that only animals can do, but fungi, bacteria and even plants, also absorb nutrients derived from other organisms; in effect, they 'eat' sugars, amino acids and other chemicals.

When the organisms that are eaten are already dead, the interaction is called scavenging or **saprotrophy**. If the food organism is still alive when the consumer encounters and eats it, the interaction is known as predation or herbivory, depending on whether the food organisms are animals or plants. Of course, to be comprehensive, it would be prudent to make separate studies of organisms that eat fungi, bacteria or micro-organisms, but traditionally, such interactions have not received much attention, and in terms of their food, these organisms are treated as being similar to plants or animals, depending on their precise characteristics.

Chapter 10 will illustrate how, in fact, it is not too problematic to lack a detailed study of organisms that eat fungi (known as fungivores or mycophages), bacteria (bacteriophages) or other micro-organisms. This is because the chapter will begin to integrate the lessons that have been learned from studying predation and herbivory, and it will be clear that the same principles apply to feeding types that we might classify in different ways. It is easy to see that once ecologists can understand the principles that govern the ecology of the interactions between insects and the plants that they eat, they will have a good appreciation of what they might find if they studied insects that happened to eat mushrooms.

Populations of organisms can obviously be strongly affected by the presence or absence of species that eat them, or of species that they consume. In the most extreme case, a population of predators might become extinct through lack of food. It is possible that many of the larger predatory dinosaurs became extinct because there were no prey left for them to eat, and it has been suggested that the New Zealand laughing owl (*Sceloglaux albifacies*) died out earlier this century because of the decline of its main prey, the Pacific rat (*Rattus exulans*). These rats were the main food of the owls, which for some reason were unable to thrive on the brown rats (*R. norvegicus*) that were accidentally introduced by humans and eventually replaced the Pacific rats.

The presence or absence of a population of predators can exert an equally strong effect on a population of prey. There are many examples of bird species that have been exterminated from oceanic islands once rats, cats, monkeys or humans have arrived. In general, it is difficult to establish the precise cause of extinction, but it is known that the bird known as the Ascension Island rail (*Atlantisia elpenor*) became extinct partly as a result of human predation, because the last known individual was recorded by a seventeenth century explorer, who caught it and ate it, reporting that its meat tasted like a 'roasting pig' (Fig. 9.1).

Fig. 9.1 The Ascension Island rail (*Atlantisia elpenor*) is unusual because we know that human predation was a significant factor in its extinction. The last person to record seeing any was an explorer called Peter Mundy, who, in 1656, killed them and ate them. Mundy also drew this sketch.

On some of the islands in New Zealand belonging to the groups known as Mercury and Hen-and-Chickens, the lizard-like reptile known as the tuatara (*Sphenodon punctatus*) suffers predation from the Pacific rat (*Rattus exulans*) (Fig. 9.2). When Cree *et al.* (1995) studied the animals, there were no juvenile tuataras at all on most of the rat-infested islands. By contrast, on those islands that were free of rats, a quarter of all tuataras were juvenile. In time, it is likely that the tuatara populations will become extinct on the islands with rats, because the heavy predation of juveniles by the rats will prevent the population from replacing old individuals as they die.

In extreme cases, a single population of predators may exterminate more than one prey population at the same time. The Nile perch (*Lates nilotica*) was introduced to Lake Victoria as a source of food for the people who live around the lake's shores in Uganda, Kenya and Tanzania. It feeds on several species of catfish, and a number of them have been exterminated. In the deeper waters of the lake, all catfish species have vanished (Goudswaard & Witte 1997).

9.2.1 Organisms that create their own organic resources

Many organisms create their own food by using the sun's energy directly to make organic nutrients from simple chemicals. The most notable group of such organisms is the group that comprises the green plants, which use chlorophyll to capture the sun's energy to make chemical energy. This chemical energy, in the form of adenosine triphosphate (ATP) is then used to drive the chemical reactions that turn carbon dioxide and water into carbohydrates. Some bacteria use the heat energy of geological features such as hot springs in a similar way. Organisms that create their own food are called **autotrophes**. Populations of such organisms have very clear indirect effects on other populations; for example, they are used as food by other organisms and

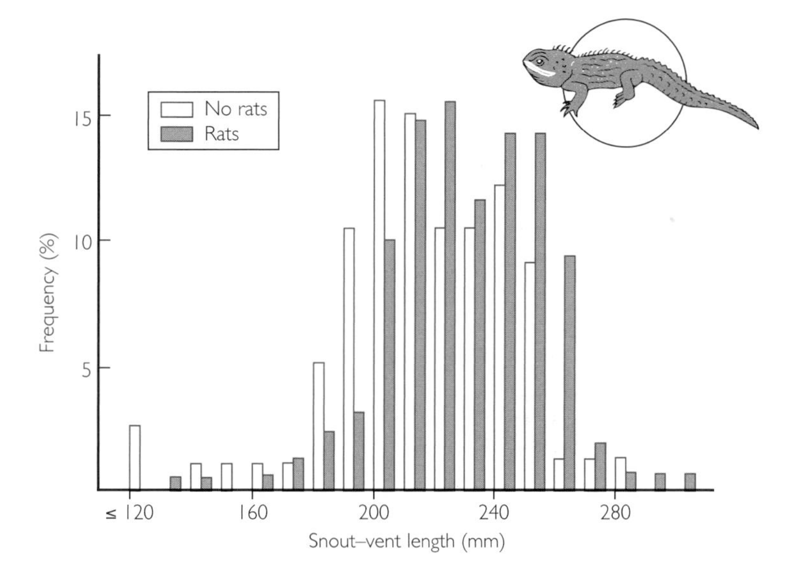

Fig. 9.2 Size distribution of tuataras (*Sphenodon punctatus*) on islands in the Mercury and Hen-and-Chickens groups. On islands where the Pacific rat (*Rattus exulans*) occurs there are almost no medium-sized tuataras and none below 120 mm in length, but on islands where the rat is absent, small and medium-sized individuals are relatively common. (From Cree *et al.* 1995.)

they sometimes generate interspecific competition for light. They are ecologically extremely significant and their importance was dealt with in detail in Chapters 4 and 5, and the dynamics of their populations were described in Chapters 6 and 7. These organisms will also be described in Chapter 12, which will investigate their role as the base of the trophic structure of ecosystems. They will not be discussed any further in this chapter because their feeding has no direct effect on other populations.

9.3 Predators and prey

Predation is the name given to killing and eating an animal. Most predators are themselves animals, and to most people the word 'predator' normally brings to mind big cats, sharks, eagles, or perhaps some of the more fearsome dinosaurs, such as *Tyrannosaurus rex* (Fig. 9.3a–c). All of these animals tend to be portrayed in films and works of fiction as violent and cruel hunters. There are, of course, many smaller carnivores, such as shrews, starfish and ladybird beetles (Fig. 9.3d,e). The animals that form the food of carnivorous animals range from large mammals to the tiniest water fleas.

In fact, not all predators are animals. There are about 400 species of plants that are at least partly carnivorous, distributed widely throughout the Americas, Eurasia, Africa and Australasia. The Venus's flytrap (*Dionaea muscipula*) is a well-known example that is native to a small area of North and South Carolina, where it inhabits wet pine woodlands and sandy bays (Fig. 9.3f). It catches insects and other small invertebrates in its traps and slowly digests them. Other carnivorous plants include the butterworts (*Pinguicularia* spp.) and the sundews (*Drosera* spp.), both of which have wide distributions. Among the most dramatic carnivorous plants are the pitcher plants (*Nepenthes* spp.) of South East Asia, which kill and digest not only tiny invertebrates but also sometimes rodents and birds.

9.3.1 Functional responses of predators

Organisms that eat animals must expend energy finding their prey, and often in chasing and killing it. Even if they do not have to chase the prey, they will use energy producing digestive enzymes, or take time choosing their food to ensure that they do not inadvertently eat something harmful, perhaps because it is toxic. Because of the high energy costs of predation, predators will thrive best when they have evolved efficient methods of finding and catching the animals that they eat.

It is especially important for predators not to waste time and energy searching for prey in places and at times when they are likely to use more energy in the search than they gain from whatever food they catch. In a similar vein, if searching, catching or handling prey results in the predator being exposed to predation itself,

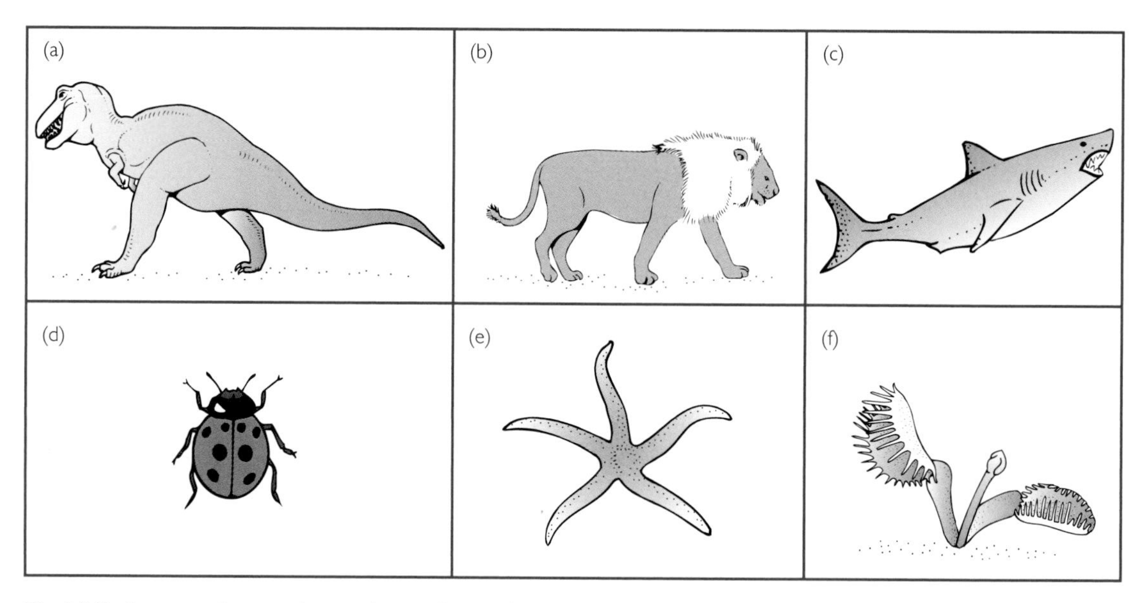

Fig. 9.3 Predatory organisms come in many shapes and sizes including: (a) reptiles such as the extinct dinosaur *Tyrannosaurus rex*, (b) mammals such as the lion (*Panthera leo*), (c) fish such as the great white shark (*Carcharodon carcharias*), (d) insects such as ladybird beetles, (e) echinoderms such as starfish, and (f) plants such as the Venus's fly trap (*Dionaea muscipula*).

there will be strong selection for the predator to minimize this exposure.

One of the most important factors that governs the rate at which a predator consumes food is the density of the food source. Imagine a southern right whale (*Balaena glacialis*) such as might be observed off the coast of Australia. The whale feeds by straining the water through the huge sieve-like plates in its mouth (known as baleen). In this way, it sieves krill (various species of Euphausiacea crustaceans) and other planktonic animals. In the few metres just below the surface, these tiny animals can live at enormous densities of more than 60 000 individuals per cubic metre of water, and when they do so, a whale could perhaps eat tens of kilograms of krill in each mouthful. Very large whales sometimes eat as much as a ton of krill in a single feeding. But if the whale found itself swimming through a part of the sea where its food existed at a lower density (for example, several metres below the ocean's surface), its food intake would be correspondingly lower. Figure 9.4 shows how, in this simplistic view of the situation, the whale's feeding rate would vary with different densities of krill. The graph assumes that the whale feeds in the same way, whatever the concentration of its food.

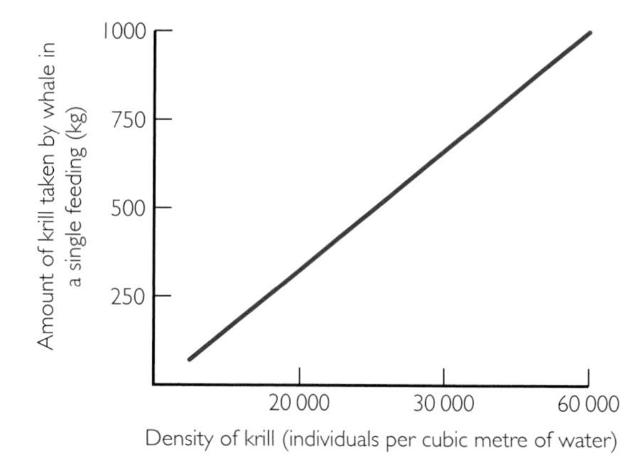

Fig. 9.4 In theory, the amount of food taken by a southern right whale will increase linearly with the density of the krill that the whale eats.

Among predators, this type of response to changes in prey density, in which food intake increases in direct proportion to food density, is uncommon, but it is not completely unknown. Sims and Quayle (1998), for example, produced some evidence that the basking shark (*Cetorhinus maximus*) probably feeds in a manner something like this. The ecologists were able to track sharks in the English Channel because the fish tend to feed near the surface, and can be easily followed. They also measured the density of planktonic animals from within 3 m of the feeding paths of the basking sharks. It turned out that basking sharks spend significantly more time in the areas of highest zooplankton density than they do in poorer areas, remaining for up to 27 h in rich patches. The ecologists were not able to measure the precise amount of zooplankton eaten by individual sharks over particular periods, but their results, coupled with what is known about how basking sharks actually catch their prey, suggest that the rate of food intake depends on the density of zooplankton in a way that is not unlike the relationship shown in Fig. 9.4.

For many predator species, there are more complex considerations that determine the rate at which they can consume prey. Imagine a spider waiting on its web to catch some flies; the denser the population of flies, the more flies the spider will catch each day, and so the more food it will eat. However, there is a maximum number of flies that the spider can eat, because however quickly it works, it must take a few minutes to eat the fly once it has caught it, and the spider will probably also need to spend time repairing the damage caused to its web by the struggling fly. The time taken to consume each prey item and prepare for the next is called the **handling time**, and it is another important factor in the rate at which predators can consume prey.

When a predator has to spend time dealing with each prey item, there is a limit to the number of prey items that can be eaten. Even if the time taken to find and kill each prey could be reduced to zero (as it is when the spider's web is catching more flies while the spider is eating one that was caught earlier), the total number of prey that can be eaten will be determined by the speed at which the predator can eat the prey once they have been killed. As prey density rises, it becomes easier for the predator to find prey, but the length of the handling time stays the same. In other words, the handling takes an increasing *proportion* of the predator's time. Thus, although the total number of prey consumed in a given time continues to grow, the *rate* of increase in food consumption slows down.

For example, Fig. 9.5 illustrates how the number of reindeer (*Rangifer tarandus*) killed by wolves (*Canis lupus*) varies with the density of reindeer in the Gates of the Arctic in Alaska (Dale *et al.* 1994). When the reindeer live at a density of about one animal per square kilometre, the wolves can easily catch and eat as much as they need (at three reindeer per wolf per month, each wolf is eating roughly twice its own weight each week). Higher densities of the prey make almost no difference

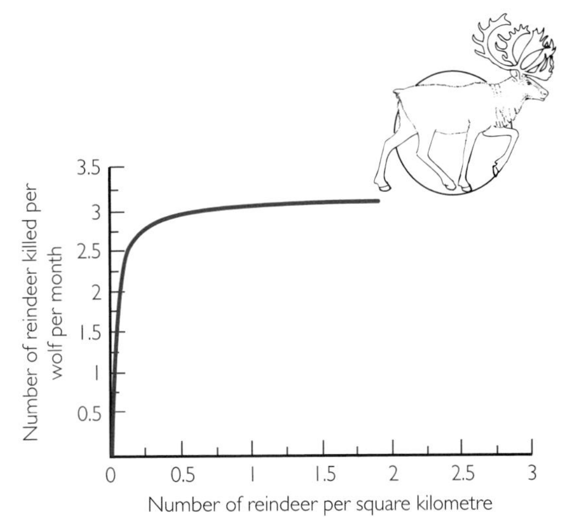

Fig. 9.5 When reindeer are very common, wolves eat about three reindeer each per month, but when reindeer are less abundant, it is more difficult for the predators to find the prey, so they consume less.

to the wolves' food intake. It may be easier to find reindeer, and the wolves may spend less time hunting, but their food needs are satiated. However much more abundant the reindeer become, there is not enough time for each wolf to eat any more each month.

However, at lower densities, that is not the case. When there is only one reindeer in each 8 km^2 (or 0.125 reindeer per square kilometre), the wolves must spend much longer looking for food, and the rate at which they can consume prey is much lower; they eat only about 1.5 reindeer per wolf per month, half as much as when the reindeer are more abundant.

Different predator species respond differently to changes in prey density, partly because of differences in handling times, and the different behaviours are known as **functional responses**. Predators that can handle their prey with a handling time that is virtually zero, like the basking shark, are said to show a type I functional response. A type II response is the name given to the behaviour of predators for which the time taken to handle each prey is not zero, and where the rate of food intake increases as prey density increases when prey are uncommon, but begins to level off as prey density becomes very large; the wolves in Fig. 9.5 show a type II response to the changes in density of the reindeer.

These descriptions of predator behaviour assume that each kind of predator eats only one kind of prey, but this is often not the case. The wolves at the Gates of

Alaska, for example, sometimes eat moose (*Alces alces*) and even snow-shoe hares (*Lepus americanus*) and beavers (*Castor canadensis*). When such a choice exists, predators may simply ignore some types of prey when they are very scarce, and concentrate their efforts on kinds of prey that are more abundant. This kind of behaviour is called **prey switching** by ecologists.

A demonstration of such behaviour comes from an experiment on the choice of food shown by the goshawk (*Accipiter gentilis*) and sparrowhawk (*A. nisus*). Both of these predatory birds eat European blackbirds (*Turdus merula*), which normally have completely dark plumage; the males are black all over and the females are dark brown. Götmark (1996) asked whether the hawks would avoid a different sort of blackbird that looked slightly different and was rarer than the normal type. He presented the hawks with stuffed blackbirds, some of which looked normal, and others with wings that had been painted red and were consequently easier to see than the normal birds. It was clear that the hawks perceived the abnormal blackbirds as suitable prey because they sometimes attacked them. But stuffed blackbirds with normal plumage (the familiar type) were attacked far more frequently.

It is entirely possible that, if red-winged European blackbirds became much more common than the all-black variety, their predators would start to ignore the all-black birds and concentrate on killing the red-winged type. This switching between food sources is exhibited by a wide range of predators, including humans. For example, in what is now Zimbabwe, in the 1880s, King Lobengula decided that he would no longer permit the hunting of hippopotamuses (*Hippopotamus amphibius*), which had been reduced in number (Mackenzie 1997). The effect was that hunters switched their efforts to other big game. This is a very unusual example, because the King recognized in advance that continued hunting of hippopotamuses would lead to significantly reduced hunting success. More normally, switching would occur because the hunters were already experiencing a reduced return. For example, in the Chesapeake bay area of eastern North America, the oyster fishermen switched to catching crabs because oyster stocks declined, and it became uneconomic to harvest them.

One of the effects of this switching between prey is that when a particular prey species is uncommon, it is very rarely to be found in the diet of the predator, and it remains so unless the population of the prey species reaches some threshold. This can lead to a third type of functional response, illustrated by Fig. 9.6, which shows the rate at which red foxes (*Vulpes vulpes*) consume rabbits (*Oryctolagus cuniculus*) at Yathong in New South Wales in Australia.

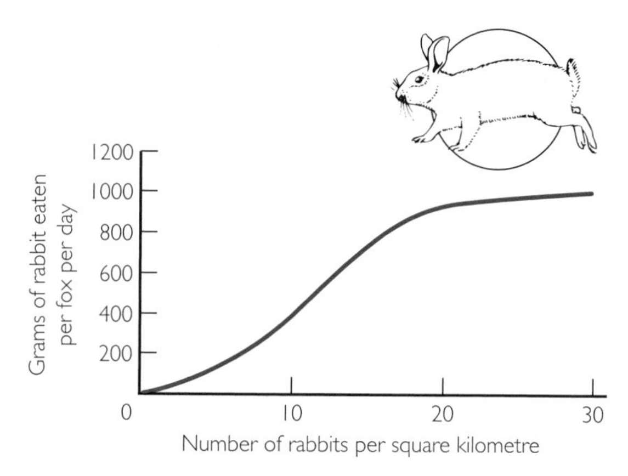

Fig. 9.6 Foxes in the Yathong area of Australia show a type III functional response to variations in the density of their rabbit prey.

In places and at times when rabbits are uncommon, foxes do not catch very many at all, and even as the rabbit population doubles from two to four individuals per square kilometre, the rate of consumption stays the same, at no more than about 120 g of rabbit per fox per day (which represents less than one rabbit per fox per week). However, by the time the rabbit population has reached 10 per square kilometre, the consumption by foxes has increased dramatically to 650 g of rabbit per fox per day (three or four rabbits per fox per week). As with a type II functional response, consumption must eventually level off, and the foxes fare little better when the rabbits reach 30 per square kilometre than they do when the rabbit population density is at half that level (Pech *et al.* 1992).

9.3.2 Mathematical models of predation systems

The fact that different predators have different functional responses to varying densities of their prey means that it could be complex to develop a single mathematical model to give an appropriate description of all examples of predation. However, ecologists can make life easy for themselves by dismissing much of the detail, and allowing mathematical models to describe the overall, general pattern.

Recall that Section 6.3 developed a model of populations that describes the rate at which the number of individuals (or their density) changes over time, and that this rate was called dN/dt. For a population of a single species, in the absence of any interactions with other species, Chapter 6 determined that this rate might depend on the reproductive rate of the population, r, the current size (or density) of the population, N, and the carrying capacity of the habitat patch, K:

$$\frac{dN}{dt} = rN\left(1 - \frac{N}{K}\right) \qquad \text{(eqn 6.12)}$$

In order to learn anything about predation from mathematical models, it is necessary to produce separate models for the population of predators and the population of prey.

9.3.2.1 *A model for the prey population*

Beginning with the population of prey, it is possible to dispense with any considerations of carrying capacity. This is because any significant effect of predation will prevent the prey population from growing until it reaches its carrying capacity. If the prey population were to reach the carrying capacity, ecologists could conclude that the predators were having no effect on the size of the prey population. That would not mean that the predators were not killing any of the prey; they could, for example, pick off individual prey that were sick and would have died anyway, and they could do so without actually affecting the total number of surviving prey.

Wild dogs (*Lycaon pictus*) and cheetahs (*Acinonyx jubatus*) in the Serengeti may, in certain circumstances, provide an example of such predators. Fitzgibbon and Lazarus (1995) believe that when they are hunting Thomson's gazelles (*Gazella tomsoni*), these predators select the old and sick, which are easier to catch. In doing so, the wild dogs and cheetahs are not reducing the overall effective size of the population of gazelles.

Once the carrying capacity has been removed from any considerations, the term $(1 - N/K)$ can be ignored in the equation because, as was clear in Section 6.3.5, it describes how the growth of a population is reduced when it becomes sufficiently large that it is close to its carrying capacity. Without this term, the model is a simple description of unlimited growth in the population of prey:

$$\frac{dN}{dt} = rN \qquad \text{(eqn 9.1)}$$

The population growth described by the equation will be mediated by the actions of the predators, so that it is necessary to subtract a term that accurately summarizes the rate at which predators kill individual members of the prey population.

The effect of a predator population in reducing the size of a prey population depends on three things. First, it is clear that predators will kill most prey when the predators themselves live at high densities; the predator density is normally called P. Second, predators will have a greater effect if the prey population is denser: when the prey are common, predators will encounter them more frequently than when the prey are at low population densities. Third, a particular population of predators will kill individual prey with a particular efficiency. This efficiency may depend on a number of factors, including the degree to which the predators concentrate exclusively on the prey in question rather than taking a more varied diet that includes other prey species; the speed at which the prey can run away; or the success of any mechanisms that the prey have evolved to provide them with camouflage. The overall capture efficiency is normally given the symbol a'.

To combine these three factors to give a single rate of predation is easy:

rate of predation = predator efficiency × predator density × prey density (eqn 9.2)

or:

$$\text{rate of predation} = a' \times P \times N \qquad \text{(eqn 9.3)}$$

which is normally written as:

$$\text{rate of predation} = a'PN \qquad \text{(eqn 9.4)}$$

This represents the number of prey individuals being removed from the population by predation, so to produce a simple model of the prey population, it is appropriate simply to subtract this rate of predation from the growth rate of the population. In other words:

$$\frac{dN}{dt} = rN - a'PN \qquad \text{(eqn 9.5)}$$

9.3.2.2 *A model for the predator population*

In the preceding section, the density of the predator population was defined as P, so that the rate of change of the population will be denoted by dP/dt.

The term $a'PN$ has already been defined as the overall rate at which a population of predators captures and kills individuals of the prey population. The growth rate of the predator population depends on the rate at which this intake of food resources can be converted (by the physiology of the predators) into reproductive success, and hence into new young predators recruited into the population. This physiological efficiency is generally denoted by the letter 'f', so that the birth rate of a predator population is given as:

$$f \times a'PN$$

or:

$$fa'PN$$

A complete model must include the death of predators from the population, and this depends partly on the existing size of the predator population, P, and partly on the death rate, called q. The overall rate at which individual predators die is therefore given as:

$$q \times P$$

which is normally written as:

$$qP.$$

The model of change in the size (or density) of predator populations is thus simply that the rate of change of the population (dP/dt) is equal to the birth rate minus the death rate, or:

$$\frac{dP}{dt} = fa'PN - qP \qquad \text{(eqn 9.6)}$$

9.4 Herbivores

In many ways, herbivory is very similar to predation, with the simple difference that the organisms being eaten are plants rather than animals. However, there is another principal difference—a herbivore usually eats only part of the individual food plant, and in many cases does not kill it. There are, of course, many situations in which predators eat only part of their prey; for example, hawks and eagles do not normally eat the feathers of the small birds that they eat. But there are very few predators that routinely leave their prey alive after eating only parts of them. It would not be impossible; a shrew could eat just half an earthworm, and if the remaining end happened to be the worm's head, the worm might survive and regenerate. However, the overwhelming majority of herbivores do not routinely kill the individual plants that they eat.

This difference could cause very significant differences in the effects on the populations of food organisms. A flock of sheep can continually graze a field of grass to the extent that few of the individual grass plants are able to produce seeds and reproduce. This can happen for months on end without necessarily affecting the overall size of the grass population.

In parts of southern Yukon in Canada, there are two main herbivores of the grasses and herbs, the snowshoe hare (*Lepus americanus*) and the ground squirrel (*Spermophila parryi*). Two of the common food plants are the anemone *Anemone parviflora* and the bluebell *Mertensia paniculata*, but John and Turkington (1997) found that the herbivores did not affect the populations of these plants. They constructed fences, and for 5 years excluded the hares and squirrels. However, the density of stems and the number and size of the leaves of the anemone and the bluebell were no different on the experimental plots to those plots that were left as controls for the herbivores to graze.

9.4.1 Mathematical models of herbivore systems

Many of the basic effects of a population of herbivores on the population of their food plants are the same as the effects of a population of predators on the population of their prey. In other words, the mathematical equations that were developed in Section 9.3.2 to describe predators and prey, will also suffice in some circumstances for the study of herbivores and the plants that they eat.

Thus, the equation:

$$\frac{dP}{dt} = fa'PN - qP \qquad \text{(eqn 9.6)}$$

can describe a population of herbivores, where P is the density of herbivores and N the density of a particular plant species used as food by the herbivore. And the equation:

$$\frac{dN}{dt} = rN - a'PN \qquad \text{(eqn 9.5)}$$

can describe the population of the plants.

However, as has already been noted above, there is a fundamental difference between most predators and many herbivores. Predators tend to kill the individual prey that they eat, while many herbivores do not, in fact, kill the plants on which they feed. When a pigeon removes some of the leaves of a large cabbage, it does not (necessarily) kill the plant.

Herbivores that eat seeds are an exception, because they almost always kill the individual seeds that they eat (animals that eat the soft parts of a fruit but do not digest the seeds are quite separate—they are actually mutualists, gaining food from the fruit but also dispersing the seeds). And, of course, many herbivores do indeed kill the plants that they eat; if the imaginary pigeon had pulled up the cabbage plant when it was very small, the cabbage plant would have died.

In circumstances where herbivores kill individual plants, the situation is exactly analogous to predators and their prey, and the mathematical models that encapsulate the details of predatory interactions are a good basis to describe herbivorous interactions. But in circumstances where the herbivores do not generally kill their food plants, ecologists must be a little more careful in developing mathematical models.

In situations where ecologists are not concerned with counting precise numbers of plants, eqns 9.5 and 9.6 may still be useful. Counting individual plants is difficult anyway, and it might instead be appropriate to measure the total mass of green leaves in a population rather than the exact number of individual plants of which they are part. In effect, this approach redefines N, so that instead of standing for the number of plants per unit of area, it stands for the mass of green leaves per unit area.

This is not simply cheating to make the models appear more suitable. It will be necessary to alter the value of a' in situations where it is appropriate to make N stand for mass of leaves rather than numbers of individuals. Recall that a' is the efficiency with which the herbivore removes food from the plant population. In the original case, where N stood for numbers of individual plants, this would be the rate at which herbivores killed whole plants, which is very low in many situations. However, once the plant population has been redefined in terms of mass of living green leaves, it is also necessary to redefine a' to stand for the rate at which the herbivores consume a given mass of leaves. This will be a much higher rate, and will be reflected in changes in the value of N. When a swarm of insects strips a plant of its leaves but does not kill it, the value of N would not change if it were defined as a measure of individual living plants, but it would change dramatically if it were defined to measure the mass of leaves in the plant population.

If, however, a particular study remains interested in the total number of individual plants in a population (as it might if it were interested in trees in a forest), it would be rash simply to accept without thought that models of predation will also serve for herbivores. But in this situation, there is much to be learned from the simple observation that herbivory is very similar to parasitism. Herbivores eat some of an individual plant without killing it, in exactly the same way that an influenza virus uses some of a person's resources without actually killing him or her. In other words, whatever mathematical models are developed for populations of parasites and their hosts will also serve for herbivores and their food plants. Parasitism is discussed in detail in Section 9.5.

9.5 Parasites

Most people usually think of parasites as being agents of disease. Human diseases such as malaria, dysentery, AIDS (acquired immune deficiency syndrome), athlete's foot, chigger sores and the common cold are all caused by parasitic bacteria, protozoa, insects, fungi or viruses (Fig. 9.7). Sometimes parasites eventually kill the sufferer; for example, most people infected with cerebral malaria, caused by the protozoan *Plasmodium falciparum*, die within a few days if they are not treated. Other parasitic diseases, such as other forms of malaria, do not necessarily kill the patient. The food poisoning bacterium *Salmonella* does not normally kill people in Western countries who are otherwise healthy, although it may cause death in the elderly, infants or patients who are already weakened by other medicial conditions. Some parasites, such as the chigger (a flea, *Tunga penetrans*), never cause death, expect perhaps by creating wounds that allow secondary infections by other pathogens or in some other extraordinary way.

Box 9.1 Ecological simulations

Ecologists often use mathematical models such as those described in this chapter to compare their theories with actual events in nature. This is especially true for predator and prey populations. For example, if we use the equation for prey population growth in the presence of a predator:

$$\frac{dN}{dt} = rN - a'PN$$

and the equation for the growth of the predator population:

$$\frac{dP}{dt} = fa'PN - qP$$

and calculate population sizes for long periods of time, we can simulate changes in the size of these populations. Then, by changing key parameters, such as the proportion of prey that are eaten by the predators (a'), the population growth rate (r) or the efficiency of conversion of prey to predator reproduction (f), we can examine the effect of these variables on the dynamics of the interaction.

Before beginning such simulations, we must choose some starting points. Generally, the more realistic we make these the better, but even unrealistic values can show us important interactions among the variables. We can start by setting the value of a' to be 0.01. This would be equivalent to one prey killed per 100 prey in the habitat. We will also set N, the initial number of prey, to be 1000 individuals, and P, the number of predators, to be 5. Next, we will set f, the conversion of prey to predator reproduction, equal to 0.01. This would mean that for every 100 prey that are eaten by the predators, the predator population produces one new offspring. Finally, we will set prey growth rate, r, equal to 0.05, and predator death rate, q, equal to 0.05.

The results of our simulation under these initial conditions are shown in Fig. 1a (labelled $a' = 1/100$). Notice that the simulation predicts population cycles to occur. Additionally, the predator population peak lags behind that of the prey. These population oscillations are typical of such models, and similar cycles have been found in nature, but are by no means universal.

Now, what happens if the predator becomes more efficient at harvesting prey? By keeping all other variables the same, and changing a' to 0.03, or three prey harvested per 100, a very interesting thing happens. The prey population goes extinct in about 37 years! Apparently, the predators are so efficient at catching and killing the prey that they reduce the prey population to extinction (Fig. 1b). This may be similar to human hunting and fishing, or to the case of a new predator being introduced into habitats containing prey that have no defences to the new predator (Section 9.2).

Of course, this is just one example of how mathematical simulations can be used to explore population interactions. Many complexities could be added to these models, such as manipulation of the prey population growth rates, the efficiency of predator utilization of prey, predator-free space (refuges) for prey, habitat heterogeneity that causes differences in predation or prey growth rates, or the addition of multiple prey populations. Regardless of the complexities of the models, they can be valuable tools in providing insights into ecological interactions.

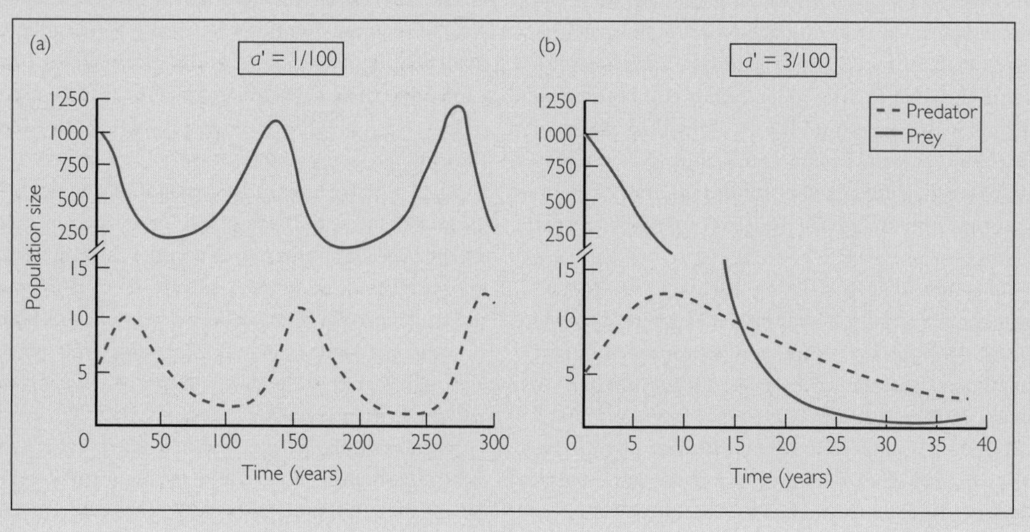

Fig. 1

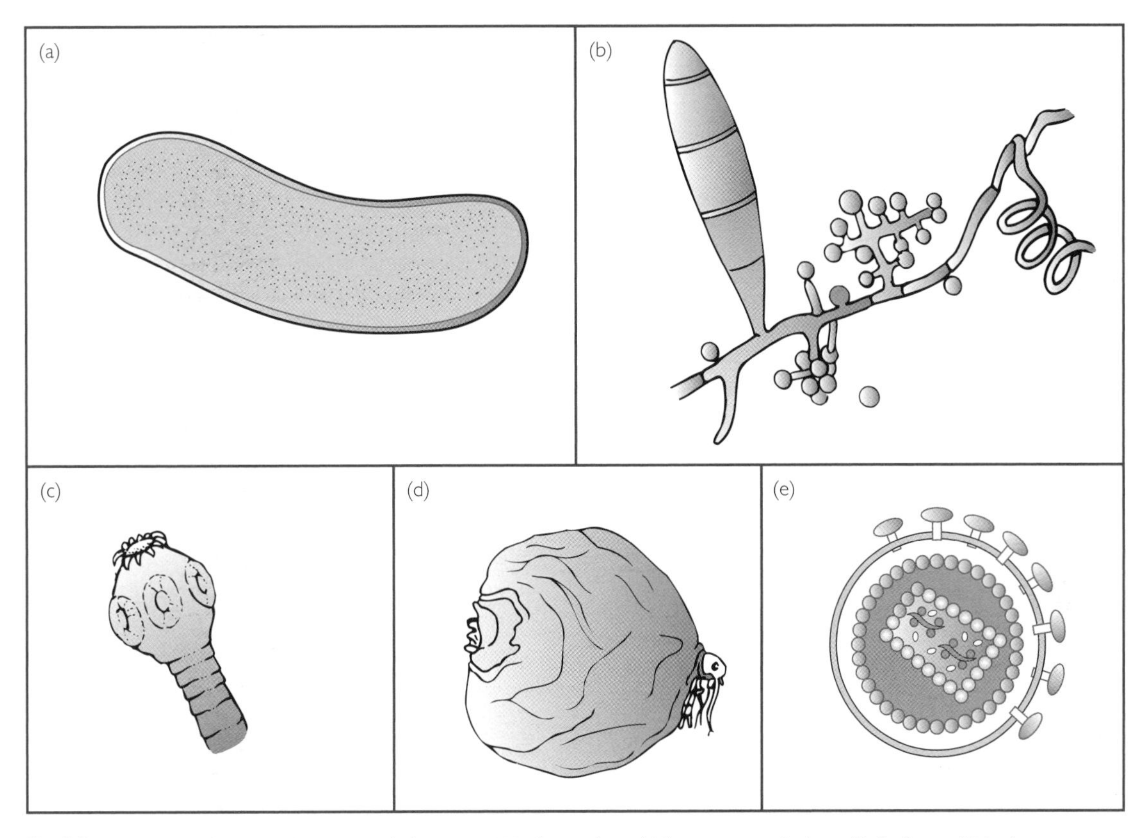

Fig. 9.7 Organisms that parasitize humans: (a) the bacterium *Helicobacter pylori*, which causes stomach ulcers; (b) the fungus *Trichophyton mentagrophytes*, which causes athlete's foot; (c) the tapeworm *Taenia solium*; (d) the chigger flea (*Tunga penetrans*), the females of which burrow under the skin; and (e) the HIV virus, which causes AIDS.

Just about all multicellular organisms can suffer from similar parasitic illnesses, although the agents are not always as small; the crustacean *Gnathia maxillaris* uses its sucking mouthparts to parasitize fish, and the tapeworm *Taenia* can grow to several metres in length inside the human gut. Indeed, the plant with the largest flowers in the world is a parasite. *Rafflesia arnoldii*, native to Borneo, produces flowers that can be as large as 1 m in diameter. Apart from the flower itself, the organism is almost unrecognizable as a plant because the cells that obtain nutrients form a mass of filaments within the roots of another plant, usually a vine.

Some organisms are completely parasitic and obtain all their nutrition from their hosts, while others are only partly parasitic, and gain some of their nutrients by other means. Entirely parasitic plants, such as *Rafflesia* and broomrapes (*Orobanche* spp.), have red or pink stems that contain no chlorophyll, and they tend to have no leaves. By contrast, although the Australian Christmas tree (*Nuytsia floribunda*) has modified roots that invade and parasitize the roots of other woody plants, it also has green leaves and can obtain some of its own nutrition by photosynthesis.

The common feature of all these parasites is that they steal some resources from the organism that they are parasitizing (known euphemistically as the **host**), and cause the sufferer to operate less efficiently than they otherwise would. With a common cold, or mumps or measles, people normally operate less efficiently because they are in pain, or have stiff joints, or feel tired or light-headed. But more generally, the reduction in efficiency caused by parasites could be manifested as a reduction in reproduction or growth, or an inability of an organism to defend itself against competitors, predators or herbivores. In other words, the evolutionary fitness of an organism is reduced by the parasite.

The rust fungus (*Uromyces rumicis*) shows very clearly how parasites reduce the resources available to their hosts. The fungus infects the broad-leaved dock (*Rumex obtusifolius*), with the infection appearing as pustules on the leaves. Hatcher *et al.* (1997) compared the resources in infected leaves with those in uninfected leaves. The leaves of healthy plants contained much higher concentrations of nitrate than the leaves of infected plants. Moreover, the effect of the infection was not limited to the leaves on which the fungus pustules occurred. The physiology of the dock plants was compromised sufficiently that even the healthy leaves of infected plants contained lower concentrations of nitrate than the leaves of uninfected plants.

The reduction in evolutionary fitness caused by parasites can also be seen dramatically in the effects of various species of nematodes, especially *Obeliscoides cuniculi*, on the snow-shoe hare (*Lepus americanus*) (Murray *et al.* 1997). In a study site in Manitoba in Canada, a sample of hares was treated with a drug that kills the parasites, which normally live in the stomach. A separate sample of hares was treated with a placebo, to ensure that any observed effects were not the result of the handling and injection procedures. Under certain conditions, the treatment with the antinematode drug had no effect on the hares, but at other times the effects were impressive. During one summer, hares with experimentally reduced parasite loads were almost two and half times as likely to survive as animals that were still carrying naturally high parasite burdens.

Parasites can also affect a population of their hosts, not by killing individual hosts, but by affecting the rate at which they can reproduce. The immune defences of the host are used in combating the parasites, thus using resources that might otherwise be used for reproduction. The protozoan parasite *Pleistophora intestinalis* reduces the reproductive success of the water flea *Daphnia magna* (a tiny freshwater crustacean) by up to 30% (Poulin 1998). In extreme cases, parasites might even render their host completely infertile. Some strains of the snail *Lymnaea peregra* do not reproduce at all when they are infected with the parasite *Diplostomum phoxini*.

9.5.1 Mathematical models of parasitic interactions

Parasitic interactions are very like predatory ones, and the various mathematical models that have been developed to describe the effects of parasites are in many cases similar to those that describe the effects of predators.

We can start by developing a mathematical description of the host population, and in doing so, it is possible to start by assuming that, initially, the hosts are all perfectly healthy and that none of them are suffering from infection by any parasites. In this situation, there is no reason in principle why ecologists cannot start with the basic logistic model of a single population, which was developed in Section 6.3.5:

$$\frac{dN}{dt} = rN\left(1 - \frac{N}{K}\right) \qquad \text{(eqn 6.12)}$$

Now the model can be altered to take account of the negative effect of a population of parasitic organisms. Recall that in Section 9.3.2, the effect of predators was incorporated by subtracting a term based on the rate at which predators killed members of the prey population. A similar term can be incorporated to take account of the rate at which parasites affect members of a host population.

However, the effect of a population of parasites is not as simple as that of a population of predators, because, as was noted in Section 9.5 above, parasites do not necessarily kill all of their hosts, and can reduce the reproductive success of others.

The model needs separate terms for the lethal and non-lethal effects of the parasites. The rate at which an individual parasite reduces the host population by killing individual hosts is known as α, while δ is the rate at which an individual parasite reduces the host population by affecting their reproduction. (In this context α is completely unrelated to the competition coefficient, described in Section 8.3.1, which is also called α.)

The total effect of the parasite population will depend on how many parasites there are, just as the effect of predators in eqn 9.4 depended on the size (or density) of the predator population. And just as the letter P was used for the density of predators, it can be used for the density of parasites.

The reduction in the size of the host population resulting from parasites killing individual hosts is therefore given by:

$$\alpha \times P$$

which is normally written as:

$$\alpha P$$

and likewise the reduction in the size of the host population resulting from impaired reproduction is given by:

$\delta \times P$

or:

δP

The total reduction in the size of the host population as a result of both effects of the parasites is simply the sum of both of these terms:

$\alpha P + \delta P$

It is now possible to take the simple course of subtracting this term from the basic logistic model of the host population to give an overall description of the host population in the presence of the population of parasites:

$$\frac{dN}{dt} = rN\left(1 - \frac{N}{K}\right) - (\alpha P + \delta P) \qquad \text{(eqn 9.7)}$$

The values of α and δ have been used to stand in for the average effects of parasites on the survival and reproduction of the hosts. In fact, ecologists who are interested in the details of how a population of parasites might affect a host population have developed much more complicated models. For example, it might be desirable to incorporate a term that describes the probability that an individual host will become infected by a parasite, and the number of other individuals in the host population that are likely to catch a disease from a single infected host. It may be necessary to divide the host population into individuals that are already infected and those that are susceptible to infection. There may even be a need for a third category for individual hosts that have recovered from the disease and have become immune to future infection, rather as children who have had measles generally do not catch the disease a second time even when exposed to fresh infection.

These kinds of mathematical models can be extremely important in studying infectious diseases. An entire discipline, called **epidemiology**, has been developed to track and forecast the spread of diseases in humans, crop species and wildlife. For example, in a study of how the measles virus (*Morbillivirus* spp.) spreads through isolated human populations, such as that of Reykjavik in Iceland, Rhodes and Anderson (1996) developed what many epidemiologists would consider a relatively simple model. It included terms for the duration of the epidemic (t), the number of people infected (s), the probability that an individual person would become infected (v) and the number of uninfected people entering the population (μ), and two different equations (with two variable terms, b and c, and a constant term, a), as well as the possibility of separate analyses for each month (denoted by τ and a different number for each month).

9.5.2 Using a model of a parasitized host population to describe herbivory

When the effects of herbivores were investigated in Section 9.4, it was noted that individual herbivores do not always kill the individual plants that form their source of food. The herbivores, however, by stealing the resources of the plants, affect the evolutionary fitness of the individual plants. In other words, many herbivores are really parasites.

This logic allows ecologists to apply the models of parasitism to herbivorous interactions. The term ($\alpha P + \delta P$) can be used to measure the effect of herbivores on a plant population; in this instance P is the density of herbivores, α is the rate at which individual herbivores kill individual plants, and δ is the rate at which individual herbivores reduce the size of the plant population by damaging individual plants sufficiently to impair their reproduction.

9.5.3 Diverse, complex life cycles make it difficult to develop models of parasite populations

Although the model encapsulated in eqn 9.7 describes populations of hosts that are suffering from the presence of a population of parasites, there remains a need to develop a mathematical model to describe the parasite population itself. As a very basic model, it is appropriate to use the mathematical description that was developed in Section 9.3.2, which investigated the dynamics of populations of predators:

$$\frac{dP}{dt} = fa'PN - qP \qquad \text{(eqn 9.6)}$$

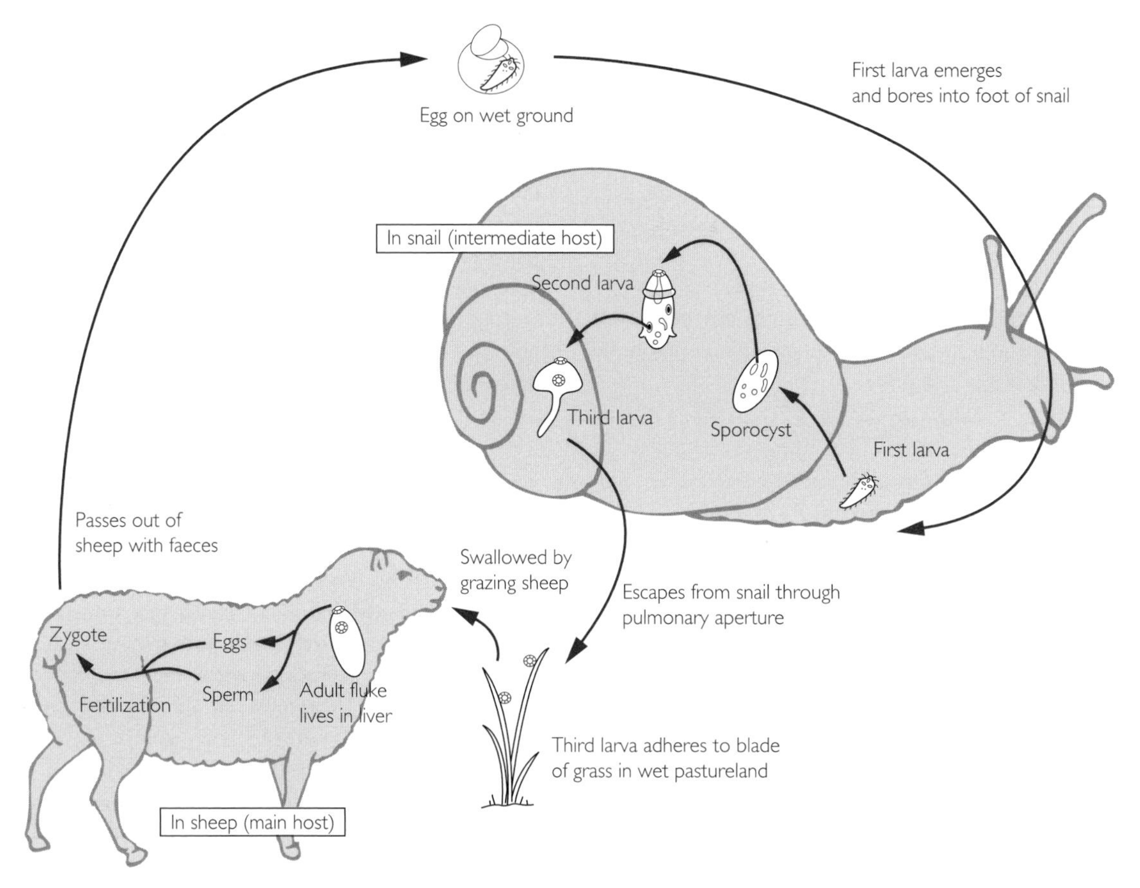

Fig. 9.8 The complex life cycle of the liver fluke *Fasciola hepatica*, which has several larval stages in two different hosts.

P now stands for the density of parasites, q is the death rate of the parasites, a' is a measure of the efficiency with which individual parasites infect new hosts, and f is a measure of how efficiently the physiology of individual parasites turns the host's resources into new parasites by reproduction.

However, it is easy to see that this model will be inadequate to describe parasite populations in many circumstances. When the model was used to describe predator populations, a' was a simple measure of the rate at which individual predators killed individual prey, but it has no such simple interpretation when applied to parasites. A parasite might infect a host that already has parasites in its tissues, or one that has previously been infected and consequently has the immune response to fight off the new infection. Such complex-ities will render the simple model effectively useless.

To produce a more sophisticated mathematical description of parasite populations, it would be necessary to use information about the number of individual parasites that infect an individual host, together with the rate at which individual parasites are passed between individual hosts, as well as many other features of the parasites' biology.

A major problem arises because different species of parasites are spread between hosts in different ways. The viruses that cause common colds might provide a relatively simple situation in which individual viruses are passed between people through personal contact, or through the air.

But the liver fluke *Fasciola hepatica*, which lives in the livers of sheep, has a much more complex life cycle (Fig. 9.8). Adult flukes live in the bile passages of the sheep, and reproduce sexually to produce eggs, which pass out of the sheep in the faeces. A larva emerges from each egg, and uses the tiny cilia on its surface to swim through water on the grass to find a particular species of snail (*Limnaea truncatula*). The larva penetrates the snail's foot, where it forms a structure called a sporocyst, in which the cells of the fluke develop into a different sort of larva, which then burrows into the snail's liver. Feeding on the snail's tissue, this larva develops into yet another form of larva, which eventually escapes from the snail via the opening through which it breathes. This third form of the larva then remains in the grass until a sheep accidentally ingests it when it eats the grass. Inside the sheep's intestine, the larva produces an immature fluke, which migrates to the sheep's liver to begin the life cycle again.

A mathematical model that accurately described the population of flukes would be complex in its form. Each larval form might be subject to different pressures; some of them would be killed if their snail was eaten by a bird, while the first larva, which locates snails by moving through water on wet grass, might suffer in a drought. Thus the model might need separate terms for the density of each of the larval forms, as well as a term for the adult flukes. Terms would also be needed for the efficiency with which each form develops into the next: What proportion of the first larval forms manage to find a snail? What proportion of the second larval forms reach the liver of a snail?

Other parasites have even more complex life cycles. Like the flukes, malarial parasites (*Plasmodium* spp.) have a number of different life stages, but unlike the fluke larva (which must find a snail of the right species), the malarial larvae can be spread by more than one species of mosquito, and in some cases can persist in a variety of different mammal species.

To develop mathematical models that would provide a sufficient description of parasite populations is therefore no easy task, and ecologists cannot simply add or subtract relatively simple terms as they can in models for predatory and herbivorous interactions, or for populations of hosts infected by parasites.

However, investigating why mathematical modellers are unable to produce such simple models of parasite populations, teaches ecologists much about the interactions in which they are involved. The need for more complex models points to the fact that many parasitic species, such as the liver fluke or malarial parasites, are affected by many features of the biological environment (such as populations of more than one species of host) and of the physical environment (such as drought).

This appreciation of the interconnectedness of a wide variety of biological populations and features of the physical environment will prove extremely useful in Chapters 11–14, which investigate how various interactions combine to form ecological communities and ecosystems.

9.6 Mutualism

Almost all of the interactions that have so far been considered have involved at least one of the populations having negative effects on other populations. Two competing species, for example, affect one another negatively. In predation and herbivory, the predators and herbivores have negative effects on the animals and plants that they eat, and parasites have a variety of harmful effects on their hosts.

In some ecological interactions, however, two populations may both benefit. For example, many herbivores have beneficial bacteria that live in their guts. Without such intestinal micro-organisms, grass-eating mammals, such as cattle and rabbits, would be unable to digest cellulose, which is the principal constituent of plant cell walls and is probably the most abundant organic chemical on Earth. In fact, enzymes for degrading cellulose are very rare in animals, and a wide range of herbivores rely on bacteria to do the job for them.

Insects as well as mammals often rely on bacteria to assist their digestion. Cockroaches have mutualistic bacteria that live inside the cells adjacent to the gut. These bacteria are necessary for the insects to obtain essential nutrients, and cockroaches that are treated with antibiotics, and which consequently have no bacteria, cannot survive on natural diets.

Thus, rabbits and cockroaches have beneficial interactions with micro-organisms that digest nutrients on their behalf. But the bacteria also benefit from the interactions. They gain a warm, protected environment, and a constant supply of chewed up organic material on which to feed. In fact, although the presence of the bacteria is essential for the proper nutrition of the herbivores, most of the simple sugars produced by the bacteria are used by the bacteria themselves.

Animal ecologists have not studied mutualism as much as competition, predation and herbivory. This is partly because it is difficult to do so (the bacteria in cockroaches' intestines are difficult to find) and partly because mutualistic interactions are often less obvious to the casual observer. Nevertheless, mutualistic interactions should not be relegated to an insignificant concept at the back of the mind, and plant ecologists have studied them more closely. Mutualists are an enormously important part of the world's ecosystems, and have played a major role in generating the vast biodiversity we see around us. Evolutionary studies on the diversification of insects and flowering plants over the past 130 million years highlight the tremendous potential of mutualism as a diversifying force. Mutualisms involving animals that pollinate plants (in return for nectar or other rewards) have had very significant roles in affecting the size, shape and colour of flowers, as well as influencing the sensory abilities of the animals.

One of the most universal kinds of mutualism is an association between fungi and the roots of green plants. In Section 2.2.2, the association, which is known as a mycorrhiza, was mentioned as being important in tropical forests. In fact, examples occur in almost every family of flowering plants, as well as in ferns, mosses and other non-flowering groups of green plants. Fossilized remains of the earliest land plants have mycorrhizae, leading to the hypothesis that this mutualism played a major role in the successful invasion of the land by plants. In the commonest form of mycorrhiza, the fungus forms a network of filaments called hyphae, which penetrate the roots of the plant spreading between the cells of the root, and forming vesicles inside some of the cells near the surface of the roots (Fig. 9.9). The hyphae also spread through the soil near the roots of the plant.

In some cases, the benefits of the interaction to the plant are very easily demonstrated. For example, Little and Maun (1996) asked whether the presence of mycorrhiza was important to the success of the Canadian marram grass (*Ammophila breviculata*). They were uncertain of the species of fungi that normally lived in a mutualistic interaction with the grass, so they isolated potential fungi from the sandy soils of the Pinery Provincial Park, where the marram grass grows naturally. They then planted marram grass seeds in pots of sterile soil, and inoculated half of the pots with the fungi. After 20 weeks, the seedlings that were growing in pots with the mycorrhizal fungi had an average leaf area of 37 cm^2 in total. But the control seedlings that were

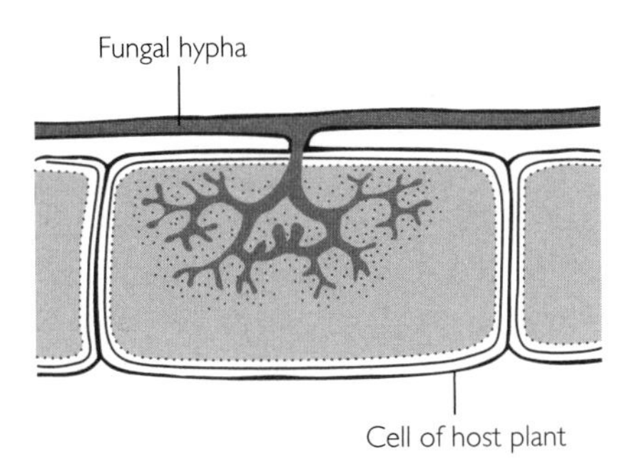

Fig. 9.9 The type of fungal structure known as a vesicular arbuscular mycorrhiza. The fungal hyphae have penetrated the wall of the host plant, and the fungus and plant are able to exchange chemicals.

deprived of the mutualistic relationship with the fungi were only about two-thirds of the size—their average leaf area was 26 cm^2.

Mycorrhizal fungi have also been shown to protect the grass known as bearded fescue (*Vulpia ciliata*) from infection by a parasitic fungus. The beneficial mycorrhizal fungus (*Glomus* spp.) suppresses the development of the harmful fungus (*Fusarium oxysporum*) in the roots of the grass (Newsham *et al.* 1995).

The fungus also appears to benefit from the mycorrhizal interaction, although the advantages have proved more difficult to demonstrate. It is generally believed that the fungi involved in mycorrhizae obtain carbon from their plant partners. There is no doubt that in some mycorrhizal interactions the plant shifts the allocation of its carbon resources, which are fixed from atmospheric carbon dioxide during photosynthesis. Carbon is moved into the fungal hyphae, where it is used in fungal respiration (Rygiewicz & Andersen 1994).

9.6.1 Mutualistic relationships can be very close

Some mutualistic interactions are so close that it can become difficult to separate the identities of the two partners. One of the most well-known examples of this situation comes in the form of lichens, which are familiar organisms in the temperate regions of Europe and North America, where they grow on the trunks of trees and on rocks and buildings. They are commonly seen on the surfaces of bricks and stones in old buildings and in graveyards, where they encrust the headstones. But lichens are hugely successful, occurring at very high altitudes in the Himalayas, in the hot deserts of

Africa and in the constantly freezing environments of the Arctic and Antarctic.

Lichens are actually mutualistic partnerships between fungi and algae (or in some cases cyanobacteria). However, each unique combination is so intimate that each one is classified as a single species. In Namibia, for example, combinations of algae and fungi are classified as *Caloplaca volkii*, *Zygophyllum stapfii* and *Lecidella crystallina*. Such lichens consist of mixtures of cells, some of which are algal and others of which are fungal. In some cases, the two sorts are mixed up at random throughout the body of the lichen. In most, however, the two partners form distinct layers, with the alga generally sandwiched between two different layers of fungal cells.

9.6.2 Some mutualists cannot survive without their partner

Although mutualisms benefit both of the populations involved, many such populations could persist without the organisms with which they interact. Most of the algae involved in forming lichens are perfectly capable of surviving without their fungal partners.

In other cases, however, the mutualism is obligatory. This appears to be true of all of the fungi that have been found in lichens, none of which has ever been found as a free-living organism.

There is even a hypothesis that some mutualisms have evolved to become so strong that one of the partners can no longer even be thought of as a separate organism. The most dramatic example is the suggestion that cell organelles, such as chloroplasts and mitochondria, evolved from single-celled organisms (such as bacteria) that lived in a close mutualism inside the cells of other organisms. The organelles now function as factories for chemical reactions in the cells of all multicellular organisms; the sun's heat and light is converted into useful chemical energy in chloroplasts, and mitochondria convert chemical energy into other useful forms. The organelles now have no independent existence of their own, and are an integral part of the cells of multicellular organisms.

The theory that organelles evolved from independent organisms such as bacteria is plausible because of the large number of similarities between modern bacteria and cell organelles. Many bacteria and most mitochondria are rod shaped, and both have an outer membrane. Some organelles even have their own DNA, separate from the genetic material in the cell's nucleus.

Mutualisms between flowering plants and the animals that pollinate them also vary from extreme specialization to loose associations. Some plants can be pollinated by many species of insect, while rewarding the pollinator with food in the form of nectar or pollen. Other plants, such as many tropical orchids and 900 species of tropical fig (*Ficus* spp.), have only one species of highly specialized pollinator. In the case of the fig and fig wasp, the male fig wasp never leaves the fig fruit throughout his lifetime.

9.6.3 Mathematical models of mutualism

Following the logic of some of the mathematical models developed earlier, it may seem simple to develop models that will provide a reasonable description of mutualism. That does not in fact prove to be the case, for reasons that turn out to be obvious once they have been encountered.

Recall that in Section 8.3.1, a term was added to a basic model of a single population, with the aim of incorporating interspecific competition into the mathematical description. In doing so, Chapter 8 described how the importance of the interaction would depend on the existing size of the two competing populations (which were called N_1 and N_2) and on the strength of the competition between them, which was described with a term called the competition coefficient, denoted as α. Because an interaction could be asymmetrical, there were two values of α. One describes the effect of an individual organism in population 1 on the size of population 2 ($\alpha_{2,1}$) and one describes the opposite effect ($\alpha_{1,2}$).

Having multiplied the size of the first population (N_1) by the size of the second population (N_2), the total was multiplied by the competition coefficient to give a term $N_1N_2\alpha_{1,2}$. This represented the overall effect of all the individuals in population 2 on the total size of population 1, and was simply subtracted from the mathematical model of a single population to account for the reduction in population 1 that is experienced because of competition with population 2.

So to account for mutualism in basic models of populations, it seems at first sight that ecologists could simply invent a similar term to α, which instead of measuring the negative effect of competition would measure the positive effect of mutualism. Rather than subtracting it from the model, modellers could add it, to account for the increase in a population of one species caused by the presence of mutualistic individuals in a population of the second species.

However, taking this approach would create a mathematical model that allowed populations to rise indefinitely. A slight increase in one of the populations would have a positive effect on the second population. This in turn would feed back into the model causing an increase in the first population, which would lead to another increase in the second population. This positive feedback would carry on forever with both populations becoming increasingly enormous. This is clearly unrealistic, because mutualists can only help organisms to obtain resources that actually exist; they cannot create resources that are not already there.

For example, a mycorrhizal fungus might assist a plant in obtaining phosphorus from the soil, such that the plant becomes extremely efficient at taking up this mineral. But there will still be a limit to the amount of water available to the plant, and however much phosphorus the plant population can obtain, it will remain restricted by the availability of water.

There are in fact different ways of developing the existing models to describe populations of mutualistic organisms, and they contain details which are unnecessary for an understanding of how mutualisms operate in ecological communities. Thus, the details will not be developed here. This does not mean, however, that it is a waste of time to consider a possible mathematical model, but end up rejecting it. Taking a model of interspecific competition and trying unsuccessfully to adapt it via a single logical step to describe interspecific mutualisms has revisited an important feature of ecological interactions. No matter how strongly a positive interaction affects a population of organisms, the population will ultimately be limited by the availability of one of the resources that form the population's ecological niche.

That is why evolution by natural selection is such an important feature of studying ecology. In a population with limited resources, it will always tend to be those individuals that are best suited to the current environment that survive to reproduce and create the next generation.

9.7 Sarcophages, detritivores, saprophytes, coprophages and scavengers

All of the methods of obtaining food that have so far been investigated involve using the resources of living organisms. Predators, herbivores, parasites and mutualists all take resources from other living organisms, and consequently have direct effects on the populations of their prey or hosts. There are many other organisms that obtain organic resources by eating other organisms only after the food is dead. Animals, that behave in such a way, especially larger ones such as crows, jackals or some crabs, are called scavengers. Other organisms, particularly microbes and many fungi that spread through dead wood, are called saprophytes, and many smaller animals and micro-organisms are called detritivores. Where the dead matter is originated as animal tissue, sarcophagy is another term for eating the dead material. However, the distinctions between the various terms are entirely subjective, and the interactions can be treated as entirely equivalent. For convenience, the name **detritivore** can be given to all organisms that eat dead matter; its literal meaning is 'eating debris', so it suits organisms that might otherwise be called sarcophages, saprophytes and scavengers (Fig. 9.10).

There is another way of obtaining organic resources from non-living material; some organisms digest faeces, and are known as **coprophages**, from the Greek word *copro* for excrement. Some animals, such as rabbits (*Oryctolagus cuniculus*) eat their own faeces in order to digest plant material that is only partly broken down by a single passage through the gut. Such organisms are mostly herbivores, and their populations are described above. Other organisms eat the faeces of other animals, and can be treated as detritivores, with a similar biology to scavengers or sarcophages. Many invertebrates derive at least some of their organic resources from faeces, but the most well-known examples are the dung beetles (of the family Scarabidae), which bury balls of elephant dung into which they lay their eggs (Fig. 9.10c). The developing larvae live entirely on the plant matter that the elephant failed to digest.

This is actually a very important interaction in many grassland habitats, which aids in the breakdown of partially digested plant material and the return of nutrients to the soil for uptake by other plants. Without the activity of dung beetles, nutrient cycling and plant

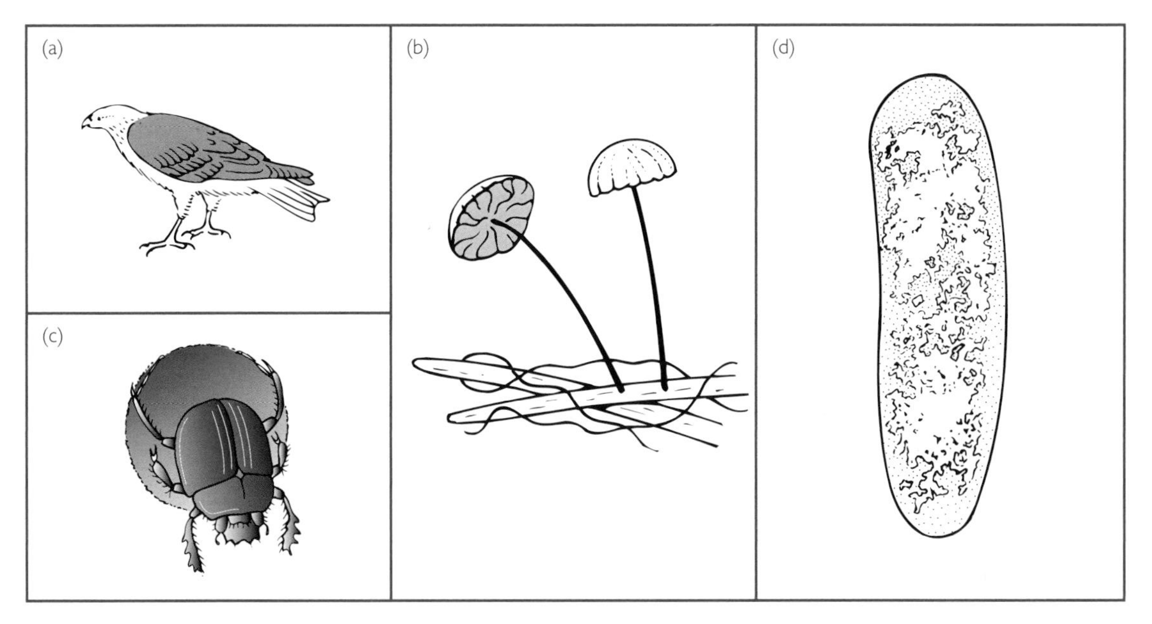

Fig. 9.10 Many different kinds of organisms feed on detritus—dead organic matter: (a) the red kite (*Milvus migrans*), which eats dead animals; (b) the fungus *Marasmius androsaceus*, which specializes in digesting decaying pine needles; (c) the dung beetle (members of the family Scarabidae), the larvae of which live on the faeces of mammals; and (d) the bacterium *Escherichia coli*, which as well as living in the human gut is also found in rotting plant material and soil.

productivity would be greatly reduced in these types of communities.

Many detritivores obtain only a small part of their organic resources from dead matter. For example, the fungus *Boletus subtomentosus* produces enzymes that digest phenols and cellulose, and can consequently obtain some of its resources by digesting dead plant material. But it is also a mutualist that forms mycorrhizae, like those that were investigated in Section 9.6, normally deriving some nutrition from pine trees (*Pinus* spp.). Similarly, many animals that are generally considered carnivorous actually make up some of their diet from organic detritus. Throughout South America, the bird known as the crested caracara (*Polyborus plancus*) catches and kills a variety of insects and small vertebrates, but also eats dead small mammals and carrion from larger ones, whenever it is available.

The detritivore populations have no direct effect on the source populations of the food they eat, because the organisms they eat have already died from some other cause. Detritivore populations are often classified as **donor-controlled**, because the size of the population is thought to be controlled by the 'donation' of dead material by other organisms. Of course, as Chapters 11 and 12 will reveal, detritivore populations can affect their 'donor' populations indirectly by affecting the recycling of nutrients back to other organisms in the ecosystem.

There is no need to design special mathematical models to describe populations of detritivores. The models that were developed in Chapters 6 and 7 are suitable for individual populations, and models of interspecific competition (see Chapter 8) describe populations where more than one species are using the same dead material.

9.8 Chapter summary

All organisms need organic resources, and they obtain these in one of two ways; either they produce their own by utilizing atmospheric carbon, which they obtain by splitting carbon dioxide using the sun's energy, or they digest other organisms, which may be dead or alive when they are eaten. As carbon atoms cycle among different individual organisms in this way, populations of different species increase and decrease in size.

Natural selection to increase the efficiency of finding, capturing and eating prey has led to numerous sensory, behavioural, and morphological adaptations in predators and herbivores. In response, plants and prey have evolved mechanisms to escape or defend themselves from being eaten. Thus, predator–prey and parasite–host interactions have large effects on community biodiversity and population growth and distribution.

Mutualisms benefit both species involved in the interaction, and can be obligatory, with one or both partners not being able to survive without the interaction, or they may be facultative, with one or both partners being able to function outside of the mutualism.

Organisms that feed on dead plant or animal tissue are fundamentally different from those feeding on live organisms. This is because their populations are 'donor-controlled', passively relying upon the supply of detritus. They have little direct effect upon the populations that they derive nutrition from, but they may have large indirect effects by playing essential roles in the breakdown of detritus and the recycling of nutrients and minerals.

Recommended reading

Brodie III, E.D. & Brodie Jr, E.D. (1999) Predator–prey arms races. *BioScience* **49**: 557–568.

McCallum, H. & Dobson, A. (1995) Detecting disease and parasite threats to endangered species and ecosystems. *Trends in Ecology and Evolution* **10**: 190–194.

Oldroyd, B.P. (1999) Coevolution while you wait: *Varroa jacobsoni*, a new parasite of western honeybees. *Trends in Ecology and Evolution* **14**: 312–315.

Selosse, M.-A. & Le Tacon, F. (1998) The land flora: a phototroph–fungus partnership? *Trends in Ecology and Evolution* **13**: 15–19.

Wall, D.H. & Moore, J.C. (1999) Interactions underground. *BioScience* **49**: 109–117.

Winfree, R. (1999) Cuckoos, cowbirds and the persistence of brood parasitism. *Trends in Ecology and Evolution* **14**: 338–342.

Chapter 10

Similarities and differences in ecological interactions

10.1 Describing interactions

From the beginning of Chapter 1, it has been clear that ecological studies must deal with the fact that there are millions of different species of organisms in the world, and must also take account of the fact that each species is divided into a number of separate populations. There are thus many millions of populations of organisms, and each population can interact with many others within the same community. The features of each of these interactions are distinct from the features of all the other billions of potential interactions that might occur.

But as scientists, whose rationale is to break problems down into manageable questions, ecologists cannot hope to study the details of more than a tiny fraction of these interactions. We can never make any progress in understanding the principles that govern ecological interactions unless we can simplify these details into a smaller number of manageable concepts and general rules. This is why ecologists break down the variety of interactions into broad categories, such as competition, predation and detritivory.

However, as has already been noted in Section 9.1, there is something artificial about the words predation, herbivory and parasitism, together with other terms used to describe different types of interaction. These words, or the roots on which they were based, were not coined to be used in a meticulous and precise way by modern ecologists, and we sometimes find that they are not very precise descriptions of the different types of interaction that we might choose to define.

The truth is that, although the terms make a helpful shorthand, and are useful for particular studies of interactions, like those that were described in Chapter 9, ecologists need to think about ecological interactions in a different way in order to gain a full understanding of the importance of these interactions in ecological communities.

10.2 Variation in ecological interactions

Section 8.5 stressed that interspecific competition is an important part of the biological environment of organisms, and described how its effects can drive evolutionary change within populations. For a complete appreciation of other ecological interactions, such as predation, detritivory and parasitism, it is necessary to fit them into the same framework. In doing so, it is essential to remove any preconceptions about the details of the different kinds of interactions. It is sometimes necessary to think of herbivory, predation, competition, mutualism and parasitism as different variations of the same basic concept.

10.2.1 Interactions with different names can be remarkably similar

To see how two different types of interaction can be ecologically equivalent, it is merely necessary to observe that corals are animals, but that they have many of the characteristics of plants—coral colonies tend to be fixed in a particular spot, for example. Some of the individual polyps develop into reproductive structures, just as certain parts of plants develop into flowers, seeds and fruits. Organisms are not concerned with whether other organisms are plants, animals or micro-organisms. Thus, animals that eat coral, such as tropical fish, behave in much the same ways as animals that eat plants, such as beetles and caterpillars. To the ecologist studying such interactions, there is no real difference between a coral colony and an oak tree.

This is manifest if one considers the similarities between organisms that consume parts of an oak tree (which we choose to call herbivores) and organisms that eat coral (which we call predators). It is clear that in these particular cases, predation and herbivory are almost identical (Fig. 10.1).

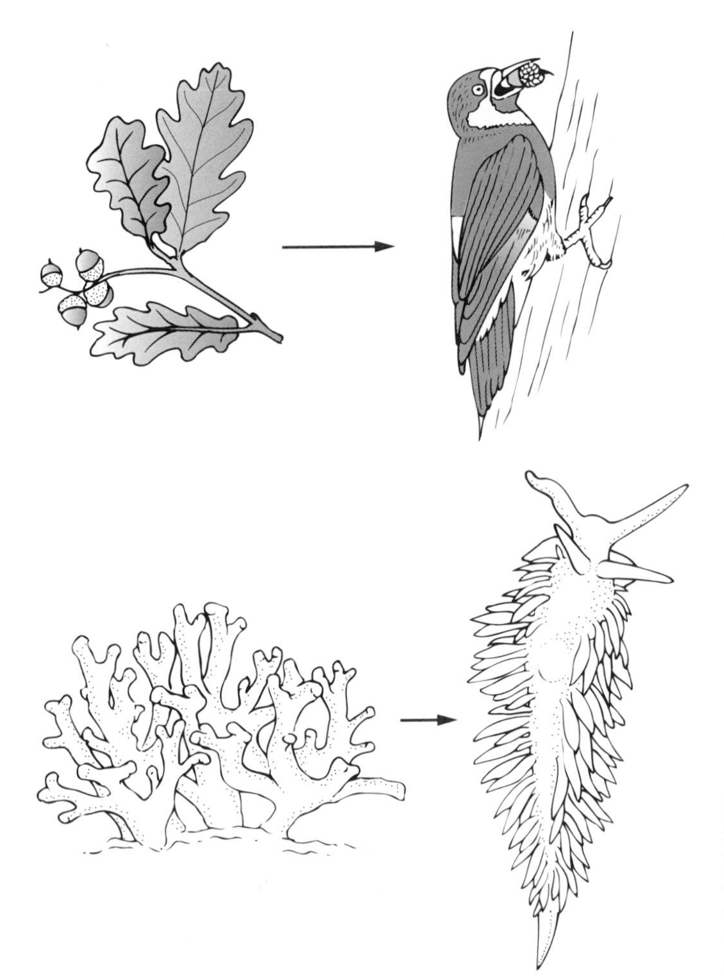

Fig. 10.1 When woodpeckers eat acorns, the interaction is known as frugivory, or fruit eating, a specialized kind of herbivory, which is the general name for eating plant material. When sea slugs eat coral polyps, the interaction is known as predation, because corals are animals. However, the overall effects of the interactions are the same.

A huge variety of consumers depend on oak trees; some eat the leaves, some the fruits and others the bark. For example, in North America, grey squirrels (*Sciurus carolinensis*) eat large quantities of acorns of the white oak (*Quercus alba*), while acorns are also the principal constituent in the diet of the acorn woodpecker (*Melanerpes formicivorus*). In other parts of the world, other birds and mammals eat the acorns of different species of oaks. The leaves of the oak are eaten by many kinds of insects, especially caterpillars of moths and butterflies; for example, in Britain there are at least eight species of moths with names that reflect the fact that either the adults or the caterpillars are associated with oak trees, including the oak beauty (*Biston strataria*), the oak eggar (*Lasiocampa quercus*) and the oak processionary (*Thaumetopoea processionea*). Many animals also consume the bark of the oak tree, including the larvae of the bark beetle *Xyleborus dryographus*, which tunnel just below the surface of the bark, and the beaver (*Castor canadensis*) and some species of deer.

The herbivores that eat oak leaves are essentially compromising the tree's ability to photosynthesize, which is its way of producing food. Those that consume the bark, among other things, affect the tree's system of protecting itself and of transporting water and nutrients, and those that eat acorns reduce the tree's rate of reproduction.

Exactly the same set of problems may face a colony of coral polyps (or other cnidarians such as sea anemones) as face the oak tree. Some predators may consume the polyps that are responsible for digestion, while others may eat those that undertake the colony's sexual reproduction. It may well be true that many coral feeders are not as specialized as the animals that eat the various parts of oak trees. However, this is not always the case. Some kinds of sea slugs, such as *Spurilla neapolitana*, tend to specialize in eating polyps that contain nematocysts, which are tiny arrow-like structures that the coral uses as a defence. The sea slugs then retain the offensive structures for their own use in repelling predators.

10.2.2 Interactions with the same name can be very different

Just as interactions with different names can be very similar, so those that are given the same name can have somewhat different effects in the real world. For example, Section 9.4.1 described how, when developing mathematical models to describe interactions, the model that was built to describe predation could be used as a description of herbivory, but only if the herbivores in question generally killed the individual plants that they ate, just as predators kill their prey. Examples of this situation might include some burrowing rodents, which destroy the tuberous roots of their food plants, such as the naked mole rat (*Heterocephalus glaber*) from Ethiopia and Kenya, which lives exclusively on roots and tubers. Similarly, the larvae of chafer beetles, such as the garden chafer (*Phyllopertha horticola*), kill their food plants by eating the roots, and are consequently pests of crop plants such as cabbages or cauliflowers (*Brassica oleracea*).

Populations of organisms that do not generally kill their food plants but eat only parts of them (such as bark beetles, sheep or rust fungi) can be described by a different mathematical model, like the one that was developed in Section 9.5.1 to describe parasitism.

It is easy to see how ecological interactions with the same name can in fact be very different in their effects by considering populations of parasites that interact with populations of humans. In each case, the interaction is known as an example of parasitism, but the effect of individual parasites on individual humans varies enormously.

In most cases, *Streptococcus* bacteria cause mild discomfort, such as a sore throat, with no serious long-term consequences. Some parasites, such as tapeworms (*Taenia* spp.), cause long-term problems but in normal circumstances do not kill the victim. Others, such as

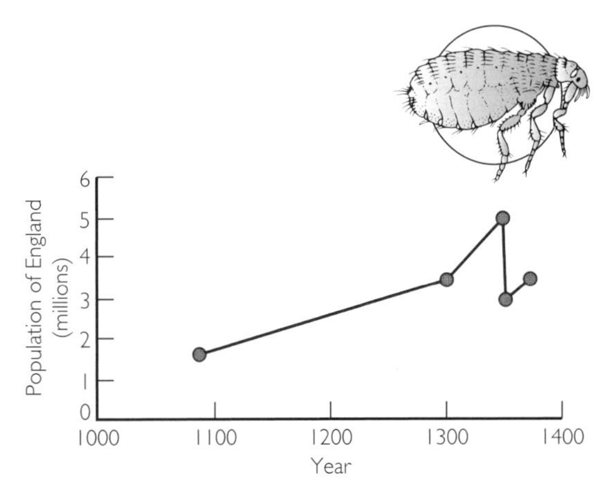

Fig. 10.2 The population of England dropped dramatically as a result of the parasitic bacterium *Yersinia pestis*, the cause of bubonic plague. The bacteria were spread by fleas, which are themselves parasites. In 4 years between 1347 and 1351, a huge epidemic of the plague, known as the Black Death, swept through Europe and Asia, killing approximately 25–35% of the population, perhaps as many as 25 million people.

HIV (human immunodeficiency virus), often debilitate the victim sufficiently that he or she may die, although the virus itself is not the cause of death. Yet more parasites are themselves deadly, and of these, some (such as the ebola virus) cause death swiftly, while others (such as relapsing malaria, *Plasmodium falciparum*) might wear the body down over many decades, but undoubtedly shorten the lives of many of their victims.

There is also great variety in the ways in which these organisms are transmitted among their human hosts. Syphilis (*Treponema pallidum*) is transferred between hosts during sexual contact, *Salmonella* bacteria are ingested with food, dengue fever is spread by secondary hosts (mosquitoes), *Staphylococcus* bacteria can enter the body through infected wounds, and the influenza virus is transmitted via droplets spread in coughs and sneezes.

Section 9.5 hinted at how the effects of these parasites on the populations of hosts can also be very different. Bubonic plague caused by the bacterium *Yersinia pestis* can wipe out whole populations of people—it killed about a third of all the people in fourteenth century England, leaving some villages empty and derelict, and forcing many families to move large distances (Fig. 10.2). But meningococcal meningitis, which is an infection also caused by a variety of bacteria (most commonly *Neisseria meningitidis*) and which can be lethal, has no overall effect on the size of human populations. Other deadly bacterial infections, such as cholera (*Vibrio cholerae*), probably reduce the overall size of human populations in places where they are endemic and have significant effects on the rate of survival among infants.

It is clear that with such a variety of different interactions between individual parasites and their hosts, and such a diversity of possible effects, it is important for us to realize that the name 'parasitism', and also the words 'herbivory', 'predation' and 'mutualism', mean nothing to the organisms involved. Each interaction between two individual organisms is unique.

10.2.3 Even interactions that appear to be identical can be very different

It is possible to see that ecological interactions show great variety by observing that the same two species can interact in two different communities, but that the effects may not always be the same. For example, a population of herbivores may reduce the size of a population of its plant food in one community, but have no effect in other places where the herbivore population is controlled by predators.

This is almost certainly true in the case of the white-tailed deer (*Odocoileus virginianus*). There is no doubt that some populations of white-tailed deer have an effect on the population abundance of some plant species, because careful experiments have shown the effects. Bowers (1997) constructed fences to exclude the deer from square plots of 36 m^2 of woodland on the west side of the Shenandoah River in Virginia, and recorded the effects on the woodland plants. There were almost no individuals of the Tatarian honeysuckle (*Lonicera tatarica*) in places where the deer were present, but the plant increased in abundance by an average of almost six-fold in the plots where the deer were kept out.

This population of white-tailed deer is probably not controlled by predators, but other populations of the same species probably are. For instance, Schmitz and Sinclair (1997) observed that controlling predators such as wolves often results in unusually high densities of large herbivores, including white-tailed deer. Observations in places where white-tailed deer have been controlled by the hunting activities of human predators (e.g. in Massachusetts) suggest that the effects of deer on plant species do indeed vary (see Sinclair 1997).

In other words, the effect of populations of white-tailed deer is in some places to reduce greatly the populations of plant species such as the Tatarian honeysuckle, but in other places, where predators are common and keep the deer population in check, the deer may have no overall effect on the abundance of their food plants.

A complete understanding of ecological interactions must accept that such variation is commonplace, and any overall theory or set of mathematical models must be flexible enough to take account of the variation, not just among different interactions, but also the disparity between separate examples of interactions between the same two species.

10.3 All interactions are unique

Section 8.4.2 described how the thorn bush *Acacia senegal* competes with its close relative *Acacia zanzibarica* for pollinators. Both need the same resource, but they cannot both have it, so the interaction was defined as competition. Parasitic tapeworms steal human resources from the gut after the humans have digested the resources. Once again, the two species, humans and tapeworms, need the same resource, but the interaction is defined as parasitism.

Section 8.2.1 investigated an intermediate situation, in which jaegers force gulls to regurgitate their meals, so that they can steal the partially digested squid and fish. It is not immediately obvious whether this is an example of competition or parasitism. If the jaegers watched while the gulls found the fish, and then caught the fish for themselves (in the way that *Acacia* bushes attract their own pollinators), then the interaction would be an example of competition. If they waited until the gulls had digested the food fully and ate the gull's living tissue (as the tapeworm does), we would have called the situation an example of parasitism. But the interaction seems to fall somewhere between the two—by simple, strict definitions, it is neither competition nor parasitism. Ecologists have been forced to invent a new term for this kind of interaction—**kleptoparasitism**.

This example shows very clearly that, in fact, interactions cannot always be easily classified into neat categories. Each interaction between two populations is unique, and to understand interactions more fully, it is essential to find a unifying way of describing them.

There is nothing theoretically to distinguish between the effects of tapeworms on humans, and the effect of grasshoppers on grass. The tapeworm parasites appropriate some of the organic tissue of the humans and use the molecules for their own growth, survival and reproduction. The effect on the humans is to make them less effective in their own growth, survival and reproduction. In exactly the same way, the grasshoppers appropriate some of the grass tissue to aid their own growth, survival and reproduction at the expense of the grass plant's growth, survival and reproduction.

In fact, all of the interactions that have been examined in Chapters 7–9 can be viewed, quite simply, as systems for moving organic, carbon-based molecules among populations. Thus, instead of partitioning the interactions into discrete units that we choose to call predation, competition, parasitism and detritivory, we could classify them according to their effects in terms of the amounts of carbon moved, and the direction in which it passes. Such a system of classifying ecological interactions—based on the functional redistribution of carbon—will ultimately be more illuminating than one based on unnecessarily precise definitions of historical terms.

10.4 More than one interaction between the same two populations

The ecological interactions that have so far been investigated tend to assume that when an individual of one species encounters an individual of a second species, there is only one possible way in which the two individual organisms can interact.

It is plain that this is not the case. Many primates, like the rhesus macaque (*Macaca mulatta*), are plagued by external parasites such as ticks and lice, but they also feed on these invertebrates, often removing them from the bodies of other members of their social group. The monkeys are thus both parasitized by and predators of the ticks.

Because two individual organisms can interact in more than one way, some pairs of populations may interact in more than one way. For example, the way in which jaegers chase gulls and force them to regurgitate their food (described in Section 8.2.1) is not the only interaction between these species (Fig. 10.3). Jaegers also catch some of their own live prey, so that they are more conventional competitors with gulls because they are using the same resources. Jaegers will also raid the nests of gulls and eat their eggs and young chicks—the parasitic jaeger (*Stercorarius parasiticus*), also known as the arctic skua, eats chicks of various gull species. But some species of gulls, including the glaucous gull (*Larus hyperboreus*), will also eat the eggs or chicks of the jaegers. So the two species are reciprocal predators on one another.

A population of jaegers will affect a population of gulls negatively through ordinary exploitation competition, through kleptoparasitism and through predation of the gulls' chicks. But the jaegers will also affect the gulls positively because the gulls will obtain food by eating the jaeger eggs.

The complex and mutually contradictory ways in which two populations can interact are well illustrated by a relative of the wild cabbage found in Spain, *Hormathophylla spinosa*, and dodder (*Cuscuta epythimum*), a plant that appears to be entirely parasitic. Dodder is not particularly specialized and can live on a variety of different species of plants, but it derives all its nutrients from its host. It has no leaves, and attaches itself to the host by means of tendrils and swollen suckers. Small threads develop and invade the stem of the host plant. These threads act as roots, and through them the dodder plant steals water and nutrients from the xylem and phloem vessels of the host.

Since dodder has no real roots and no leaves, it seems impossible to escape the conclusion that its effect on the host is negative; host populations that suffer parasitism from dodder must be expected to perform worse than host populations that are free of the parasites.

But in the high mountains of Sierra Nevada in southern Spain, where dodder affects a number of species of green plants, its effects on host populations do not appear to be universally negative. Populations of the host plant *Hormathophylla spinosa* vary in the degree to which they suffer from dodder infestation: between 0% and 20% of the plants are parasitized. Although the direct effects of the parasitism have not been measured, it is inconceivable that the dodder is anything other than a serious parasite in terms of the nutrients that it extracts from the hosts. However, the overall effects of the dodder are very close to neutral, because of the indirect effects brought about by the dodder. For example, wild ibex (*Capra pyrenaica*) actively avoid eating parasitized plants. Since ibexes are large herbivores (they can weigh up to 80 kg), they consume large quantities of herbaceous plants, and it is considerably

Fig. 10.3 Jaegars and gulls of various species (*Larus* spp.) interact in a variety of ways: (a) jaegers chase gulls and force them to regurgitate their food; (b) jaegers and gulls eat some of the same foods and hence are in exploitation competition; (c) jaegers will eat the chicks of gulls; and (d) gulls will eat the eggs of jaegers.

to the advantage of the *Hormathophylla* plants that the ibexes avoid dodder-infested plants, even when they are less than a metre away from uninfested, heavily grazed plants of the same species. Furthermore, being parasitized by dodder also protects plants from attack by the weevil *Ceutirhynchus*, which eats the seeds of *Hormathophylla* (Gomez 1994).

Thus, although dodder is unequivocally a 'parasite' in that it directly steals the resources of the *Hormathophylla*, the overall effect of the dodder population on the *Hormathophylla* population is neutral.

In other cases, however, one population can have more than one mutually reinforcing effect on another. An extreme example of this comes in the effects that the slug *Deroceras hilbrandi* has on the butterwort *Pinguicula vallisneriifolia* in southern Spain (Zamora & Gomez 1996). The plant is partly carnivorous and has sticky leaves that trap a variety of insects and other invertebrates, but the leaves are green, and the plant derives some of its nutrition from photosynthesis in the normal way. The slug obtains some of its organic nutrition by herbivory. It eats the leaves of the butterwort, and in wet areas up to 18% of the leaf area can be consumed by the slugs in this way. But the slug is partly a scavenger, and will eat the carcasses of flies and other insects. A principal source of such carcasses is the sticky surface of the carnivorous plant, and some plants could suffer a theft rate of up to 80% of the

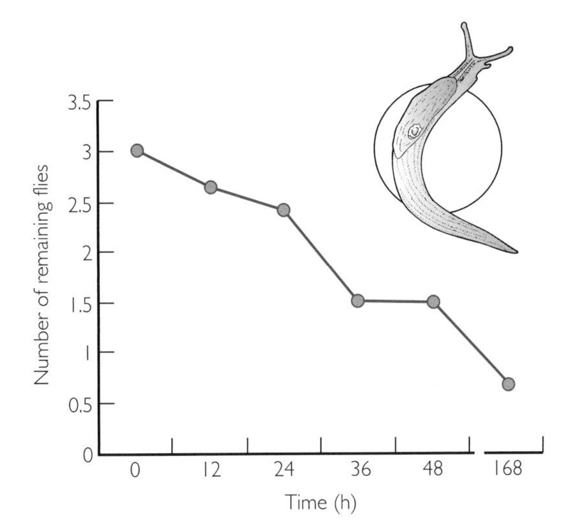

Fig. 10.4 The average number of flies remaining on the sticky leaves of the butterwort (*Pinguicula vallisneriifolia*) over a period of 168 h. The loss of flies, which the plant would otherwise digest, is a result of theft by the slug *Deroceras hilbrandi*. The slug has two negative effects on the plant because as well as taking the flies, it also eats the leaves of the butterwort. (From Zamora & Gomez 1996.)

insects they have trapped. Thus, the slug is both a kleptoparasite and a herbivore on the butterwort, and affects the plants entirely negatively in two quite separate ways (Fig. 10.4).

10.5 All interactions have common elements

Having recognized that all ecological interactions are unique, and that some of them have multiple effects on the populations involved, it would be possible to believe that it is difficult to have a real understanding of how interactions affect communities of organisms. No single person could study the details of all the many thousands of interactions that take place within a single habitat.

It is possible, and sometimes profitable, to ignore the precise details of interactions and produce mathematical descriptions of whole classes of interactions at the same time. These mathematical descriptions are only possible because of the fundamental principle that interactions are all about how organisms gain organic resources, sometimes at the expense of other organisms.

Thus, it is feasible to produce a single understanding of ecological interactions by temporarily forgetting about individual organisms and their behaviour, but focusing instead on how organic resources are moved among populations. Chapter 12 will return to the details of the ways in which carbon-based molecules are moved around ecosystems.

10.6 Interaction coefficients

One way in which it is possible to summarize the effects of one population is to use a quantitative measure of the average effect that individuals of one species have on a population of a second species. This approach has already been adopted in Section 8.3.1, which defined the competition coefficient—a measure of the amount of organic resources that organisms in one population lose when they are competing with organisms in a second population, relative to the organic resources they could obtain in the absence of the second population.

Recall that $\alpha_{1,2}$ is the overall effect on the size of population 1 when population 2 is increased by one individual, and that the effect is caused by competition between organisms in the two populations of different species.

There is no reason why this logic should be restricted to competitive interactions. In theory it would be possible to define predation coefficients, parasitism coefficients, herbivory coefficients and so on. But Section 9.1 has already observed that dividing interactions into these categories is really only for the convenience of ecologists. So in theory, it would be possible simply to expand the use of the term $\alpha_{1,2}$ to measure the effect of extra individuals in one population on the size of a second population, irrespective of the precise nature of the interaction between the two populations.

In practice there are problems with this approach, as was revealed in Section 9.6.3, which developed a mathematical model to describe populations of mutualists. Because two mutualistic populations have positive effects on one another, mathematical models that simply adopt the concept of α would provide unrealistic results in which the mutualistic populations could keep growing forever. An increase in the first population would cause an increase in the second population, which would cause another increase in the first population and so on.

Thus, it is not possible merely to transfer the concept of α to ecological interactions other than interspecific competition. However, in order to understand real ecological interactions in real ecological communities, it is not necessary to worry too much about the theoretical details of particular mathematical models.

It remains true that, within the limits set by real situations, it is possible to measure the effects of a population of one species on a population of another, and quantify it. This might be done by removing individuals of one species, and observing the effects on the number of individuals in a second species. Additionally, an experimenter could import individuals of one species to increase its abundance, and count the change in the numbers of individuals of another species.

Such measurements can be called **interaction coefficients**, provided it is recognized that, if a mathematical model of a population is to be built, these coefficients must be treated with great care, because there is no guarantee that the same numerical effects will apply at different population densities. The numerical effects might also differ in different habitats or different populations, as was clear in the example of the white-tailed deer and Tatarian honeysuckle described in Section 10.2.3.

Such interaction coefficients have one significant practical advantage. If, as in the case of dodder and *Hormathophylla* in Section 10.4, there are a number of different types of interaction taking place between the same two populations, the details can be ignored, and a study could simply measure the average overall effect of individuals of species 1 on the population of species 2.

As was noted in Section 8.3.1, with reference to competition coefficients, real estimates of interaction coefficients are difficult to obtain because they involve complex, detailed and potentially lengthy experiments. But that does not detract from the overall value of the concept.

10.6.1 Interaction coefficients tell us nothing about the ecology of individuals

Interaction coefficients give information about the effects of ecological interactions on populations of organisms, and thus they help ecologists to understand their subject in terms of the definition in Section 1.5, in which ecological studies were said to be about the distribution and abundance of species and populations.

It is important to return to the central idea that ecology is also about the interaction between individual organisms and their biological and physical environments. It must never be forgotten that interaction coefficients are mathematical constructs that help ecologists to reduce complex sequences of events into summaries that they find manageable. Interaction coefficients do not give any information at all about the complex of interactions between individual organisms that combine to produce the overall effect of one population on another.

Indeed, it is perfectly possible for the individuals of a population of one species to have negative effects on the individuals of a different population, but for the interaction coefficient between the two populations to be zero. Such a situation is illustrated at its most extreme by the interaction between people and their farm animals.

Domestic chickens (*Gallus gallus*) are kept in all sorts of ways by farmers around the world, both for eggs and as a source of meat. In some places, they are packed in large numbers into battery sheds, while in many others they are kept in small numbers by rural families. One uniform feature of all the populations of chickens, however, is that many (if not most) of the animals are killed before they have the opportunity to breed. Most males are killed for food, and even though a female may lay many eggs during her lifetime, many farming practices will not allow them to be fertilized. But the population of chickens remains at a fairly constant size that is determined by the available resources (most of which happen also to be provided by humans). Thus, the interaction between individual humans and individual chickens is highly negative for most of the individual chickens, but can be seen as effectively neutral (or even positive) on the number of individuals in the population.

Exactly analogous situations occur in thousands of ecological interactions, including many examples of herbivory. Beetles eating the leaves of a large tree, or antelope browsing in the African bush, have negative effects on the individual trees and bushes, but they do not necessarily have any effect on the size of the local population of trees or bushes.

It is difficult to imagine situations in which the individual organisms in a population are positively affected by an interaction, but where the population itself is negatively affected. However, it would certainly be possible for positive effects on individuals to translate into a neutral effect on the population. Imagine a situation in which the size of a predator population is limited by the presence of an infectious disease, but where there is more than enough food for the predators, consisting of two different, very abundant prey species. Individual predators benefit from the interaction with each of the prey populations, but the overall population of predators would remain at the same size even if one of the prey species became extinct. The predator population would still be limited by the infectious disease, and there would still be plenty of food in the form of the remaining prey species.

This example returns to a central theme in ecology; populations of organisms are more than the sum of the individuals of which they consist. Thus, the mathematical construct of an interaction coefficient is an extremely useful way to summarize the effect of one population on another, but it tells ecologists absolutely nothing about the effect of one organism on another.

10.7 The importance of individual organisms

Discussions in earlier chapters have several times returned to one of the central dilemmas for ecologists —the distinction between individuals and populations. Ecologists can rarely hope to learn much about the overall distribution and abundance of populations if they spend too much time studying the details of a small number of individual organisms, but they can never convince themselves that their understanding of populations is very deep if they fail to appreciate how that understanding reflects the lives of individuals.

When ecologists investigate interaction coefficients, they are generally choosing to summarize the effects of whole populations on one another. However, they cannot always ignore the details of how one individual organism affects another. Part of this detail is studied by **behavioural ecologists**, whose aim is to understand the 'survival value of behaviour'. Behaviour can be defined rather loosely so that it not only includes traditional ideas of 'animal behaviour', but also certain aspects of the biology of plants. This area of investigation is called behavioural ecology because the ways in which behaviour affects survival are dependent on the ecology of the individual organisms involved.

10.7.1 Why behaviour ecology matters when studying ecological interactions

Because behavioural ecologists study survival and reproductive success, their investigations are of direct relevance in understanding population ecology. Chapters 6–9 identified that survival is a key element in determining the size and density of populations. In Section 6.3.3, developing a simple model of an imaginary population of frogs uncovered an appreciation of how the details were changed when the models started to include realistic assumptions about the likely survival rate of the frogs. In Section 9.3.2, which examined models of predators and prey populations, it was helpful to include a factor to measure the effectiveness of individual predators in catching individual prey. This is exactly the kind of detail that behavioural ecology seeks to explain.

The kinds of activity that are generally studied by behavioural ecologists include questions such as: How do organisms choose their mates? How do they optimize their feeding efficiency when they are foraging? When does it pay to live in a group rather than as a single organism or a pair?

Section 9.3.1 investigated three functional responses: the different ways in which predators and herbivores respond to changes in the population density of the organisms that make up their food. The discussion investigated the effects of such responses by individual herbivores on the size of the populations of their food plants, and on the potential size of the herbivore population. Behavioural ecologists might study the effect of such responses on the survival of individual herbivores.

The integral nature of the various strands of ecological study can be seen in a comprehensive study of the chimpanzee (*Pan troglodytes*) and the red colobus monkey (*Colobus badius*) in Tanzania (Stanford 1998). The subtitle of the book is *The Ecology of Predator and Prey*, and the study deals in depth with the standard ecological issues such as the effect of the predator population on the prey population. However, it also deals with the ways in which the predatory interaction affects, and is affected by, the behaviour and social structure of the chimpanzees.

10.7.2 Behavioural ecology helps us to understand biological diversity

The study of social insects, such as termites, ants, bees and wasps, illustrates very clearly the close relationship between behaviour and ecological diversity. These invertebrates are important parts of many terrestrial ecological communities, and have direct and indirect interactions with a wide variety of plants, animals, fungi and micro-organisms.

Recall that Chapter 1 explored the concept of biodiversity, including the genetic make-up of a population. Part of the reason that social insects make such good objects of study is that many of them (including all the ants, bees and wasps, but not the termites) have unusual genetics. Just like most multicellular organisms, female ants, bees and wasps have a full complement of chromosomes, half of which are inherited from their mothers and half from their fathers. But males of these groups are produced from unfertilized eggs, which contain only half a set of genetic material, all of it provided by the mother; in other words male ants, bees and wasps have no father.

This unusual phenomenon allows ecologists to learn about the ecology of social insects by using one of the powerful concepts that we developed in Chapter 1: the value of comparison. Ecologists can use the fact that ants have a different genetic system from other insects as a tool in understanding their behaviour, and its effects on their ecology. Indeed, Bourke and Franks (1995) examined the links between the social behaviour of ant species and a number of the concepts that have been of interest to us in previous chapters: the interactions between genes and populations, the diversity of ants, and the importance of the reproductive rate (*r*) and the carrying capacity (*K*).

Another area that has proved helpful in trying to understand biodiversity is the study of mate choice. There are many examples of animal species in which the females choose the individual males with which they will mate. Peahens (*Pavo cristatus*) prefer to mate with peacocks that have longer, more elaborate tail feathers (Fig. 10.5) (Petrie *et al.* 1991). The evolution of mate choice in animals has helped ecologists to understand the ways in which biological diversity is maintained. For example, among the different species of cichlid fish in Lake Victoria, the males are often brightly coloured and the females of each species tend to prefer males of a particular colour. Early in the evolution of closely related species, such preferences may have prevented hybridization, and thus have been crucial in maintaining the identity of the different species.

Mate choice is also a good example of a 'behaviour' that may in fact apply to organisms other than animals. For example, there is good evidence that the female organs of some plants exert a 'choice' over which pollen grain eventually fertilizes the egg. The mechanisms by which this might occur are far less obvious than they are in animals, but at least two different methods are reasonably clear, although the precise details may not be understood.

The first mechanism is delayed fertilization, in which the female organs wait for some time after the first pollen reaches them, allowing a number of pollen to build up, before any of the pollen can fertilize the eggs. In the sweet gum (*Liquidambar styraciflua*), the delay is between 1 and 3 weeks, while in some species of oak (*Quercus* spp.) the delay can be longer than a year. Willson and Burley (1983), who examined this process, believed that delayed fertilization may be a female tactic to increase the greatest number of potential mates, from which the female organs can in some sense 'choose' the best.

The second way in which the female organs of plants might exercise a choice over which male will fertilize the egg is by producing more than one embryo, but aborting most of them and allowing only one to develop. The carpels of most kinds of oak trees (*Quercus* spp.) typically produce six potential eggs, and several

Fig. 10.5 The behaviour of individual organisms is very important in the study of ecology, and the field of behavioural ecology studies the adaptive value of behaviour. Although behavioural ecologists often study subjects such as mate choice in animals, the scientific questions are equally important in other kinds of organisms. In peacocks (*Pavo cristatus*), it is relatively obvious how a female chooses the male with which it will mate. In fact, similar 'choice' occurs in oak trees (*Quercus* spp.), in which only about one-sixth of potential eggs develop into acorns, and in some sense the tree is 'choosing' which of the pollen grains that fertilizes the eggs will have its genes represented in the offspring.

can be fertilized by different pollen grains, but only one matures, so that at some level, the plant is deciding which of several pollen is allowed to fertilize the successful seed.

Despite the obvious importance of behavioural ecology, it will not be treated in any depth in this book because it is not of prime relevance to the definition of ecological study that was developed in Section 1.5. This definition is concerned with how the interactions of individual organisms with their environment affect the distributions and abundances of populations and species. However, ecologists should always remember that interactions between individual organisms are extremely important, and should always keep them in the back of their minds when they are studying the kind of ecology that fits within the slightly more restricted definition that was developed in Chapter 1.

10.8 Interactions where organic resources are not transferred between populations

So far, ecological interactions have been described in the context of carbon-based organic resources being transferred between two different organisms. A herbivorous organism eats plants and then uses the plants' resources to build and maintain its own existence. Likewise a parasite co-opts the organic resources of its host.

However, individual interactions between two organisms do not always result in the transfer of resources. An Amazonian skunk (*Mephitis semistriatus*) is clearly interacting with a predatory Argentine grey fox (*Dusicyon griseus*) when it repels the predator with its sulphurous musky spray, but no organic resources are transferred between the two organisms.

An interaction does take place when a person is stung by a stinging nettle (*Urtica dioica*), even though the transfer of tiny amounts of acid does not really fit a definition of ecological interactions that is couched in terms of the movement of carbon among organisms within an ecosystem.

These interactions between individual organisms occur only because one of the organisms has evolved a defence mechanism to reduce the chance of its being killed or harmed in one of the ecological interactions that we have already discussed. In spraying the fox, or stinging the person, the skunk and nettle are reducing the chances of predation or herbivory occurring. And, of course, in doing so, they are themselves using their own organic resources—they must manufacture the sulphurous repellant liquid and stinging acid.

There are many, diverse examples of interactions between individual organisms where there is no direct cycling of carbon-based organic resources. As in the case of skunks and foxes, or nettles and people, these interactions are generally by-products of those interactions that were examined in Chapters 6–9. They are, in fact, responses that have evolved by natural selection in response to selection pressures exerted by predators, herbivores, parasites or mutualists.

These responses generally involve the use of resources, just as the nettle and skunk use resources in producing unpleasant repellant liquids. In fact, it may be very costly for plants like nettles to produce the chemicals (known as secondary metabolites) that are used in defending themselves against herbivores. In some cases, using resources to produce these chemicals appears to be so costly that the plants do not produce them unless they are attacked by herbivores. When the herbivorous Colorado beetle (*Leptinotarsa decemlineata*) feeds on the leaves of potato (*Solanum tuberosum*) or tomato (*Lycopersicum esculentum*) plants, it can cause the rapid accumulation of chemicals that inhibit enzymes that break down proteins (Harborne 1997). In the absence of the herbivores, the plants produce little or none of these proteinase inhibitors. Since these chemicals interfere with the digestive system of the beetle, they deter the insects from feeding on the tomato or potato plants.

Inevitably, such an interaction reduces the amount of resources that are available for the individual prey or plants to put into growth and reproduction. At least in theory, the interaction can thus affect the size or distribution of populations, even if the repellant is so successful that the predators or herbivores never actually succeed in killing or damaging the individual prey or plants.

In fact, in the case of plants defending themselves against herbivores with secondary metabolites, the evidence of a trade-off between defence and growth or reproduction is rather weak. There is no good evidence that the kind of defensive interactions shown by individual nettles and tomatoes have any strong effects on the populations to which those individuals belong.

It would be foolish, however, to pretend that a lack of evidence for such effects was proof that no such effects ever occur. The probability that they do occasionally occur is suggested by circumstantial evidence that, to produce poisons and repellants, organisms require a flow of resources diverted from primary functions, and that some plants do not go to the expense of producing secondary metabolites until they are stimulated to do so by attack from herbivores.

10.9 The strength of interactions can vary in different environments

Interaction coefficients are not fixed quantities that are always the same, wherever populations of the same two species happen to coexist. The quality and strength of interactions change according to the enormous number of factors that can vary in the biological and physical environments. This is clear in the case of white-tailed deer, described in Section 10.2.3, which have a strong negative effect on some of their food plants in areas where the deer are common, but not in places where predators regulate the size of the deer population. A more precise example came from the laboratory experiment described in Section 8.4.1, which noted that competition between two species of seaweed flies is very delicately balanced. In the laboratory, when the flies are fed on chopped seaweed, a population of *Coelopa pilipes* has a positive effect on a population of *C. frigida*. But this positive effect disappears with a simple change in the environment—feeding the flies with minced instead of chopped seaweed (Hodge & Arthur 1997).

A more complex example comes from the negative interactions of a parasite and a herbivore on the broad-leaved dock (*Rumex obtusifolius*). The plants suffer herbivory from the beetle *Gastrophysa viridula*, with a typical rate of consumption of about 3 cm^2 of leaf per beetle each day (Hatcher *et al.* 1997). However, when the physical environment is altered by the addition of nitrate or ammonium fertilization, the interaction has a much less strongly negative effect on the plants. Under these conditions, each beetle consumes only 1.5 cm^2 every day. The fungus *Uromyces rumicis* infests the leaves of the broad-leaved dock, but again, increasing the level of nitrate available to the plant reduces the level of infection by as much as 71% (Fig. 10.6).

It is not clear how these interactions are mediated by the availability of nitrate, nor is it clear how strong the interaction coefficients actually are. The amount of damage to individual broad-leaved dock plants is reduced by an increase in nitrate, but it is not obvious what effect this would have in the long term on the overall population of the docks. Small increases in nitrate can bring about large differences in the amount of damage caused by the beetles and fungus. Thus, it is reasonable to suppose that a single, small change in the

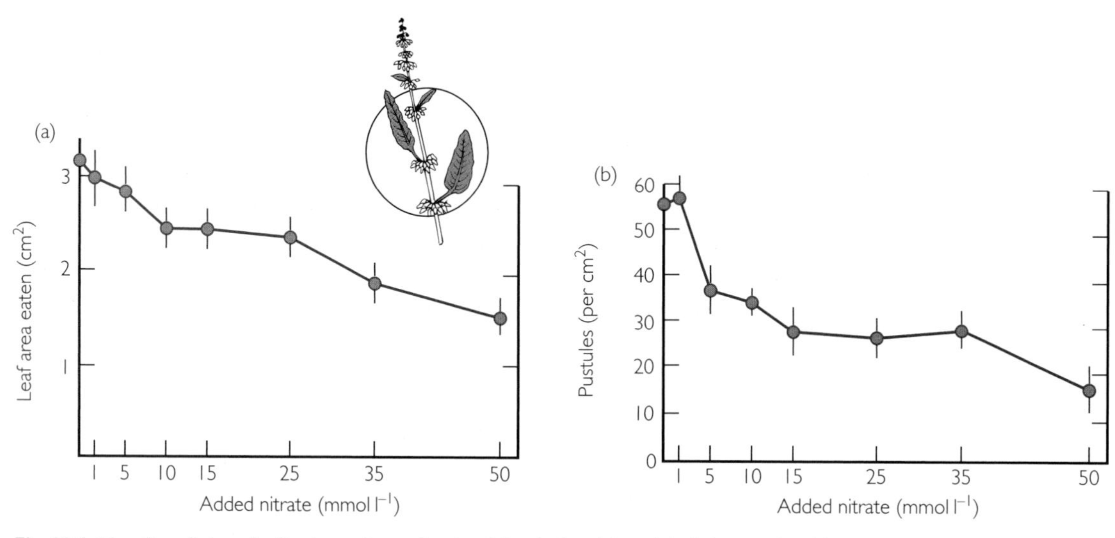

Fig. 10.6 The effect of nitrate fertilization on interactions involving the broad-leaved dock (*Rumex obtusifolius*). (a) The average area of dock leaf eaten each day by each individual of the beetle *Gastrophysa viridula*, and (b) the average number of pustules of the fungus *Uromyces rumicis* per square centimetre of leaf area 8 days after inoculation. In both cases, changing the physical environment by adding large doses of nitrate fertilizer causes a dramatic change in the effects that the herbivore and parasite have on the plants. (Adapted from Hatcher *et al.* 1997.)

physical environment has a significant effect on the magnitudes of the interaction coefficients.

10.10 More than two populations

So far, interaction coefficients have been considered in the context of two interacting populations. But Section 10.1 noted that in real ecological communities it is necessary to consider more complex sets of interactions, where many populations may have very many mutual interactions. For a full understanding of the importance of ecological communities, it is essential to appreciate how all of these interactions fit together. Chapter 11 will begin to do this in detail, as it describes the nature of ecological communities. But it is important to remember that there is nothing inherently special that differentiates community interactions from those interactions that have been described and investigated in this chapter and Chapters 6–9. Communities are simply collections of populations that interact with one another, either directly or indirectly.

Importantly, before gaining an appreciation of the structure of ecological communities, it is imperative to understand that, just as the physical environment can affect the strength of an interaction coefficient, so can the biological environment.

In other words, the strength of an interaction between two populations can vary with the presence or absence of other populations. This is true of the interaction between the Indian meal moth (*Plodia interpunctella*) and the wasp *Venturia canescens*, in the presence of a granulosus virus, known as GV.

The wasp is a parasitoid of the moth—the female wasps lay their eggs inside the body of the moth's caterpillars, and the larval wasps eat the larval moths alive, starting from the inside. By a general definition, this can be viewed as an example of predation because the wasp larvae invariably kill the caterpillars by eating them. However, the female wasps can detect whether or not a caterpillar is infected with the granulosus virus (Sait *et al.* 1996). If the infection is present, the wasp lays fewer eggs than she would in a healthy caterpillar. Wasp larvae are less likely to survive in caterpillars that are heavily infected with the virus than in those caterpillars that are either healthy or have only a light virus infection.

The moth population may be less strongly affected by the wasps in the presence of the virus, because the wasps are avoiding infected moths. Whether or not there is any overall effect on the moth population is unclear—those that escape parasitism by the wasps may all be killed by the virus.

10.10.1 Complex sets of interactions

Sections 8.4.3 and 10.4 described how one population can affect another in indirect ways, and Section 10.10 above illustrated that interactions can become more complex when three or more species are involved. Added together, these two facts mean that some sets of interactions are difficult to track.

This can be seen clearly in the interactions between the butterfly *Maculinea rebeli*, the flowering plant known as the gentian (*Gentiana cruciata*) and the red ant (*Myrmica schenki*) (Fig. 10.7). The caterpillars of the butterfly start life by eating the flowers of the gentian. Eventually, they fall to the ground, where they are found by the red ants. The ants mistake the caterpillars for their own larvae, and carry them to the colony, where the caterpillars continue to be fed by the worker ants.

Taken individually, there are two obvious interactions: the butterfly larvae are herbivores on the gentian, and can also be considered as parasites on the ant colony, since they steal the food that the worker ants bring, intending to feed it to the real ant larvae.

Thomas *et al.* (1997) looked at the effects of the gentian population on that of the ants. They asked whether the red ant colonies that were near to gentians were more or less successful than colonies that were more distant from the flowering plants. One of the measures of success that Thomas *et al.* used was the number of worker ants in the colonies. They found that ant colonies whose centre was within 2.5 m of a gentian plant had an average of 291 workers. Colonies situated at a greater distance from a gentian had an average of 606 workers, more than twice as many as those close to the gentians.

In other words, the presence of a population of gentian plants has a negative effect on a population of red ants, because the plants allow the butterfly population to persist. The strength of the interaction is likely to depend on the population density of the butterfly population.

In effect, this is an example of the apparent, or 'predator-mediated', competition discussed in Section

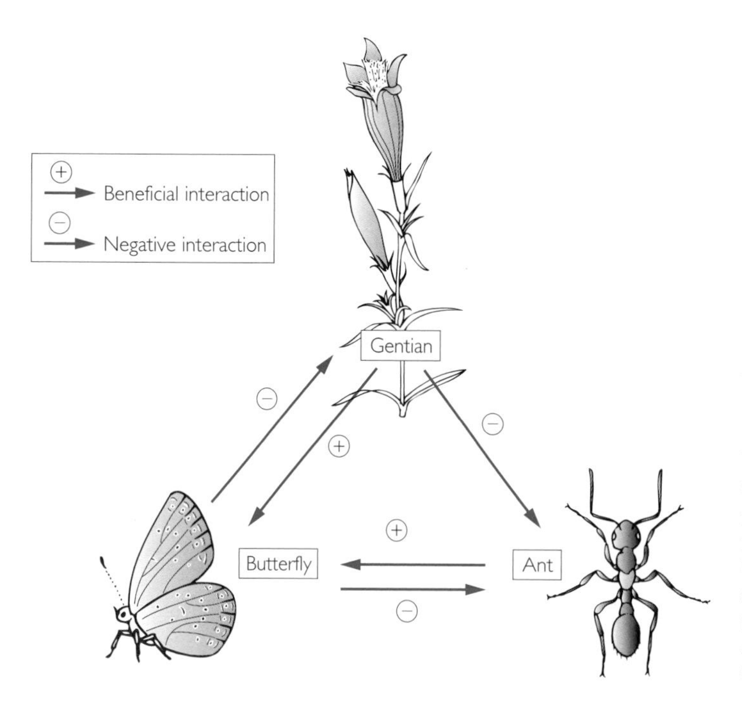

Fig. 10.7 In the complex set of interactions between the gentian (*Gentiana cruciata*), the red ant (*Myrmica schenki*) and the blue butterfly (*Maculinea rebeli*), the larvae of the butterflies eat the gentian but are also parasites in the ant colony, where they mimic the ant larvae and are fed by the ants. Thus, populations of both the gentian and the ant have positive effects on the population of butterflies, but the gentians have a negative effect on the ants, because their presence allows the butterfly population to persist. Ant colonies are more successful, in terms of number of individuals, in places where the gentian is absent.

8.4.3, although it would not be normal use of ecological terminology to describe the butterfly as a predator.

10.10.2 Many, diffuse interactions in the same place at the same time

It has proved simplest to concentrate on relatively clear sets of complex interactions, but Section 8.7 observed that there can be hundreds of populations in the same community, many of which could potentially affect one another.

In many cases, it is simply not practical to elucidate the effect of each species on each other. This is evident if one considers the many micro-organisms that occur in every community and ecosystem. Bever *et al.* (1997) noted that the dynamics of plant populations were affected by a host of such micro-organisms and fungi in the soil. There are bacteria that promote the solubility of minerals in the soil, bacteria that fix nitrogen, bacteria that produce hormones or suppress pathogens, as well as pathogenic fungi, bacteria and nematodes, and mycorrhizal fungi. The net effect of all these populations of soil organisms on all the populations of plants growing in the soil could be vast.

In theory, it would be possible to study the individual effect of each of the populations of soil organisms on the population of plants, but it would also be essential to study the effect of each population of soil organisms on each of the other soil organism populations. For example, Section 9.6 observed that the mycorrhizal fungus *Glomus* competes with, and affects the population of, the pathogenic fungus *Fusarium oxysporum*, and that this interaction affects the population of the bearded fescue grass (*Vulpia ciliata*).

Even if ecologists ignore most of the organisms in a community, and define a subset based on their evolutionary relationships, there will still be a great many potential interactions to study. When Cambefort (1994) investigated 11 different communities of dung beetles (Scarabaeidae) in the Ivory Coast in West Africa, the communities contained an average of 111 beetle species. If Cambefort had chosen to study all of the potential interactions among the dung beetles, ignoring all other organisms, he would have had to estimate an average of more than 12 000 different interaction coefficients in each community

In theory, this huge number of potential interactions within each community is no barrier to a full understanding of the functioning of ecological communities. While it may be more time-consuming to perform the calculations, there is no obvious reason why, given enough computing power, an ecologist could not

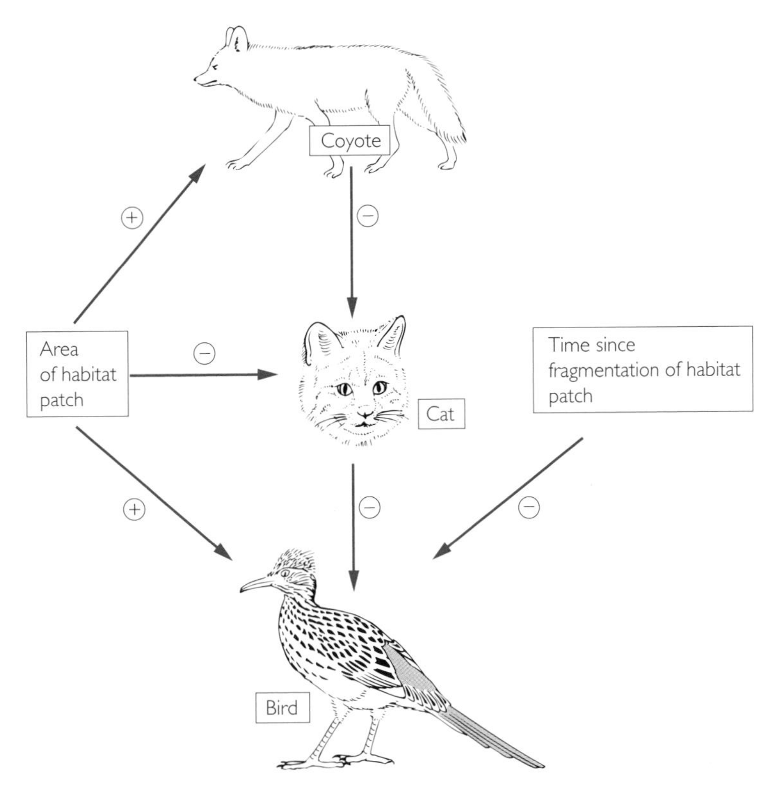

Fig. 10.8 The complex set of interactions in an ecological community in southern California cannot be seen simply as the sum of all the individual interactions. The birds are eaten by both cats (*Felis sylvestris*) and coyotes (*Canis latrans*), and the population of cats is negatively affected by the population of coyotes. The overall effect on the birds depends on whether one or both predators are present in a patch of habitat. The situation is further complicated by the fact that larger patches of habitat tend to have higher densities of birds and coyotes but lower densities of cats, and patches that have been isolated for long periods have lower bird densities. (Reprinted with permission from *Nature* **400**: 563–566. Copyright (1999) Macmillan Magazines Limited.)

calculate the overall effect of all the interactions in a community.

In practice, there are considerable difficulties to this approach. First, there is no community for which anyone has ever made an attempt to measure even a fraction of the interaction coefficients. There is not even a subset of five or 10 species within a community for which all the coefficients have been estimated.

Second, even in relatively simple systems, it appears that it is not viable to make the simple assumption that all the interactions together are simply the sum of their individual parts. Crooks and Soulé (1999) were interested in the difference between patches of habitat that are home to coyotes (*Canis latrans*) and patches from which they are absent. The patches of scrub in southern California are also the habitat of five other common carnivorous or omnivorous mammals: the grey fox (*Vulpes cinereoargenteus*), the domestic cat (*Felis sylvestris*), the skunk (*Mephitis mephitis*), the raccoon (*Procyon lotor*) and the opossum (*Didelphis virginianus*). The presence or absence of coyotes seems to have a significant effect on the interactions among these predators and at least some of their prey. For example, the interactions between coyotes and cats appear to have a very strong impact on the abundance of various scrub-breeding birds, such as the Bewick's wren (*Thryomanes bewickii*), the spotted towhee (*Pipilo erythrophthalmus*) and the greater roadrunner (*Geococcyx californianus*). In other words, the effects of interactions among coyotes, cats and small birds cannot simply be calculated by studying each individual interaction in isolation (Fig. 10.8). The interactions between cats and small birds and between coyotes and small birds could be studied separately, but in the real world there are other effects that depend on whether both coyotes and cats are present, or whether the small birds experience predation from just one or the other.

10.11 Chapter summary

There is much to be learned by studying the details of examples of competition, predation, herbivory, mutualism and parasitism, but it is unhelpful for ecologists to believe that these terms represent rigid descriptions of clearly defined and separate types of ecological interactions.

In fact, the ecologist who wants to understand fully the ways in which interactions between organisms combine to form ecological communities must recognize that each interaction is unique, but that every example involves a common element: the movement of organic resources. The functioning of ecological communities, which will be examined in Chapters 11–14, depends on this common element.

Recommended reading

Hudson, P. & Greenman, J. (1998) Competition mediated by parasites: biological and theoretical progress. *Trends in Ecology and Evolution* **13**: 387–390.

Johnson, C.N. (1996) Interactions between mammals and ectomycorrhizal fungi. *Trends in Ecology and Evolution* **11**: 503–507.

Roy, B.A. (1994) The use and abuse of pollinators by fungi. *Trends in Ecology and Evolution* **9**: 335–338.

Takabayashi, J. & Dicke, M. (1996) Plant–carnivore mutualism through herbivore-induced carnivore attractants. *Trends in Plant Science* **1**: 109–113.

Chapter 11

Ecological communities

11.1 Properties of ecological communities

Earlier in this book, communities were defined as assemblages of species that co-occur at the same time and place, and this definition is explored further in Box 11.1. This broad definition can be applied at many different scales. For example, ecologists may study the insect community in the bark of a tree, or the entire community of trees and animals in Costa Rica. The choice of scale in community studies is dependent upon the interests of the ecologist, and the questions being asked (Box 11.2).

Community ecologists are interested in the patterns of species occurrence and the interactions among different species in a community. They ask questions such as: Why do some communities have more species than others? How quickly can a community recover from disturbance? Are there general patterns in the kinds of species that occur together in a community? What types of interactions do we see among species in a community, and how do they affect the species composition of the community?

Some ecologists view communities as a collection of individual species, each having similar environmental requirements and therefore occurring in the same area. Consequently, a forest may consist of many plant species, each adapted to the same physical environment. These trees, shrubs and herbs are eaten by insects and other animals that are also able to live in the climate of the area. A community in this view is simply the sum of its component parts. Other ecologists view communities as being unique assemblages with characteristic **emergent** properties, such as the number of species or the variability or stability of a particular community.

Historical views of communities were often dominated by subjective observations. The similarity in overall appearance and presence of certain dominant species in different patches of vegetation caused the American community ecologist Frederick Clements to refer metaphorically to communities as **superorganisms** (Clements 1916). What Clements meant was that each community had a specific make-up, its own typical history, and its component species played specific roles in that community. The whole was greater than the sum of its parts. One of the predictions made by Clements was that the species in a community would have similar distributions, evolutionary histories and

Box 11.1 Community definitions

Ecologists generally define communities as all the organisms living together in the same place at the same time. Because communities are difficult to study, ecologists tend to use this definition flexibly. For example, many people restrict their study to a single taxonomic group of organisms. Thus, some studies talk about the 'bird community' or 'plant community' occupying a site. Other studies might define a subset of populations according to their ecological interactions. Such studies might define a 'predator community' or 'herbivore community'.

In an effort to reduce the uncertainty created by such flexible definitions, some ecologists use separate terms for different subsets of the populations occupying a community. For example, all the predators occupying an area might be termed the 'predator **guild**', while all the birds might be named the 'bird **assemblage**'.

Although these definitions can be useful, the truth is that most of the investigations that use these terms are trying to reveal the same kinds of ecological understanding. The definitions create distinctions, but the studies are all motivated by the same set of questions about ecological communities.

Box 11.2 Issues of scale

For their own convenience, ecologists define the scale at which they study communities.

Mammal ecologists might study very small communities, such as all the squirrels in a single park or garden. Morley (1998) studied the small mammal community of the Mkomazi Game Reserve in Tanzania, while Eltringham *et al.* (1998) investigated the large mammals of the same area. Alternatively, ecologists may use boundaries created by humans, such as the political borders between countries. Greenwood *et al.* (1996), for example, were interested in the mammal community of Great Britain. Pagel *et al.* (1991) studied the mammal community of the entire North American continent, while Letcher and Harvey (1994) chose to study the mammals of an even larger area—the entire Palaearctic region, which covers Europe, Asia, North Africa and Arabia.

The scale of each of these studies was defined partly for the convenience of the investigators, but each shed light on the ways in which mammal populations coexist and interact within communities. So long as ecologists are conscious of the restrictions created by the scale of a particular study, they can investigate community ecology at any scale they choose.

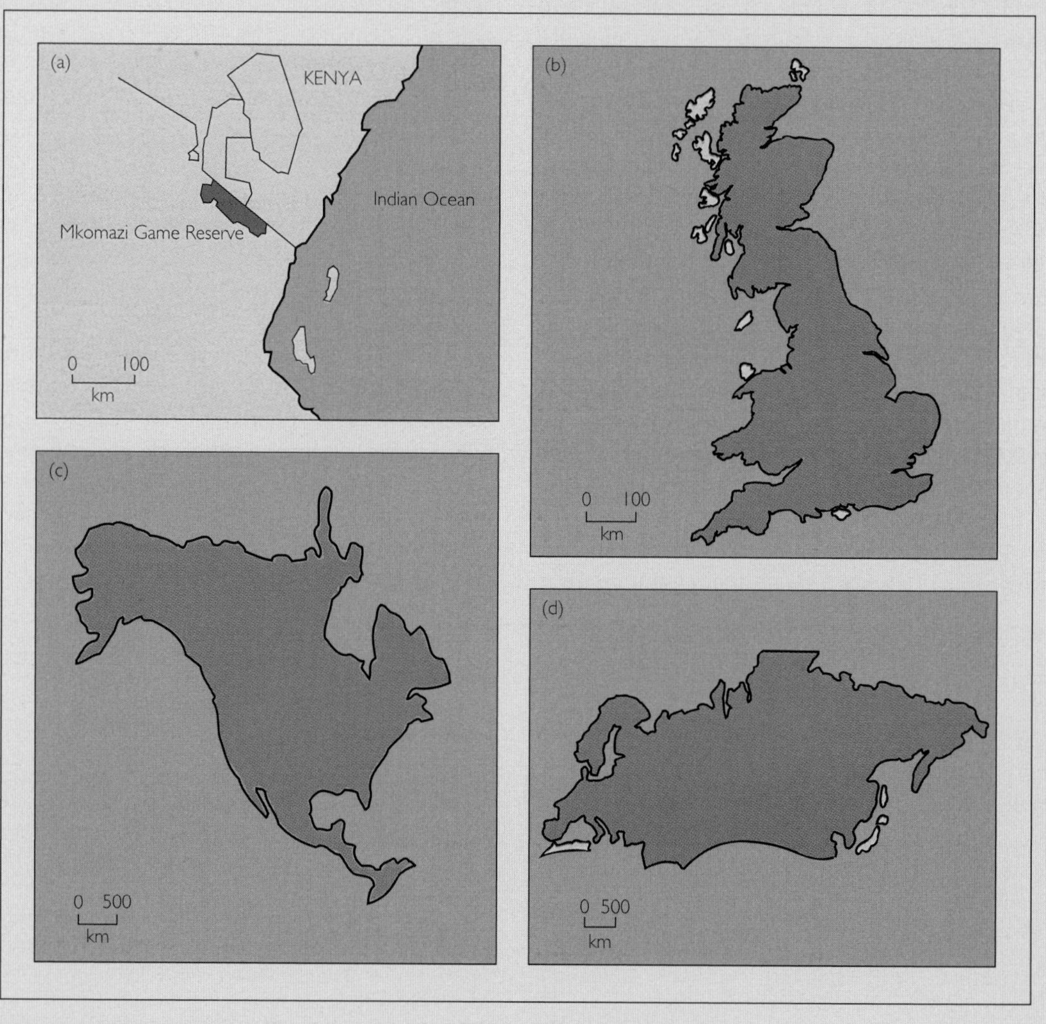

Fig. 1 Different scales can be used depending on the focus of the study, for example: (a) the Mkomazi Game Reserve in Tanzania, (b) Great Britain, (c) North America, including Mexico, and (d) the Palaearctic.

strong interactions with one another. There would be sharp, distinct boundaries between different communities, as we often see when a forest lies next to grassland. Additionally, if the community were disturbed by an event such as fire, it would redevelop along a specified path into the same climax community that was there before the disturbance. A **climax** community is defined as a stable, long-lasting community that represents the final stage of change in a community.

Henry Gleason (1926) did not agree with Clements' view of communities. Gleason was a proponent of the individualistic concept of communities. Gleason developed his ideas by sampling forest vegetation along a north–south transect in the midwestern United States. He concluded that changes in species abundance and presence occurred so gradually that it was not really practical to divide vegetation into specific associations or communities. He also believed that each individual species had its own distribution determined by its responses to the environment. If some species were often found together, this simply reflected similar adaptation to the physical environment. Between 1930 and 1960, results similar to Gleason's were obtained by other ecologists in different forest types, deserts and montane communities.

The true nature of communities probably lies somewhere in the middle of these two views. There is a great deal of chance involved in determining which particular species are present in a community, and there is also variation in the species making up communities in similar environments. Some communities have more integration and interaction among species than others. On the other hand, there do appear to be some properties of communities that stem from species interactions and which go beyond the simple sum of the species present.

11.1.1 Emergent properties

There are properties of groups that come from analysing them as a whole, rather than as individuals. For communities, we call these traits emergent properties. Emergent properties include such characters as:

1 **Physiognomy**: the outward, external appearance of the dominant vegetation. For example, we can describe a forest in Europe as having a physiognomy of tall, deciduous-leaved trees, with two or three subcanopy layers of smaller trees and shrubs.

2 **Species richness**: the number of species in a community. Species richness by itself may not tell us much about the community, because we have no idea of the distribution of individuals among species.

3 **Species evenness**: the number of individual organisms in each species, relative to the total number of individuals in the community.

4 **Species diversity**: a combination of species richness and species evenness. A community with 10 species, each with 100 individuals, is more diverse than a community with 10 species, five of which are represented by 95 individuals and the other five of which are represented by 10 individuals.

5 **Species guilds**: these are groups of species that are similar in their growth habit, in the way they acquire food or in the ways they live. In other words, they have similar niches. For example, we could classify all of the insects that eat seeds in the community into an insect seed-eating guild. We could say that all of the grass species make up a guild of narrow-leaved, herbaceous plants, or that all the predatory vertebrates form a predator guild.

11.2 How are communities constructed?

What factors are involved in determining the identity and number of species in a community? Populations of animals and plants are determined locally by survival, reproduction and dispersal of individuals. These processes are affected by the ability of the organism to tolerate the physical environment, the availability of essential resources and interactions with other organisms. It is hard to separate out the effects of each of these factors, because they are not independent of each other. For example, changes in the climate and the presence of competing species can both affect the availability of resources. Climate conditions and the presence of diseases or parasites can also change the requirements for resources. One way ecologists can separate out the effects of each of these variables is to conduct manipulative experiments where only one factor is varied and all other conditions are held as constant as possible.

11.2.1 The role of biological interactions in community structure

What biotic factors do ecologists think are important in structuring communities? Some of the answers have already been given in Chapters 8 and 9, which examined the effects of interactions on population regulation. Any interaction that affects the population size or distribution of a species has the potential to affect community structure. Hence, ecologists have focused on competition, predation, mutualism and pathogens as biotic forces in communities.

11.2.1.1 The role of competition

One way in which ecologists examine the importance of competition in structuring communities is the so-called **ecomorphological approach**. This approach assumes that if competition is important, then we should see greater morphological differences among species in a community than would be predicted by chance. In effect, it is an extension of the idea described in Section 8.6.3 that there might be limits on how similar two species can be if they are to coexist. However, in studying communities, the idea becomes more complex, dealing not just with differences between two species, but with differences among a group of coexisting species.

For example, we might expect greater differences in the beak sizes of the insectivorous birds in a single community than we find for all of the insect-eating birds in the surrounding region. One of the challenges in this approach is determining what group of organisms make up a random assemblage from the surrounding region. In other words, what is the proper 'species pool'? If we compared beak sizes of birds in a small 1 ha patch of forest with a random pool drawn from all the bird species in the world, then our species pool might include species that would never have had a chance to colonize the small patch of forest. If the community was in Papua New Guinea, there would be no point in including the lesser rhea (*Pterocnemia pennata*), which is restricted to the southern tip of South America and is flightless.

An example of a study that overcame the problem of defining an appropriate pool of species is one that examined introduced species of birds in the Pacific islands of Oahu and Tahiti (Lockwood *et al.* 1993).

Because historical records of bird introductions were available, ecologists were able to compare the body size and beak dimensions of all bird species introduced to these islands over the last 100 years with those that were currently present. The body size and beak size of the surviving birds showed greater differences than would be expected from a random choice of all of the introduced birds. The assumption in this work is that bird species that are similar in body size and beak type eat similar foods and exploit similar habitats—in other words, they will compete with each other. This competition may have forced inferior competitors into extinction, leaving a community of birds with fewer similarities than the original pool of introduced species.

One of the communities that has received a great deal of attention by ecologists is that of the plants, ants and rodents in the southwestern deserts of the United States. The system consists of annual plants that grow after rainfall in either the winter or the summer. These plants grow for only a short time, flower, set seed and then die. The seeds provide a valuable food source for ants and rodents. Ants in this system generally eat smaller seeds than the rodents do, but the two groups overlap in the seeds that they eat. By removing the ant species, rodent species, or both, ecologists have been able to examine interactions between these groups. These manipulations have been made possible by constructing a series of fences around small plots in the landscape. Holes of different sizes in these fences allow different animals to be excluded from the plots once they have been removed. If the ants are removed, the rodents increase in number. If the rodents are removed, the ants become more common. If both the ants and rodents are removed, the seeds become denser, and there are changes in the number and types of plant species in the community. Thus, seed predation by ants and rodents affects the plant community, and competition between ants and rodents influences the animal community.

There are about 16 species of small rodents in these desert communities. About half of these species eat plant seeds, the rest eat insects, other small animals or plant leaves. In investigations aimed at studying just the guild of seed-eating rodents, Brown and Munger (1985) manipulated the amount of seeds in plots and the presence of the most important seed-eaters, kangaroo rats. There are three kangaroo rat species, the largest of which is the bannertail kangaroo rat (*Dipodomys spectabilis*), which eats up to 53% of the seeds on a site. Two other smaller kangaroo rats are the merriam kangaroo rat (*D. merriami*) and the ord kangaroo rat (*D. ordi*). There are five species of smaller seed-eating rodents, the desert pocket mouse (*Perognathus penicillatus*), the silky pocket mouse (*Pg. flavou*), the deer mouse

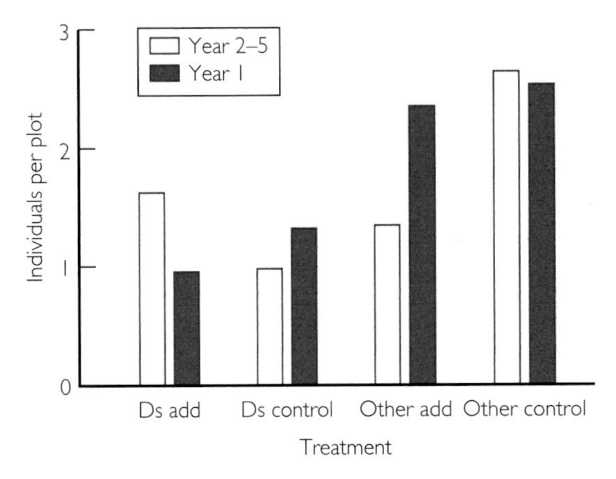

Fig. 11.1 The addition of seeds to the plots did not affect different species of kangaroo rat similarly. Seed addition allows the dominant kangaroo rat, *Dipodomys spectabilis*, to increase, while other kangaroo rats decrease. Ds add, number of *D. spectabilis* with seed addition; Ds control, number of *D. spectabilis* with no seed addition; Other add, number of two other kangaroo rat species with seed added; Other control, number of two other kangaroo rats with no seed addition. (Data from Brown & Munger 1985.)

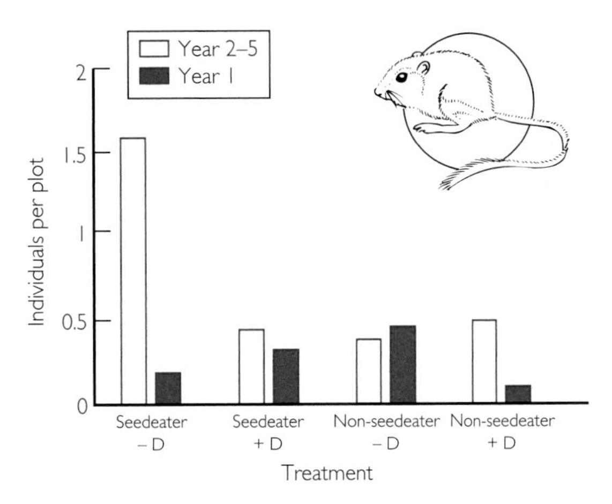

Fig. 11.2 The removal of kangaroo rats (*Dipodomys* spp.) from the plots had varying effects on other small rodents. Seed-eating rodents increased in density, while non-seed-eating rodents did not increase. Seedeater – D, seed-eating rodents minus *Dipodomys*; Seedeater + D, seed-eating rodents with *Dipodomys* present; Non-seedeater – D, non-seed-eating rodents minus *Dipodomys*; Non-seedeater + D, non-seed-eating rodents with *Dipodomys* present. (Data from Brown & Munger 1985.)

(*Peromyscus maniculatus*), the cactus mouse (*P. eremicus*) and the western harvest mouse (*Reithrodontomys megalotis*). Non-seed-eating rodents included the insect-eating northern grasshopper mouse (*Onychomys leucogaster*), the southern grasshopper mouse (*O. torridus*), and the leaf-eating whitethroat woodrat (*Neotoma albigula*). Interestingly, when the seed supply on the plots was doubled, not all rodent species increased. Instead, the largest species, the bannertail, increased by 66%, while the number of merriam and ord kangaroo rats decreased by 58% (Fig. 11.1). The extra food was dominated by the superior competitor, the bannertail, at the expense of inferior competitors.

When all individuals of the three kangaroo rat species were removed from the plots, the five other species of seed-eating rodents increased in density (Fig. 11.2). However, the insect-eating and leaf-eating rodents did not increase with the removal of kangaroo rats, which is exactly what would be expected if competition for limited seed supplies was the reason that the other seed-eating rodents were kept in check by the kangaroo rats. These experiments led Brown and Munger to conclude that limited food resources and interspecific competition play major roles in regulating the density of rodent populations and determining the organization of desert rodent communities.

11.2.1.2 The role of predation and herbivory

Predation and herbivory may exert a strong effect on the population sizes and species found in certain communities.

Power *et al.* (1985) examined the interactions between small, algae-grazing minnows (*Campostoma anomalum*), algae growing on rocks (primarily *Spirogyra* and *Rhizoclonium* spp.) and predatory fish such as bass (*Micropterus salmoides* and *M. punctulatus*) in small streams in south-central Oklahoma. These streams are characterized by shallow riffles (2–3 cm deep) separating small pools of 30–70 cm depth. In 14 pools along a 1 km stretch of Brier Creek, bass and minnows were rarely found together. When they did co-occur, it was usually after a flood. When bass were added to pools containing only minnows, the bass preyed upon the minnows. This caused minnows to reduce the time they spent feeding on algae, to move from deeper areas of the pool to shallower areas where the bass could not go, and to emigrate to other pools. Wherever the minnow occurred without bass, algae populations were grazed intensively, algal growth was reduced, and *Spirogyra* was eliminated. Thus, community composition in these pools was determined by local interactions between minnows, bass and algae, punctuated by floods that allow larger fish to move from pool to pool.

11.2.1.3 The role of mutualism

As described in Chapter 9, mutualistic interactions benefit both of the species involved in the interaction. The mutualisms can be either obligate or facultative. Any species involved in an obligate mutualism cannot invade a new community without its partner. For example, there are over 900 species of fig tree in the tropical regions of the world. Each of these fig species has one unique wasp species that carries pollen from one flower to the stigma of another flower. Without its specific wasp species, a fig tree cannot produce seeds and would soon disappear from a location.

Although they are not essential for the growth and reproduction of the organisms involved, facultative mutualisms also have an influence on the distribution of species, and hence on community structure. Perhaps the best example of this is seen in the associations between plants and their pollinators. Plants have evolved many ways to transmit pollen from one individual to another, with pollination systems based on wind, water and animals. Having an animal that acts as a pollinator carrying pollen faithfully to other plants of the same species can make a large difference in the success of a population.

The association between hummingbirds and plants is especially interesting in the tropics, and can have large effects on community structure. Many individuals of tropical plants are long distances from the nearest individual of the same species. Thus, they depend on reliable pollinators that can carry pollen over these distances. Hummingbirds are ideal for this. For example, in southwestern Puerto Rico there are three species of hummingbirds and at least 13 species of hummingbird-pollinated plants spread out along an altitudinal gradient from the sea up to montane forest above 750 m in elevation. The hummingbird and plant communities are influenced by a mixture of mutualism between birds and plants, and competition between the birds. There are two distinct classes of hummingbird and plant.

In open, drier areas there are five species of short-lived, shrubby plants that have tube-shaped red or yellow flowers. These plants are pollinated primarily by the smallest of the three hummingbirds, the Puerto Rican emerald (*Chlorostilbon maugaeus*). In forested, wetter areas there are at least eight species of tree, vine or epiphyte that produce tube-shaped red or yellow flowers. These plants are pollinated by the two larger hummingbirds, the Antillean mango (*Anthracathorax dominicus*) and the green mango (*A. viridis*). The mangos have bills greater than 22 mm long, while the bill of the emerald is less than 14 mm. There are corresponding differences in the length of the floral tubes of plants pollinated by these species (Fig. 11.3). The effect of mutualism on the communities is seen in the tight correlation between the abundance of plants and of the hummingbirds that pollinate their flowers. On adjacent islands without hummingbirds, none of the 13 species of plant occurs. When flowers are absent from the plants, hummingbirds have to move to other areas or they will starve. In return for pollination, the plants provide the hummingbirds with nectar rich in energy. Without this food, the birds cannot maintain their high metabolism.

Competition among the hummingbirds appears to be responsible for dividing the birds and plants into two groups. Habitat segregation is evident between the two large hummingbirds. The green mango occurs in habitats above 500 m in elevation. When individual green mangos encounter other species, they chase them away aggressively. This restricts the Antillean mango to large flowers at lower elevations. Ecologists might ask why the larger hummingbirds do not exclude the small hummingbird species altogether. The answer appears to be that the flowers pollinated by the emerald are so small that they produce less nectar than is required by larger hummingbird species like the mango. For these larger species, it is not worth the effort to visit the small flowers.

Beneficial interactions among organisms can influence community structure through more subtle ways than seen in pollinator–plant interactions. Sponges that grow on coral reefs are simple animals that filter the water surrounding them for microscopic food particles. They grow anchored on the substrate, so we might expect competition for space to be the primary force determining species composition of sponge communities. However, in many diverse coral reef systems, we find different sponge species growing over and intertwined with one another, without the loss of species. In coral reefs of the Caribbean, sponges actually adhere to one another. In one study of 13 species, 61% of the

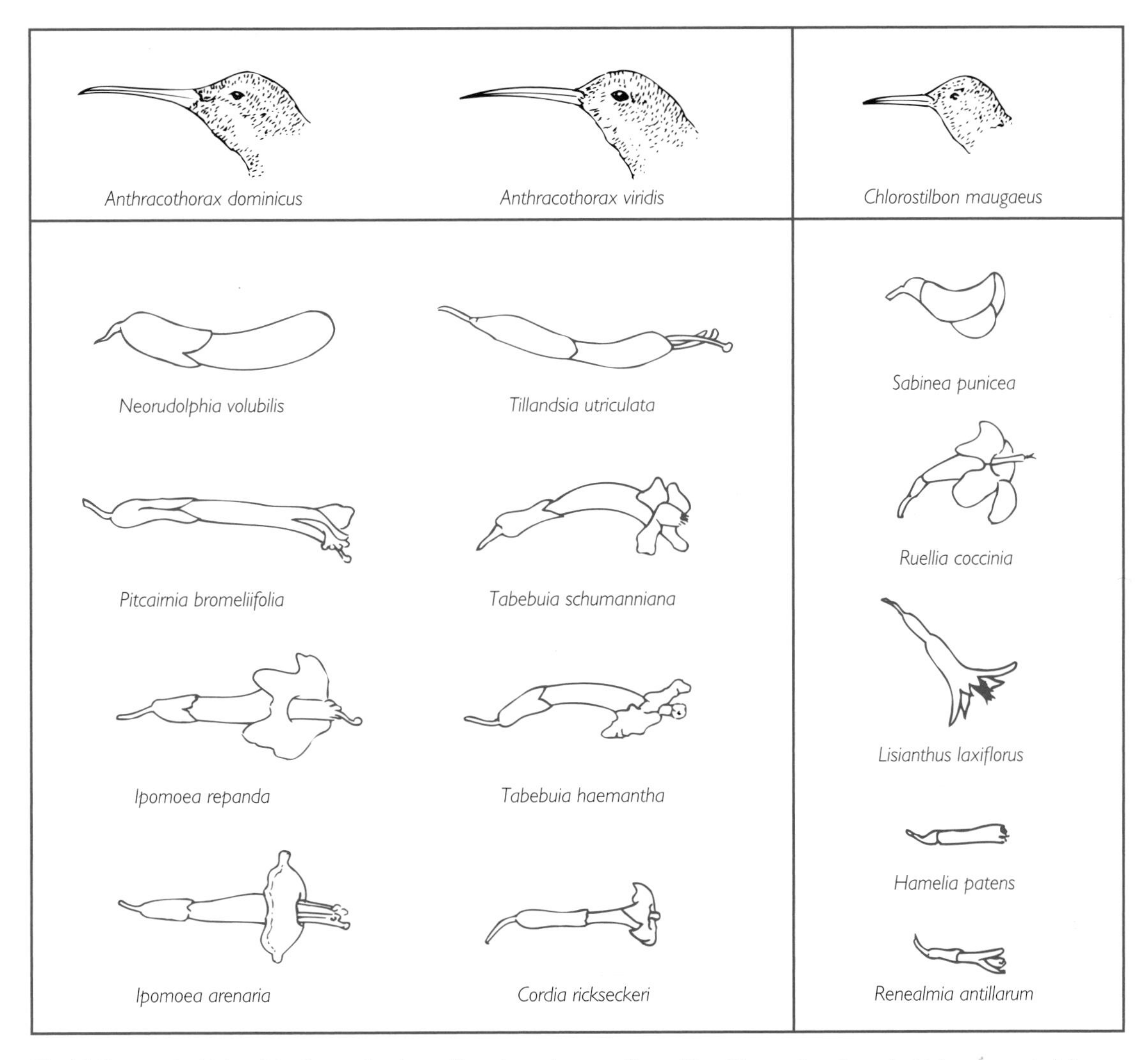

Fig. 11.3 Hummingbirds and the flowers that they pollinate in southwestern Puerto Rico. The two large hummingbird species on the left are found at different altitudes and rarely visit the short-tubed flowers on the right. The small hummingbird on the right does not have a long enough beak to pollinate the flowers on the left. The birds and flowers are drawn to scale from Kodric-Brown *et al.* (1984).

individuals were adherent to at least one sponge of another species (Wulff 1997). Growth and persistence were increased by adherence to an individual of another species in at least three of the most common sponge species in this system, the green finger sponge (*Iotrochota birotulata*), the erect rope sponge (*Amphimedon rubens*) and the yellow tube sponge (*Aplysina fulva*).

The mutualistic relationship is partly explained by the observation that there are several factors that can lead to breakage, fragmentation or death of a sponge. Although these animals can regenerate from fragments, 70% of fragments die, so there is significant mortality of sponges broken from their mooring. Hurricanes and other storms generate waves that can break sponges on shallow reefs, many fish and starfish eat sponges, sediment stirred up from storms may bury sponges, or pathogens might infect and kill a sponge. The three common sponges differ in their susceptibility to breakage from environmental factors or predation. For example, the green finger sponge is susceptible to predation by starfish and to breakage by storms. However, the rope sponge is highly resistant to predation from starfish and is resistant to storm breakage. Thus, by adhering to individuals of the rope sponge, individuals of the finger sponge appear to gain some resistance to these factors. A final piece of the puzzle is that diseases infecting these sponges are highly specific. They do not spread from one species to another, so there is little risk of an individual sponge acquiring pathogens by associating with individuals of a different species. Species diversity in these sponge assemblages may actually be enhanced by these mutualistic interactions.

11.2.1.4 The role of diseases

Parasites and pathogens can have large effects on population growth, and hence on community structure. The effects of pathogens can be direct—by causing the death of individuals; or they can be indirect—by causing reductions in the numbers of predators or competitors of a species. The structure of forests in the United States, Western Europe, Australia and East Africa has been dramatically altered by pathogens. The chestnut blight, caused by a fungus (*Cryponectria parasitica*) virtually eliminated chestnut trees (*Castanea dentata*) from North America. Similar fungus blights have affected elms in Britain and Western Europe, and hemlock in the Pacific Northwest of North America.

In tropical rainforest communities, many tree species do not reproduce well in their own shade. Carol Augspurger (1984a,b) found that for six of the nine tree species she examined, pathogens killed 30–91% of the seedlings, and that the proportion of pathogen-induced deaths was higher in the shade than in forest gaps, and highest where seedlings were very dense. Initial seedling density was highest near parental trees because most seeds were dispersed only short distances. However, because of the density-dependent response of the pathogens, most successful recruitment occurred outside the pathogen shadow surrounding the parent trees. This had two effects on the community. First, it created a greater distance between successful tree seedlings and their parents, and second, the overall species diversity of the community was increased.

Two good examples of the indirect effects of pathogens come from England and Africa. In England, rabbits (*Oryctolagus cuniculus*) were introduced by the Normans in the eleventh century to provide a source of meat throughout the year.

The rabbits escaped captivity and existed in low numbers until agricultural changes in the eighteenth century provided larger areas of suitable habitat. Their populations then exploded, and by the early 1950s they were serious pests; they prevented trees from regenerating in open areas, and reduced winter cereal harvests and production from forest plantation. The myxoma virus, native to Brazil, was introduced into rabbit populations in England in 1953. Rabbit populations crashed within 2 years, and remained low for 15 years. During this time, oak forests regenerated, most notably the remnants of the royal forest at Silwood Park, near Ascot. After the rabbits almost disappeared, the virus also became rare. The rabbits that were left were more resistant to the virus, and the virus itself changed to a less virulent strain. By the 1970s, rabbits were again inhibiting forest regeneration.

A scenario similar to rabbits in England has been seen in East Africa. Outbreaks of anthrax, a bacterial disease of large grazing mammals, greatly reduced the numbers of impala (*Aepyceros malampus*) in northern Tanzania in 1977 and 1983. This allowed the acacia woodland (*Acacia tirtukus*) (small, thorny trees) to regenerate. Impalas browse on *Acacia* seedlings, killing them before they can mature. Therefore, when there are large populations of impala, there is little regeneration of acacia woodland. The woodland is now characterized by even-aged stands of trees, dating from the periods when impala populations were low.

Despite some well-researched examples of the effects of pathogens, ecologists do not know enough about subtle effects related to disease-causing organisms. We do not know how many cases there are of pathogens changing fecundity or mortality by only a few per cent. Even such a small change could make the difference between persistence or exclusion of a species in a given community. There are probably many examples of hidden diseases that alter the competitive abilities of species and affect their distribution and position in communities.

11.2.2 Keystone species

One reason ecologists have focused so much attention on competition and predation as forces structuring communities is the presence of **keystone species**. A keystone species is defined as a species that plays a larger role in the community than would be predicted based upon its numbers or biomass.

Ecologists often label top predators as keystone species. Nevertheless, even the effects of these prominent animals on the community are not usually appreciated until they are lost. An example is that of the Pacific Ocean sea otter (*Enhydra lutris*) occurring along the northwestern coasts of North America. The sea otter is a voracious eater of abalones and crabs, both of which are commercially important. Early in the twentieth century, commercial fishermen killed sea otters whenever possible because they believed that sea otters were reducing their catch of valuable abalones. Sea otters were also hunted for their fur, and were on the brink of extinction in 1911. As sea otter numbers decreased, kelp, fish, seals, hawks and eagles also disappeared from coastal areas. Eventually the sea otter was protected from hunting, and populations began to recover. As sea otter numbers increased, kelp beds were restored, more fish appeared and seals and eagles came back. What did the sea otters do to affect this community so strongly? The answer was in one of the sea otter's food sources. Sea otters eat many sea urchins, which are major consumers of kelp and sea grass. Unchecked, sea urchins will destroy kelp beds. Kelp and other seaweeds are essential habitats for many species of fish, providing them with breeding areas and a refuge from predation. Seals and eagles feed on the fish. Without sea otters to control sea urchins, the entire community changes dramatically.

11.2.3 Interactions between abiotic and biotic forces

Physical, abiotic conditions can modify the biological interactions among species in a community. One approach to discover how these modifications act is to examine the variation in biotic interactions along gradients where abiotic conditions change. For example, harsh environmental conditions can change the effects of predation by restricting foraging time, foraging efficiency or by causing predator mortality. The fjords of New Zealand are characterized by steep, rocky walls, harsh wave action and large variations in the runoff of freshwater from the surrounding land. Organisms such as mussels, barnacles, sponges and sea anemones cement themselves to the walls and filter particles of food from the surrounding seawater. These stationary animals are preyed upon by animals, such as sea stars, lobsters, sea urchins, crabs and octopuses. Mussels are generally the dominant animals at depths of less than 6 m, sometimes covering over 80% of the rocky walls of the fjord. The mussels peak in abundance at about 3 m in depth. This depth coincides with a layer of less salty water caused by runoff of freshwater from the surrounding land. Most predators of mussels cannot tolerate low salinity, and are therefore restricted from shallower depths. They reach their peak below 6 m, where mussels cannot persist. In this community, predation limits how deep mussels can occur, and a surface layer that is low in salinity provides a refuge from predators. However, these depths are not constant. In dry months, low rainfall results in reduced runoff, which in turn reduces the depth of low-salinity water. This allows predators to move up and eat more of the mussels at shallower depths, reducing mussel numbers. The interaction between salinity, predation and rainfall results in a dynamic, complex community changing over time and space.

11.2.4 Interactions between dissimilar species

As discussed earlier in this chapter, host–pathogen interactions and many mutualisms involve species that are from different kingdoms, or of vastly different sizes. However, interactions among vastly dissimilar organisms are not limited to these two categories. For example, large grazing mammals in the savanna of East Africa can alter the communities of small rodents that co-occur with them. In central Kenya, there is a diverse community of at least eight species of small rodent. Most of these species are small, mouse-like rodents that eat plants or insects. The seed-eating species are important agricultural pests, and can serve as reservoirs for diseases affecting humans. Population outbreaks of these rodents are of major concern in this part of Africa. The most common of the rodents is a medium-sized species called the pouched mouse (*Saccostomus mearnsi*).

There is also a diverse community of large, ungulate herbivores, such as the elephant (*Loxodonta africana*), giraffe (*Giraffa camelopardalis*), zebra (*Equus grevyi* and *E. burchelli*), buffalo (*Syncerus cafer*), eland (*Taurotragus oryx*), Grant's gazelle (*Gazella granti*) and domestic cattle. In a manipulative experiment designed to test the effects of these animals on each other, large plots of 4 ha (200 m × 200 m) were fenced to exclude ungulates. Results indicated that the pouched mouse greatly increased its abundance in these fenced areas (Fig. 11.4).

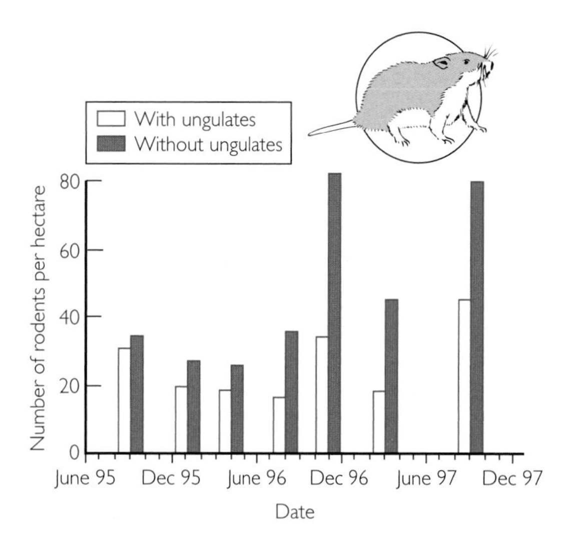

Fig. 11.4 Animals of different sizes can have large effects on each other. Here, the exclusion of large, grazing animals like antelope resulted in the increase of the small rodent, *Saccostomus mearnsi*. The antelopes and rodents compete for food, and if antelopes are excluded the habitat is improved for the rodents. As large animals disappear in Africa, rodents may increase, causing problems for human populations by eating food stores and spreading disease. (Data from Keesing, 1998.)

Male mice were significantly heavier in fenced plots, and mice from surrounding areas moved into these plots. Apparently, excluding large ungulates improved the habitat for the pouched mouse. Over 2 years, rodent diversity remained relatively constant in the fenced plots. However, in unfenced areas there was a period of a few months when the number of rodent species decreased dramatically. Thus, large grazing mammals compete with and help control rodent populations in the savanna. This is an important interaction, considering that many areas of Africa are losing native ungulate species due to land-use changes and growing human populations. The growth of rodent populations in the absence of large ungulates will increase losses of grain and the spread of disease in human populations.

11.2.5 The role of random events, history, dispersal and disturbance

Studies of sessile organisms like marine barnacles, marine algae and terrestrial plants have shown that the timing and spatial aspects of physical disturbance, offspring production and offspring dispersal play major roles in colonization and community structure. In these communities, the first seeds or larvae that encounter an open space can dominate the resulting community, and the composition of the community may depend on which species have seeds or larvae that are particularly good at dispersing. Likewise, other unique events, such as a fire, a blizzard, human activity or a drought, might affect the composition.

Thus, the unique history of a site or the chance occurrence of a nearby population of a certain species can influence the community that develops. The degree to which chance affects community structure will be dealt with in Chapter 13.

11.3 Species assemblage rules

After many years of studying bird communities around New Guinea, the ecologist J.M. Diamond (1975) asked: What do we need to know in order to predict how communities are assembled from a common pool of species? There have been many different answers to this central question of community ecology. Diamond was asking if there are rules that determine the presence or absence of species in a community. These '**rules of assembly**' are made up of generalized restrictions on species presence or abundance, and should not be confused with the hard and fast rules that govern aspects of physics or chemistry. Rather, they are general patterns in community structure that may allow ecologists to predict the types and numbers of species that may inhabit new environments, or that may move in during the recovery of a former community from disturbance. As such, these so-called rules may help in managing areas after human disturbance, or they may help ecologists to predict the effects of human activities on community structure before the effects have had time to occur. These rules do not have anything to say about the response of individual organisms to the environment. They represent ecologists' efforts to go beyond simply showing that a pattern in species composition exists. To increase predictive power, assembly rules are based on species traits, not species lists. This allows rules from one system to be applied to another.

How might an assembly rule be formulated? Imagine that there were a finite number of available niches in a community. Then, an assembly rule may be that only a limited number of species could coexist locally. A species invading a patch with a high number of species present would have a lower chance of becoming established than a species invading a patch with a low number of species. Because of their similarities in resource use, a species that is invading a community with many similar competing species may have greater difficulty becoming established than would a species invading a guild that is relatively under-represented in the existing community.

11.3.1 Studies seeking assembly rules

The number of ecological studies looking for assembly rules in communities is steadily increasing. For example, Wilson and Whittaker (1995) examined the plant community in a saltmarsh on the fringe of Church Island in North Wales. They compared the actual species composition of small plots in this saltmarsh to idealized communities constructed by randomly choosing from among all the species in the marsh. If there were no real assembly rules, the presence of a particular species would not increase or decrease the chance of any other species being present. In fact, these ecologists found a non-random structure in the species make-up of the community. Variation in species richness of the plots was only two-thirds of what was expected. This was partly because real communities generally contained representatives of several different guilds. They were less likely to contain many species from the same guild, suggesting that there was more competition between members of the same guild than between species from different guilds. Plant species with narrow leaves were unlikely to be found together, and the same is true of broad-leaved plants. This suggested two different guilds—narrow-leaved species and broad-leaved species. This might occur if there were interactions, such as competition for light, that were especially strong between plants with similar structures.

A similar study of the botany lawn of the University of Otago in New Zealand indicated that plant species were divided into narrow- or broad-leaved guilds, and into upper- or lower-canopy guilds (Wilson & Roxburgh 1994). There was a higher probability for a new species to invade this community if it was from a guild different to those already represented in the community. This resulted in a nearly equal number of species in all four guilds. Again, the mechanism behind this observation might have been strong canopy interactions among leaves of similar width and height above the ground.

In the desert rodent communities of southwestern North America that were discussed in Section 11.2.1.1, Fox and Brown (1993) gathered information on local assemblages of rodent species. For 115 sites in Nevada, and 202 other sites throughout the desert, there appeared to be nested subsets of species. At each site with a higher number of species, all of the species from more species-poor communities would be present, plus some additional species. The goal of Fox and Brown's study was to test whether each new species entering a community would be drawn from a different functional group until each functional group was equally represented. **Functional groups** are formed of those species in the same guild, in this case the seed-eating rodent guild, that have similar ways of gathering food. Those factors that affect seed gathering might include activity times (day or night), means of movement (two legs or four legs), size of preferred seeds, places where animals look for seeds (under or away from shrubs) and relatedness, or common ancestry. Fox and Brown identified five functional groups in the rodents of southwestern North America based upon their foraging strategies. Three groups ate primarily plant seeds. The first consisted of three species of bipedal kangaroo rats in the same genus, the merrian, ord and desert kangaroo rats (*Dipodomys merriami, D. ordi* and *D. deserti*). The second consisted of three mouse species that used all four legs for locomotion, the longtail, great basin and little pocket mice (*Perognathus formosus, Pg. parvus*, and *Pg. longimembris*). The third group was made up of generalist seed-eaters, including three deer mouse species, the common, canyon and piñon deer mice (*Peromyscus maniculatus, P. crinitis* and *P. truei*), the western harvest mouse (*Reithrodontomys megalotis*) and the whitetail antelope squirrel (*Ammospermophilus leucurus*). The remaining two functional groups had species that did not eat seeds. The results showed that local communities differed significantly from communities made up of a random assemblage of all possible species. There were significantly more communities in which the most species-rich functional group contained just one more species than the most species-poor functional group, than there were in which the most species-rich group was represented by two or even three more species than the least speciose group. Thus, Fox and Brown's original hypothesis—that functional groups should be equally represented in the community—was supported.

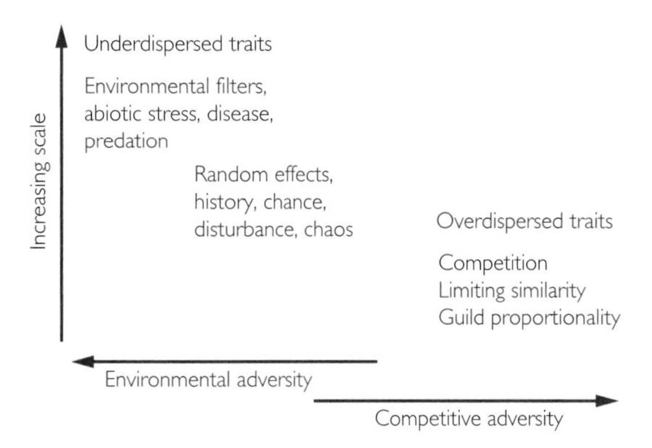

Fig. 11.5 A visual model of the forces shaping community structure. Along the x-axis there is a gradient from stressful abiotic environments on the left to stressful biotic environments on the right. On the y-axis the scale of the community is represented. Studies of regional communities or biomes fall on the high end of this axis, while local studies fall on the low end. (Adapted from Weiher & Keddy 1995.)

11.4 An integrative model of community structure

How can all of the factors discussed so far be brought together into a model of community structure? A visual attempt at such a model is shown in Fig. 11.5. On the x-axis, there is a gradient from high environmental stress on the left to high competitive stress on the right. At the left side of this axis, abiotic stress, such as drought, salinity or frequent disturbance, may limit the types of traits species possess. Only those species capable of handling the environment will persist in the community. The associations between life forms and environment outlined in Chapter 4 are examples of this. Similar species, adapted to specific environmental conditions, may be found together because of selection by the environment and convergent evolution towards similar ways of dealing with the environment.

On the right side of the x-axis, dissimilar species may occur together, because competition has limited the similarity among competing members of the community. In the middle of the x-axis is a range where chance effects or initial conditions may influence community assembly, and a random association of traits may be found. The y-axis corresponds to the scale of study. This affects the species pool being considered. Most studies of limiting similarity have looked at small scales, members of the same genus, or guilds in a very restricted area. Studies of associations between environment and traits are usually at the larger scale, looking at regional assemblages.

A final point about this model is that the factors structuring a community are not mutually exclusive. Arraying environmental filters and competitive forces at different ends of the x-axis is a little misleading. Environmental filters can constrain the traits present in species within a community, similar to the air pressure outside a balloon that restricts its expansion. Intracommunity forces like competition can then act simultaneously to limit similarity among species within a community, analogous to the air within a balloon, pushing out on the sides and causing the balloon to grow.

11.5 Ecological succession

Communities are not static systems. There are constant changes in the occurrence and abundance of the species in each community. Most communities experience some type of disturbance, such as human activities, fires, volcanoes, storms, landslides or glaciers. What patterns do ecologists see in community recovery from disturbance? This is an important question, both for natural systems and for managed systems. For example, conservation biology is often concerned with how quickly natural systems can recover from past disturbances caused by humans. The changing nature of communities, and their ability to recover from disturbance, are known as **ecological succession**, and will be dealt with in more detail in Chapter 13.

11.6 Community responses to global changes

Global changes caused by human activities—patterns of land use, atmospheric composition, disposal of human wastes and harvesting of plants and animals—will affect communities of organisms in profound ways. Already, reductions in the biodiversity of natural animal and plant assemblages and changes in the species composition of communities have been documented. Changes in land use are probably the most important causes for changes in community composition. Simply put, as habitat is altered for human use, some species are displaced from their ideal habitat. This alters species interactions and species composition of communities. If these species are keystone species, or even if they are involved in interactions with only one or two other species in the community, the loss of a single species often has ramifications beyond a localized area.

One area where humans have had a large impact is that of atmospheric composition. Atmospheric concentrations of gases such as carbon dioxide, methane, sulphur oxides and nitrogen oxides have been increasing since the early years of the Industrial Revolution in the nineteenth century. These gases have several effects, some act directly on organisms, some indirectly, through changes in climate.

11.6.1 Direct effects of increased carbon dioxide

Plants take atmospheric carbon dioxide in through their stomata (pores in the leaf surface) and use enzymes to fix it into carbohydrates. These building blocks are then used by the plant for metabolism and growth. There are three different photosynthetic pathways that plants use for fixing carbon dioxide, but the most common responds positively to increased carbon dioxide concentrations by increasing photosynthesis. However, several studies have shown that there is substantial variation in the response of different species to elevated carbon dioxide. It is sometimes difficult to predict which species will respond positively. Plant responses in calcareous grasslands are an example. These grasslands are interesting because they contain many rare species, and are very diverse. When model communities of six species native to the calcareous grasslands of Europe were constructed and grown in three concentrations of carbon dioxide (330, 500 and 660 ppm), increases in leaf and litter biomass for the entire community were observed. However, there were significant differences among species in their response to increasing carbon dioxide concentrations. One species had a negative response, two had positive responses, one had a barely positive response, and two did not have any response at all (Fig. 11.6). In addition to this variation among species, variations in genetic make-up within a species affected plants' responses (Fig. 11.7). Not only will the differences in response among species lead to community changes in future environments but these changes may differ depending on the particular genotypes in the community.

Plants commonly respond to higher carbon dioxide concentrations by increasing the ratio of carbon to nitrogen (C : N) in their tissues. Leaves of plants in

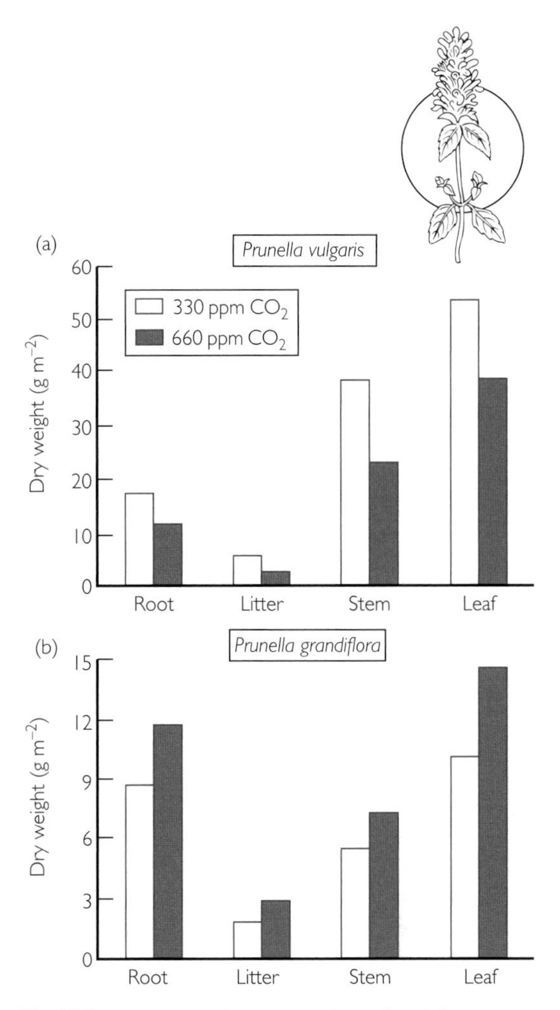

Fig. 11.6 The response of two species of *Prunella* to different CO_2 concentrations during growth. (a) *Prunella vulgaris* actually decreased growth in response to higher CO_2 concentrations, while *P. grandiflora* (b) increased growth with increasing CO_2 concentration. Distinct responses such as these, from different species may lead to changes in community composition. (Data from Leadley & Stocklin 1996.)

atmospheres with high levels of carbon dioxide may also be thicker, and have more starch and more carbon-based defences against herbivores. Insects that eat plants are often limited by nitrogen supply. How will insects respond to plants with higher C : N ratios in high carbon dioxide environments? Most feeding studies have shown that insects consume more plant biomass when carbon dioxide levels are high, in order to meet their need for nitrogen. Some insects have also grown more slowly, and developed into different pupal stages more

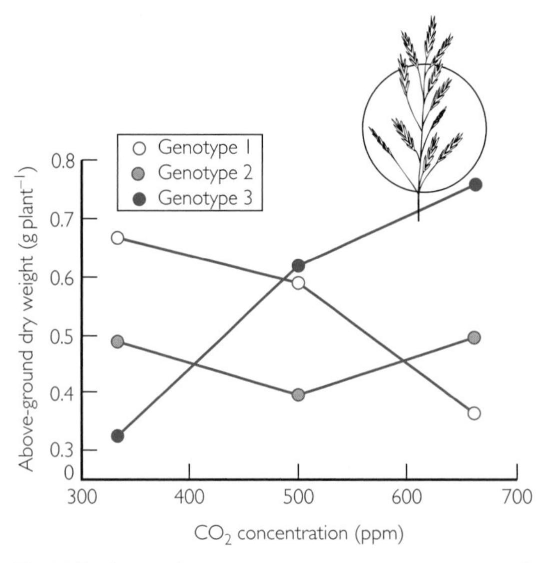

Fig. 11.7 The growth response to increasing CO_2 concentration of different genotypes within the same species may differ. Here, distinct genotypes of *Bromus erectus* did not behave similarly: some increased growth, some did not change, and some decreased growth over a range of CO_2 concentrations from 330 to 660 ppm. (Data from Leadley & Stocklin 1996.)

slowly. Little is yet known about how high levels of carbon dioxide will change insect food preferences, either for different species or for different tissues on the same plant. However, alterations in feeding behaviour and performance may lead to alterations in insect herbivore communities, and in the plants on which they feed.

11.6.2 Indirect effects of increased carbon dioxide

In addition to its use by plants in photosynthesis, carbon dioxide in the atmosphere also absorbs infrared radiation. As carbon dioxide concentrations increase, more of the infrared radiation given off from the Earth's surface will be absorbed by the atmosphere. Absorbing this radiation heats the carbon dioxide molecules, and they give off infrared radiation themselves, back to the Earth's surface. This is the foundation for the so-called greenhouse effect. The most common projections of the consequences of a doubling in carbon dioxide concentration are increases in atmospheric temperatures of 2–4°C at the equator, and 6–9°C at the poles. At present rates of carbon dioxide build up in the atmosphere, carbon dioxide should double in concentration within 100 years.

The potential biological changes that may occur with global warming will be due not only to increased temperature, but also to the rate at which temperatures will increase. The projected increases may be too rapid to be tracked by evolutionary processes such as natural selection. The capacity of species to adapt may be inhibited, increasing the probability of extinction for some species.

Many tree species take hundreds to thousands of years to disperse substantial distances. One of the most rapid rates has been found for white spruce (*Picea glauca*), which expanded its range by only 20 km for every 10 years after the last ice age 10 000 years ago. This rate of dispersal may be too slow by a factor of 10 if the species is to spread quickly enough to track projected temperature changes. Land use by humans and habitat fragmentation will slow this dispersal even more. Present-day communities will be dramatically altered because of differences in species movement rates.

An example of some of the interactions that make projections about community response to global change difficult can be seen in studies on birds. Understanding the effects of climate change on birds requires an understanding of its effects on plants. For example, the red-cockaded woodpecker (*Picoides borealis*) is endemic to mature pine and pine–oak forests in the United States. Individuals of this bird species need mature, living trees for nesting sites. If a change in climate kills mature trees, then the woodpecker could be limited by the availability of nesting sites. Kirtland's warbler (*Dendroica kirtlandii*) in northern Michigan, is restricted to a narrow area of jack pine trees (*Pinus banksiana*) that grow in sandy areas. Since this tree species is adapted to a narrow climatic range, a small change in climate may force the range of this tree to move northwards. However, north of the present distribution, there are none of the rapidly draining, sandy soils that the ground-nesting birds need to raise their young successfully. Kirtland's warbler may become extinct because of a lack of suitable habitat in the direction that climate change will force it.

In addition to vegetation, the distribution of birds is often associated with environmental factors such as the frost-free period, minimal winter temperatures and maximal summer temperatures. These responses are highly specific to a particular environment and there are likely to be community disruptions as large as those that occurred during the warming period after the last ice age. The difference will be that they will occur 10 times as rapidly.

11.6.3 Exotic species and global change

A major cause of the loss of biodiversity has been the introduction of non-native species to new environments by humans. This is often inadvertent, but sometimes the introductions are made purposely. Many of these introduced species become successful, spreading rapidly and displacing native species. Global changes in land use have favoured many of these species by opening up habitats for their establishment. Climate change may also aid in their spread, especially for introduced weedy species that can exploit the changed conditions. In the southeastern United States, two weedy vines introduced about 120 years ago are especially successful. These Asian species are kudzu (*Pueraria lobata*) and Japanese honeysuckle (*Lonicera japonica*). These two species have several properties that may allow them to spread even more rapidly in future environments. Since they are not native to North America, they have few natural enemies. They grow very fast, rapidly colonizing and dominating open habitats. As vines, they allocate a large portion of new growth to stems and leaves. This allows them to overtop other vegetation, or, in the case of Japanese honeysuckle, rapidly twine around stems of other plants and girdle them.

These species have been found to respond strongly to increased levels of carbon dioxide with increased elongation and above-ground growth. Presently, their northern distribution limits are set by the low temperatures of winter and the occurrence of early autumn frosts. As temperatures and carbon dioxide levels rise, they may become even more aggressive competitors in their spread northwards. Competition between these weedy, aggressive exotics and rare, native species (that are already endangered and placed under stress by high temperatures and increased periods of drought) may further reduce the biodiversity of native communities.

11.7 Landscape ecology

As the human population grows, it is changing the land cover of the Earth. Land cover is defined as the vegetation or habitat type found in a parcel of land. Changes in land cover affect the distribution of plants and animals, the production and storage of biomass, and the cycling of essential elements. In addition, natural disturbances such as fire, hurricanes, windstorms and pest outbreaks create patchiness in the landscape. The area of ecology that studies land cover is called **landscape ecology**.

This rapidly evolving discipline bridges the gap between community ecology and ecosystem ecology. Landscape ecologists study the causes and ecological consequences of spatial patterns in the environment. Their work is becoming an increasingly important tool in land-use planning and in conservation biology. The scale of a particular landscape study is dependent on what is being investigated. A scale of several thousand square kilometres might be appropriate for a study of the suitable habitat for the Asian tiger (*Panthera tigris*). A study of suitable habitat patches for a small land snail might encompass only a few square metres.

Landscape ecologists ask questions such as: How does the pattern of habitat types affect the ability of animals to move from one patch to another? How many patches, and of what size, are needed to maintain a population of a certain species? How does the arrangement of patches of different habitats affect the flow of nutrients from a farmer's field to neighbouring streams?

There are three important aspects of the landscape: structure, function and change. **Structure** is the spatial pattern of land cover—what is there and how it is arranged. **Function** is the interaction among spatial elements, the movement of organisms, materials and energy. **Change** looks at the structure and function of land cover over time. Also important are the causative factors that generated the landscape. These might include historical factors (e.g. past logging), species interactions, variability in abiotic conditions (e.g. topography, differences in soil type, drainage patterns), disturbance (fires or storms), activities of organisms that can change land cover (e.g. beavers felling trees and damming rivers) and human activities. Humans have influenced landscapes by changing the relative abundance of different species, extending or reducing the distributions of species, providing opportunities for weedy species to invade, altering the nutrient status of land and water, and changing land use.

In order to document the structure, function and change occurring in a landscape, ecologists have enlisted sophisticated mapping and data storage systems. The development of geographic information systems (GIS) have paralleled the development of landscape ecology. These systems allow data on landscape patterns to be collected and analysed. They usually consist of a set of computer tools for collecting, storing, retrieving, transforming and displaying spatial data. Along with the location of a site (latitude and longitude), data on soil type, vegetation cover, species occurrence and spatial interrelationships of landscape components are included in GIS. Sources for these data come from aerial photography, digital remote sensing, airborne imaging scanners, published data and censuses, survey data and photographs taken at different times.

Organisms live, eat, reproduce, move and die in landscapes made up of different habitat or patch types. How are these organismal processes affected by the landscape? To answer this question, ecologists determine food types, nutrient availability, water conditions, nesting sites, soil type, etc. for different patches. Often the decision on whether or not a patch type is suitable for a species is made by looking at where individuals of the species occur freely and reproduce the best. For example, a bird species that nests and rears its young in the interior of forests may not occur in small patches of forest. By examining current patches where the bird occurs, landscape ecologists might predict the minimal patch size of forest needed for this species to reproduce. The size and arrangement of forest patches in a sea of agricultural land may have a large impact on the success of such a species. Habitat fragmentation generally decreases the size of suitable patches, and increases the edge/interior ratio of the remaining patches. This might alter the microclimate of the remaining patch, raising temperatures and wind speed, and lowering humidities compared to larger forest patches. Edges have been shown to provide greater access for predators to interior forest areas. In North America, roads and pathways created to allow access to power-lines have provided domestic cats with access to forest interiors where songbirds nest, resulting in increased predation on the songbirds.

11.7.1 Metapopulations

The landscape is a primary determiner of the **metapopulation** dynamics of a species. A metapopulation is defined as a series of populations of the same species occurring in different patches. These populations can be divided into two different categories—source and sink populations. Source populations produce more offspring than are needed to replace the population. These offspring may then disperse to other populations. Sink populations produce fewer offspring than are needed to replace the current population. Without immigration from source populations, sink populations will go extinct. The connection between suitable patches is important to the continued success of a metapopulation in a fragmented landscape. The connectivity provided by migration corridors can maintain overall populations better than when connectivity is low. The level of connectivity that is needed to support a series of separated populations depends on the movement abilities of the organism. For example, snails need more sheltered, shorter corridors than birds.

Prime examples of habitat types that provide important corridors for forest and woodland species are hedgerows. Many forest animals do not cross open habitats, and if they do, they are often exposed to increased risks. Fencerows in Europe often form connecting corridors between forest patches for mice, birds and other animals to travel along. In addition to serving this migration corridor function, these same fencerows serve as a habitat patch in their own right. Insect diversities are especially high in these fencerows. Because they attract many insect predators, these hedgerows have reduced the need for agricultural pesticides. Populations of natural enemies of agricultural pests, such as the Russian wheat aphid (*Diuraphis noxia*), are maintained in these hedgerows, providing an important service to humans.

An example where landscape ecology approaches have been applied is in the examination of the metapopulation of tigers (*Panthera tigris*) in Nepal. Tigers were once widespread across the riverine grasslands and forests of Asia, from India to Siberia. As recently as the early 1900s, 40 000 tigers were estimated to live on the Indian subcontinent. However, increased human population growth and changes in land use have resulted in large decreases in suitable habitat for tigers. Tiger populations are now reduced to small, isolated forest remnants, with sizes of 20–200 breeding animals. These small population sizes can lead to loss of genetic variability, reductions in gene flow among populations and increased chances that an ecological catastrophe will wipe out local populations. Many scientists think the extinction of small tiger populations will become

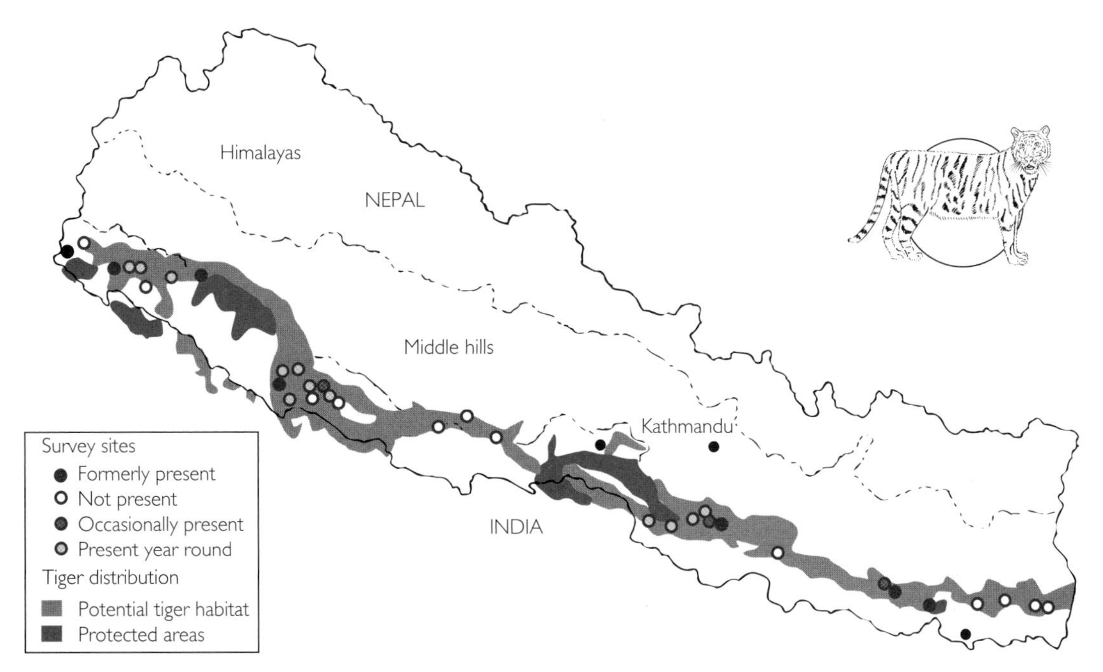

Fig. 11.8 A map of Nepal showing the current distribution of tiger populations. The potential habitat is much greater than that presently exploited. This is because much of the forest is fragmented into parcels too small to sustain tigers. Land-use change since the 1940s has resulted in large reductions in the range and population size of tigers. (From Smith *et al.* 1998a.)

more common owing to their small, isolated nature, the presence of populations in several jurisdictional areas (making management difficult) and the increased threat of poaching.

Through a combination of remote sensing, GIS and rapid assessment techniques, including field surveys and interviews with local residents, Smith *et al.* (1998a) evaluated the metapopulation of tigers in Nepal. They found that tigers were restricted to lowland forest and grasslands below 1000 m elevation, south of the Himalayas (Fig. 11.8). At higher elevations, intense agriculture excludes tigers, and south of this zone, extensive human populations in India limit the tiger's distribution. Four separate breeding populations were located during this study. They were usually separated from each other by agricultural or urban development. Habitat type was very important to tiger use. Forests, tall grassland and dry river courses sustained tigers, while denuded hills, agricultural fields, pasture and degraded forest did not. The amount of good habitat in the range of the tigers determined the success of the population (Fig. 11.9). The largest populations were centred on national parks and reserves.

Even here, some tigers occurred outside the protected lands where they were exposed to increased levels of poaching, and other interactions with humans. This study highlighted the need to expand management plans beyond park borders, to include community forests and other lands, to maximize the tiger's habitat while still providing for the needs of local people.

Landscape patterns not only affect plant and animal populations, but also the movement of energy and matter in ecosystems. For example, the presence of a forest buffer between farm fields and streams will greatly reduce the flow of nitrogen and phosphorus into the streams. In some parts laws govern the presence and size of forest buffers constructed around large estuaries, such as Chesapeake Bay on the Atlantic coast of the United States, to control the flow of pollutants and nutrients into the estuaries. Lakes surrounded by urban development or housing usually have higher levels of eutrophication than lakes surrounded by forests. We will examine the effects of global change on the fluxes of nutrients and other ecosystem processes in Chapter 12.

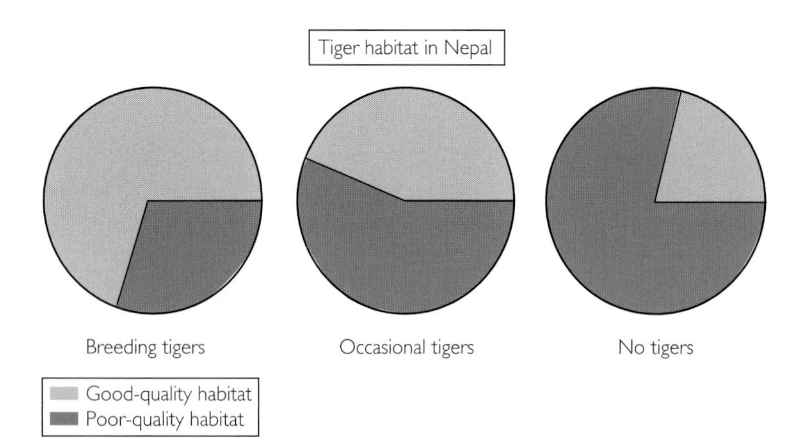

Fig. 11.9 The per cent of total habitat that is classified as 'good-quality tiger habitat' varies in different forest units in Nepal. Good-quality tiger habitat consists of forest, tall grass or dry stream valleys in areas below 1000 m elevation. Poor-quality tiger habitat includes denuded hills, agricultural land, pastures or degraded forest. These habitats usually occur in areas with high human impact. There is a critical ratio of at least 50% good- to poor-quality habitat that is needed to support breeding populations of tiger. In this figure, breeding tigers were found in forest units where over 71% of the habitat was defined as 'good quality', tigers were found occasionally in forest units where 45% of the habitat was defined as 'good quality', and no tigers were found in forest units where only 22% of the area was defined as 'good quality'. (From Smith *et al.* 1998a.)

11.8 Chapter summary

Ecological communities are sets of species that occur together in the same area at the same time. Communities have their own properties that make them more than the sum of their parts. For convenience, ecologists tend to define communities on a geographical scale that suits the questions that are being asked. Physical factors together with ecological interactions combine to determine the composition of communities.

Climate change will affect ecological communities in complex and unpredictable ways, and ecologists are increasing the effort that is put into studying landscape ecology—the ecology of how areas of land are used and what habitats cover them.

Recommended reading

Hughes, L. (2000) Biological consequences of global warming: is the signal already apparent? *Trends in Ecology and Evolution* **15**: 56–61.

Keesing, F. (2000) Cryptic consumers and the ecology of an African savanna. *BioScience* **50**: 205–215.

Laurance, W.F. (1998) A crisis in the making: responses of Amazonian forests to land use and climate change. *Trends in Ecology and Evolution* **13**: 411–414.

Lima, S.L. & Zolner, P.A. (1996) Towards a behavioral ecology of ecological landscapes. *Trends in Ecology and Evolution* **11**: 131–135.

Van Valkenburgh, B. (1995) Tracking ecology over geological time: evolution within guilds of vertebrates. *Trends in Ecology and Evolution* **10**: 71–76.

Weiher, E. & Keddy, P. (eds) (1999) *Ecological Assembly Rules. Perspectives, Advances, Retreats*. Cambridge University Press, Cambridge, UK.

Chapter 12

Ecosystems: the flux of energy and matter

12.1 Ecosystem ecology

Ecosystem ecology is the study of the interactions of organisms with the transport and flow of energy and matter. Ecologists who study ecosystems ask questions such as: What are the feeding relationships among the organisms in an ecosystem? How many different types of feeding relationships can be supported in a system? Why are some systems more productive than others? How much carbon and nitrogen are stored in the plants in an ecosystem? How rapidly do nutrients cycle through the living organisms in an ecosystem? How much of a particular nutrient is lost from the system each year?

12.2 The trophic–dynamic concept of ecosystem structure

A conceptual framework developed to explain the dynamics of energy and matter flow in aquatic ecosystems was developed in 1942 by Raymond Lindemann at the University of Minnesota. This model is called the trophic–dynamic model, and it describes the relationships between different organisms in an ecosystem by following feeding relationships among them. The organisms in an ecosystem are divided into different trophic levels, with the first trophic level containing the primary producers, the second trophic level containing the primary consumers, the third trophic level containing the secondary consumers, and on to the top consumers in the system. The model is an oversimplification, but it serves to illustrate the flow of energy and matter in ecosystems (Fig. 12.1).

Energy flows one way in ecosystems, with energy input from the sun being captured by primary producers, and large losses of energy between each trophic level, owing to respiration and inefficient energy transfer (Fig. 12.1).

Because of losses to respiration, and inefficiencies in harvesting, assimilation and digestion, the amount of energy in any trophic level is not completely available to the next highest trophic level. This observation has been used to explain the limits to trophic levels in ecosystems. Rarely are there more than four or five trophic levels in an ecosystem, and the number of individuals in top trophic levels is usually limited. Ecologists often illustrate this pattern through the use of pyramid plots, where the total amount of energy in each trophic level is plotted in a series of stacked boxes, starting with the first trophic level on the bottom (Fig. 12.2). A trophic pyramid constructed using energy is never inverted, consumers cannot use more energy than is available in their food. However, pyramids constructed with biomass can sometimes be inverted, and

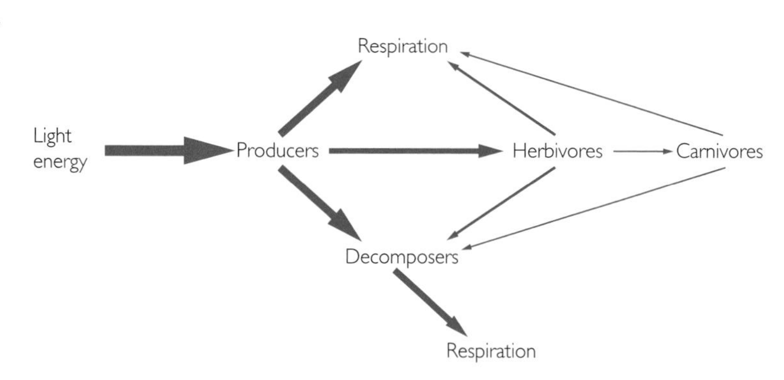

Fig. 12.1 A diagrammatic representation of the trophic–dynamic concept of ecosystem structure. Light energy is captured by plants and used to build their bodies. Much of this energy is lost to respiration, but a portion is passed onto the next trophic level, represented by herbivores. A portion of the energy in herbivores is then passed on to carnivores. When individuals in any trophic level die, they are broken down by decomposers, which recycle nutrients back into the system for uptake by other organisms.

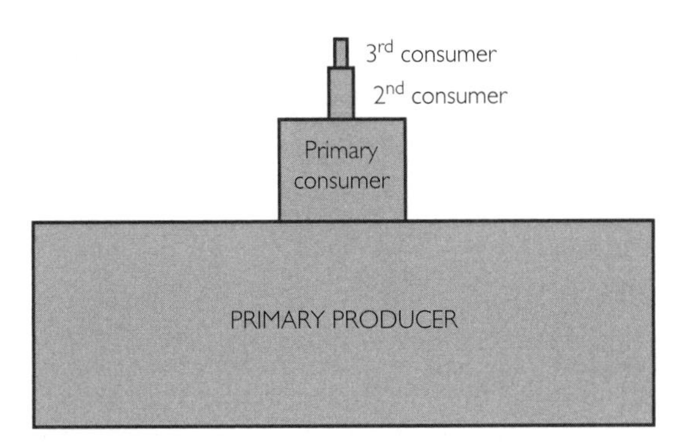

Fig. 12.2 A diagram of the amount of energy in four trophic levels of an idealized ecosystem. Each higher trophic level contains less energy due to losses to respiration within each level and inefficient transfer of energy between levels.

those constructed using numbers of organisms are often inverted—for example, there are many insect herbivores on a single oak tree.

12.3 Differences in efficiency of energy transfer among ecosystems

Once solar energy is used to fix atmospheric carbon dioxide into plant material, there are differences among ecosystems in its availability to higher trophic levels. This is because primary producers can take many different forms, from algae to trees, and have differences in the allocation of carbon to structures, which differ in digestibility. For example, woody plants allocate a significant proportion of their carbon to woody stems. Wood is not easily digested by animals, usually requiring that they have a symbiotic association with micro-organisms capable of digesting wood. In contrast, phytoplankton in aquatic systems can be largely digested by zooplankton grazing on them. In addition, in forest ecosystems, the age of a forest, or successional stage, will affect the proportion of biomass allocated to wood, leaves or other parts of the plants. During earlier stages of succession, the community will have a higher number of herbaceous species, and allocation to leaf material will be high. As a forest stand matures, a higher percentage of carbon will be tied up in woody material, and there is a rise in the respiration costs of supporting this tissue.

After allocation by plants, carbon and nutrients may be harvested by herbivores. The **exploitation efficiency** of these herbivores will be determined by the ratio of the amount of plant production they ingest to the total plant production. This ratio will be dependent on the plant life form. It could be as low as a few per cent in northern evergreen forests, but as high as 60% in African grasslands. Of the total ingested, not all will be digested. Plant material that is high in tannins, phenolics, hemicelluloses or lignin will be harder to assimilate than tissues low in these compounds. Once assimilated, the conversion into consumer biomass will depend on the respiratory demands of the consumer. Thus, endothermic consumers spend a much greater proportion of their assimilated energy on respiration than do ectothermic consumers. An endothermic animal eating low-quality plant food may have a net **production efficiency** of less than 1%, while an ectothermic carnivore that eats a high-quality animal prey may have a value as high as 30%.

12.4 Nutrient cycling and decomposer trophic levels

Detritus, or dead organic matter, is another major energy pathway in most ecosystems. Breakdown of **detritus** is a central process in the nutrient cycles of ecosystems. The productivity of open ocean habitats is often tied to the amount of decomposition and release of nutrients by bacteria in the upper layers of the water, before dead organisms can sink below the zone of light penetration. The turnover rate of detritus is often used as a measure of nutrient cycling, and forms the basis for the high productivity of some tropical communities found on nutrient-poor soils (Table 12.1). The turnover rate will depend on the structure and chemical make-up of the detritus, temperature, moisture and decomposer community. Often, large arthropods are needed to break dead plant material (litter) apart, allowing bacteria and fungi greater surface areas to attack the tissue.

As a rule of thumb, the pattern and rate of nutrient movement are more important to ecosystem function than are the absolute amounts of nutrients in the system. Nutrient movement between plants, animals, decomposers, detritus and soil will depend on the different storage pools of nutrients, and the turnover rates of these pools. Nutrient storage pools include the living organisms in an ecosystem, detritus and soil. Because decomposer activity is affected strongly by temperature and humidity, moist, warm areas generally have faster

Table 12.1 The half-life (time taken for half the material to be broken down) and rate of production of litter for three different forest communities.

Community	Half-life (years)	Litter production rate (g m^{-2} $year^{-1}$)
Tropical rainforest	0.1	1200
Temperate deciduous forest	1	800
Northern coniferous forest	7	600

turnover rates than colder, drier ecosystems. However, leaf structure (evergreen or deciduous), plant life form (woody or herbaceous), and soil structure also influence nutrient cycling. Leaf tissues that are high in lignin content may be hard to break down, and nutrients may not be released from this pool very fast. Many evergreen trees keep their leaves for several years, maintaining nutrients in their biomass and not releasing them to the system. The soil pool is influenced by soil structural properties such as the proportion of clay, sand and silt particles. Water in sandy soils drains easily, leaching nutrients away with it. High clay and organic contents increase the water and nutrient-holding capacity of soil. The presence of peat bogs, coal swamps and fossil fuel deposits shows that decomposition has lagged behind productivity many times and in many ecosystems during the Earth's history. The result is that much dead organic material has remained undecomposed for thousands, or even millions, of years.

12.5 Food chains and food webs

The path of energy and matter from one trophic level to another is often outlined by constructing food chains. A food chain is simply a listing of which organism eats a different organism in the ecosystem. Food chains are greatly simplified, however, and usually the eating relationships are more accurately depicted as a food web (Box 12.1). An ancient mathematical problem from Egypt, dating back to approximately 1700 BC, sums up the effect of a food web on the rate of energy and matter transfer in an ecosystem. This problem is stated as: In each of seven houses are seven cats; each cat kills seven mice; each mouse would have eaten seven ears of wheat; each ear of wheat would have produced 3.5 pecks of grain. How much grain is saved by the seven houses' cats? The cats would have saved over 8000 pecks of grain, equivalent to about 16 tonnes. But the cats weigh only a few kilograms. We can account for the difference by observing that at every stage in the food chain, the transfer of energy is not completely efficient. For example, when the mice eat the grain, some of the energy contained in the wheat is dissipated as heat (Fig. 12.2).

12.6 Productivity, species richness and disturbance

Chapter 2 discussed the distribution and primary productivity of the major biomes in the world. Productivity in different biomes ranged widely, from less than 50 g to over 3000 g of carbon per square metre per year. There is a broad correlation between productivity and the availability of resources (Fig. 12.3), both within ecosystems and among ecosystems. Generally, as resources such as light and nutrients increase, plant productivity increases. This, in turn supports greater productivity in higher trophic levels.

The relationship between resource availability and species richness is not always positive. Studies from a variety of ecosystems have shown a peak in species richness at intermediate levels of resource availability. What are the bases of this relationship? Many ecologists think that at low levels of resources, only a few species can survive, so species richness is low. These are usually stressful environments, where adaptations to the physical environment are most important. Only a few species have these adaptations, leading to lower species richness in low-resource environments. As the level of limiting resources increases, more species can survive and species richness increases. However, with high levels of resources, competition becomes an important force, and competitive dominants may eliminate less competitive species, resulting in a decrease in species richness.

Predation and herbivory often interact with competition and resources to determine the final biodiversity in an ecosystem. If a predator or herbivore can reduce

Box 12.1 The food web in the pitcher of a carnivorous plant, *Sarracenia purpurea*

The pitcher plant, *Sarracenia purpurea*, occurs in low-nutrient bogs. It derives its name from the vase-shaped pitcher that it forms from modified leaves. The pitcher contains rainwater that some flies, mosquitoes, mites and midges are able to live in, while other insects are trapped in the rainwater and drown. The plant is able to absorb nitrogenous compounds that are released from the breakdown of drowned insects. Insect detritus from the feeding of the larvae of the midge *Metriocnemus knabi* and the small fly *Blaesoxipha fletcheri* is broken down by bacteria and yeast in the pitcher. Algae, yeast and bacteria in the pitcher are then fed upon by various protozoa and the rotifer *Habrotrocha rosa*. The top trophic level is represented by larvae of the pitcher plant mosquito, *Wyeomyia smithii*, which feeds on algae, bacteria and yeast as well as protozoa and rotifers. This example shows that interactions among microbial community members are just as complex as those observed in plant and animal communities (Cochran-Stafira & von Ende 1998).

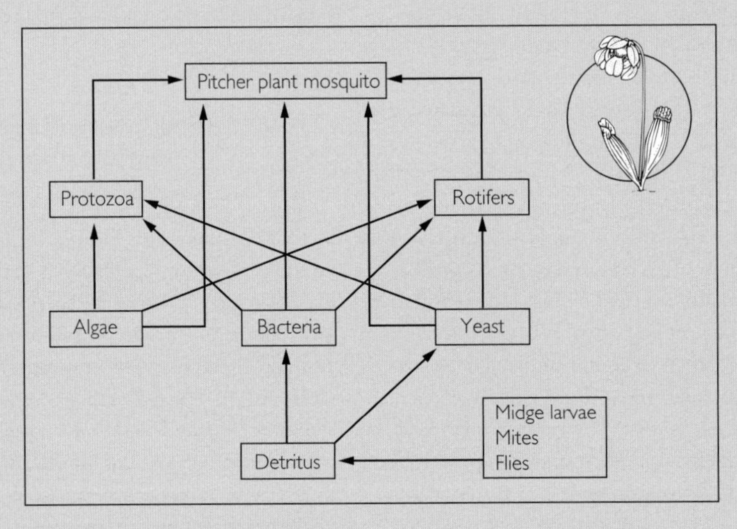

Fig. 1 The food web found in the pitcher of *Sarracenia purpurea*. (From Cochran-Stafira & von Ende 1998.)

the population of a dominant competitor, weaker competitors may be able to coexist, increasing species diversity. The interaction between nutrients, grazing and plant species richness was reviewed by Proulx and Mazumder (1998). They looked at 30 different studies containing 44 comparisons of plant species richness, nutrient availability and grazing in lake, stream, marine, grassland and forest ecosystems (Fig. 12.4). They found that all 19 comparisons from nutrient-poor ecosystems showed lower plant species richness under high grazing compared to low grazing. Only two of the 25 comparisons from nutrient-rich ecosystems showed a decrease in plant species richness when levels of grazing were high, while 14 showed increases in plant species richness. The authors concluded that high levels of grazing reduce species richness in nutrient-poor ecosystems, because resources are so limiting that plants cannot regrow fast enough to replace tissues lost to herbivory. Moreover, in nutrient-rich ecosystems, a high level of grazing allows the increase of less palatable plants, which may have been competitively excluded under low grazing pressure.

12.7 Top-down versus bottom-up control of trophic levels

Ecologists have debated whether, and under what conditions, one trophic level might control the numbers and diversity of organisms in other trophic levels. One viewpoint is that bottom-up control, or the supply of resources, is the dominant factor. This view holds that the amount of resources available to a trophic level

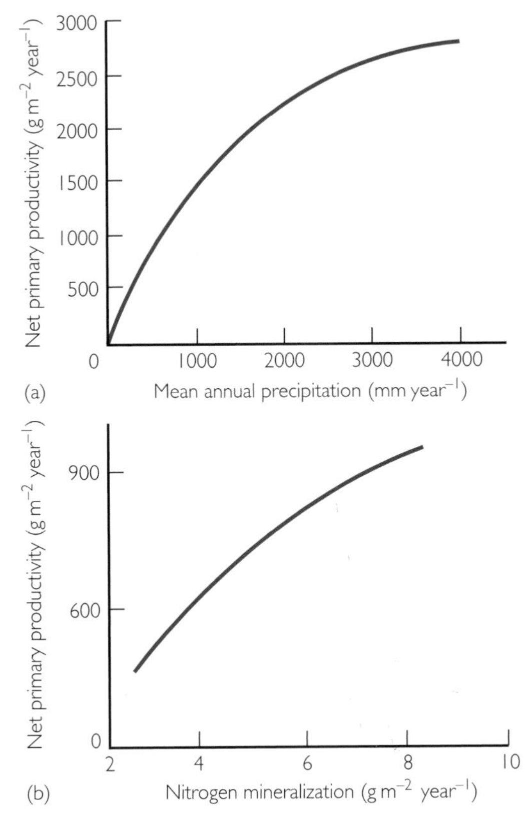

Fig. 12.3 (a) Patterns of terrestrial net productivity as a function of mean annual precipitation. (Adapted from Lieth 1973.)
(b) Relationship between productivity and nitrogen mineralization for different ecosystems of Blackhawk Island, Wisconsin, USA. (Adapted from Pastor *et al.* 1984.)

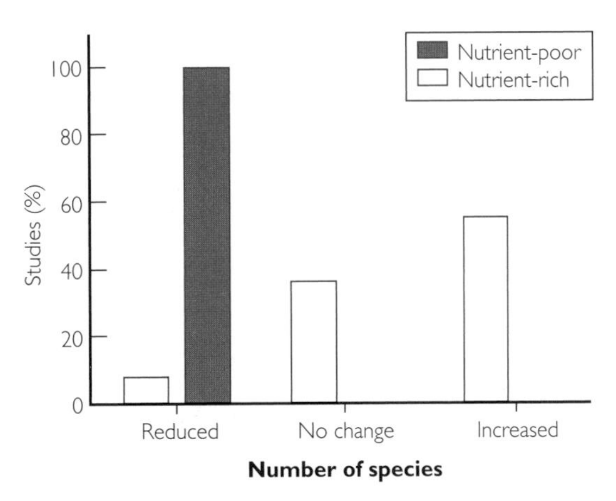

Fig. 12.4 The per cent of studies showing a decrease, an increase or no change in the number of plant species under high grazing compared to low grazing. The responses of nutrient-poor ecosystems differ compared with those of nutrient-rich ecosystems. (Data from Proulx & Mazumder 1998.)

controls the productivity of that trophic level. Because of the over-riding importance of resources, the next highest trophic level has little influence. For example, the amount of nitrogen in the soil may control plant productivity in a particular ecosystem, and the amount of herbivory by consumers does not significantly reduce plant productivity. In this view, nutrient-rich ecosystems will sustain high plant productivity, which will in turn support a large number of herbivores.

Other ecologists maintain that consumption by a higher trophic level (top-down control) is the dominant factor determining the productivity and diversity of the next lowest trophic level. For example, predators may eat so many herbivores that the herbivores never develop high enough population levels to have significant effects on the plants in a system. Top-down control of the ecosystem structure is called a **trophic cascade**, because the effects extend through several trophic levels. In this view, trophic levels alternate between limitation by consumers and limitation by resources. For example, in a three-level trophic system, predators control herbivore populations, allowing plants to be controlled by resource levels such as water or nitrogen. In a four-level system, the fourth trophic level (top consumers) would control populations of the secondary consumer to the extent that the secondary consumer would not exert control over primary consumers. Because of less control by secondary consumers, primary consumers would then develop large populations that would limit the primary producers (Fig. 12.5).

There is evidence to support both views, and aquatic systems may be fundamentally different from terrestrial systems. Rates of herbivory are generally greater in aquatic ecosystems. For example, aquatic herbivores can remove about 51% of primary productivity, three times the average amount removed by terrestrial herbivores.

Thus, aquatic herbivores have a greater ability to control primary productivity, unless they themselves are kept in check by their predators. Several manipulative studies in temperate lakes have shown cascading effects of top trophic levels on lower trophic levels (Carpenter & Kitchell 1988). Generally, these studies have shown that both abiotic and biotic factors are

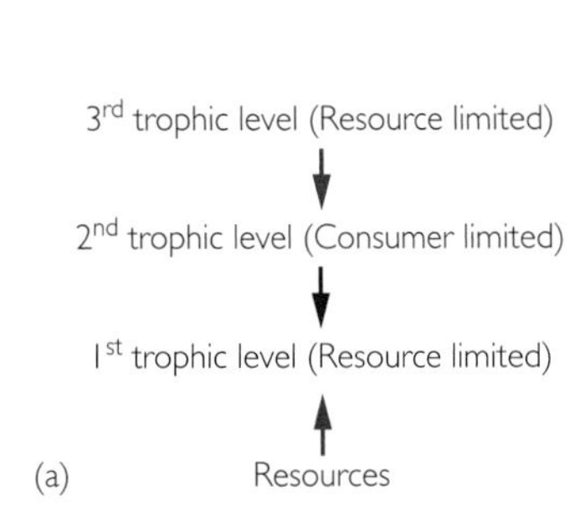

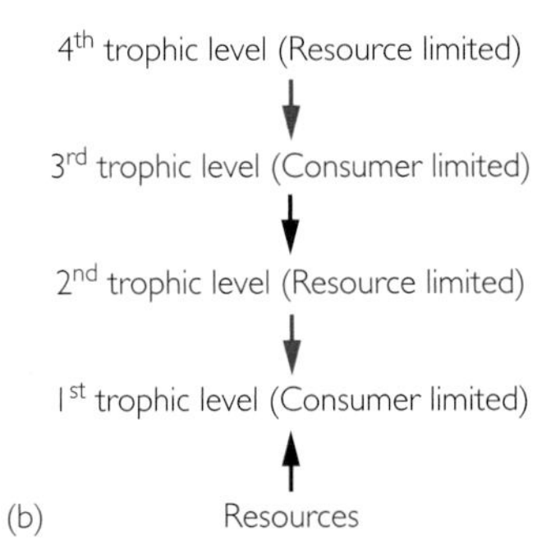

Fig. 12.5 Two trophic cascades, differing in the number of trophic levels. (a) The controls operating on different trophic levels for a system with three trophic levels. (b) A four-level trophic system. Purple arrows indicate that one trophic level exerts a dominant force on another, and black arrows indicate a lack of control.

important in a lake's productivity. The availability of nitrogen, or phosphorus, or both, will set the maximum potential level of productivity, but then the food web composition of the particular lake will set the actual amount of productivity. For example, in a lake with three trophic levels, the productivity of phytoplankton (algae), zooplankton (small animals that graze on the algae) and planktivores (small fish that eat the zooplankton) is near the limit set by nutrients. This is because the planktivores keep zooplankton populations low, which allows phytoplankton populations to increase. In contrast, in a lake with similar nutrient levels, but with four trophic levels, phytoplankton, zooplankton, planktivores and piscivores (larger fish that eat planktivores), primary productivity is usually further from the limit set by nutrients. This is because

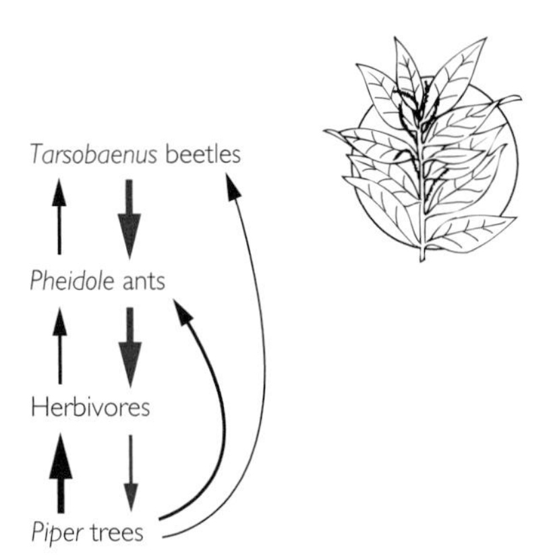

Fig. 12.6 Illustration of the feeding relationships in a tropical forest system consisting of a small tree, *Piper cenocladum*, its herbivores and their predators. Black arrows represent the effects of lower trophic levels on the levels above them, and purple arrows represent the level of control exerted by higher trophic levels on lower levels. Because the plant provides nest sites and food bodies for both ants and beetles, arrows are also drawn between these organisms. (From Letourneau & Dyer 1998.)

of the control of planktivores by piscivores, which allows larger zooplankton populations, which reduces phytoplankton numbers through grazing.

There are fewer studies of top-down effects in terrestrial systems. Often, it is hard to manipulate large terrestrial systems and exclude top predators. Another complication is that top predators often feed from more than one trophic level. However, Letourneau and Dyer (1998) examined a system with four trophic levels in a lowland tropical forest in Costa Rica. This system was made up of a small tree, *Piper cenocladum*, and insects in three trophic levels associated with this tree. There were a variety of insect herbivores in the second trophic level, such as moth larvae, weevils, leaf-cutting ants and beetles, that fed on the tree's fruits, leaves and branches. The third trophic level consisted of predators of these insects. The study focused on an ant species, *Pheidole bicornis*, that lives on the tree. These ants live in hollow petioles and stems, feeding on food bodies produced by the tree, as well as on other insects. The relationship between the tree and the ant is mutualistic: the ant provides herbivore protection for the tree, and the tree provides a home and food for the ant. The fourth trophic level had many members, but the study focused on a predatory beetle, *Tarsobaenus letourneae*, that specializes on the ants. Beetle larvae live inside hollow petioles with the ants. They kill adult ants and feed on both ant larvae and plant food bodies (Fig. 12.6).

To manipulate resource levels, Letourneau and Dyer transplanted cuttings of *Piper* trees to forest plots differing in light and nutrient levels. They also manipulated the abundance of the top predator. If top-down effects were important, then high numbers of *Tarsobaenus* beetles should reduce the number of *Pheidole* ants, which should raise the number of herbivores, which would reduce the growth of the trees. If bottom-up forces were dominant, then high light and nutrients should increase plant productivity, which would increase herbivore, ant and beetle numbers.

Observations from the different treatments showed that additions of beetles reduced the abundance of ants by five times, compared with systems without beetles. This resulted in a three-fold increase in herbivory on *Piper* leaves, and a two-fold reduction in leaf area of the plants. Thus, the fourth trophic level had a strong cascading effect on levels below it. Not all of the effects were immediate. While ant numbers were reduced within a few months of the addition of beetles, it took over a year before effects on the plants could be seen.

Light and soil nutrients had only small effects in these experiments, demonstrating that bottom-up effects were mitigated by the strong effects of the top predator. This study and others have now shown that the important question in bottom-up versus top-down control of trophic structure is not: Which control is most important? but rather: Under what circumstances is one more important than the other?

12.8 Ecosystem engineers

In 1997, C.G. Jones, J.H. Lawton and M. Shachak coined the term '**ecosystem engineer**' to describe organisms that directly or indirectly control the availability of resources to other organisms by causing physical state changes in the environment. It is important to distinguish ecosystem engineers from the trophic and biotic interactions that occur in ecosystems. The utilization of living or dead tissues as food by a consumer or decomposer, or the direct uptake and utilization of an abiotic resource (light, water, nutrients) by an organism, are not engineering. The large effects caused by a keystone species (which influences community structure through competition or predation, see Section 11.2.2) are also not included in the concept of an ecosystem engineer. Rather an ecosystem engineer, by its physical presence or activity, determines the nature of the ecosystem.

For example, trees in a forest are ecosystem engineers. Their tissues are eaten by animals and microbes, and trees compete with each other and other plants for light, water and nutrients. Nevertheless, these interactions are not what make them engineers. Rather, it is their physical structure, including branches, bark, roots and leaf surfaces that provide shelter, resting locations and living space for myriad other organisms. Small ponds full of organisms form where water gets channelled into crotches between branches, soil cavities form as roots grow and die, and leaves and branches cast shade, reducing the impact of rain and wind, moderating temperature and increasing humidity. Dead leaves fall to the forest floor, altering raindrop impact, water runoff and heat and gas exchange. In the soil, dead leaves make barriers or protect seeds, seedlings, animals and microbes. Dead trunks, branches and leaves fall into forest streams, creating debris dams and ponds. Roots bind around rocks, stabilizing substrates and moderating the effects of storms. More species are probably affected by these processes than a tree affects directly through food or competition.

Many other species are also ecosystem engineers. In fact, ecosystem engineers occur in virtually all ecosystems. Examples include organisms that make large physical structures like corals, sphagnum bogs, kelp forests or a sea grass flat. Other, more subtle examples include microphytic crusts in deserts made up of cyanobacteria and fungi that secrete polysaccharides. These chemicals bind the desert soil and stabilize it, preventing erosion, changing runoff patterns and controlling site availability for seed germination. Other examples of engineers include animals that build or destroy massive, persistent structures; or animals that build burrows or disturb soil, such as earthworms, gophers, pack rats, mole rats, alligators, termites and tilefish. Puffins that burrow on islands can create erosion of the soil surface. Large animals like elephants, bison and grazing ungulates may serve as engineers. Not all species in the area benefit from this engineering. With the damming and flooding of areas by beavers, some terrestrial organisms are eliminated, and stream organisms that are not adapted to the still water of the pond may be displaced.

Food webs and engineers can interact to change the physical state of the environment. For example, diatoms in the Bay of Fundy secrete carbohydrates that bind sand particles. This stabilizes sand movement. Diatoms are thus engineers. They are grazed upon by small animals called amphipods. If grazing is too intense, diatom populations decrease, and sand stabilization is reduced. Sandpipers are shore birds that are predators on amphipod crustaceans. When sandpipers are present, they can control amphipod populations, which in turn increases the diatom population and sand is stabilized. When sandpipers are not present, amphipod populations increase, diatoms decrease and the sand is not stabilized.

12.9 Biogeochemical cycles

The cycling of nutrients from non-living to living components and back is one of the most important of ecosystem functions. Because this cycling involves living organisms (biology), geological processes (water movement, mountain building, erosion and soil properties) and chemical transformations, these cycles are termed **biogeochemical** cycles. Biogeochemical cycles can be divided into atmospheric or sedimentary cycles, depending upon whether the atmosphere and oceans are the primary storage pools for the nutrients, or whether sedimentary strata and other rocks are the major repository.

12.9.1 The hydrological cycle

The hydrological cycle involves the movement of water. Without the hydrological cycle, other biogeochemical cycles could not exist, ecosystems could not function, and life on Earth could not be maintained. Water moves either as vapour or as liquid in the atmosphere, oceans, streams and rivers. Energy from the sun powers the evaporation of water into the atmosphere (Fig. 12.7). This atmospheric water turns over rapidly (every 10–11 days), condensing on dust or other particles in the air and falling as precipitation over the land or ocean. Although 84% of the total evaporation of water vapour into the atmosphere occurs from ocean surfaces, only 77% of the total annual precipitation occurs over oceans. The 7% difference between ocean evaporation and precipitation is made up by river, stream and overland flow of water back to the ocean.

After precipitation falls on land, part of it is intercepted by vegetation and other surfaces. This water usually evaporates back into the atmosphere. Water that is not intercepted falls on soil surfaces, where it either infiltrates into the soil, or runs off the surface of the soil. The infiltration rate is affected by the soil type, degree of slope, vegetation and characteristics of the precipitation. A short, intense rainfall will not penetrate into the soil as effectively as a longer, less intense rainfall. Once water infiltrates into the soil, some will be held in the air pores between soil particles and some will percolate down into groundwater. Groundwater can be held in deep aquifers below the soil, or it may make its way to streams and rivers, to be carried away from the ecosystem. The water held in the air pores of the soil will be

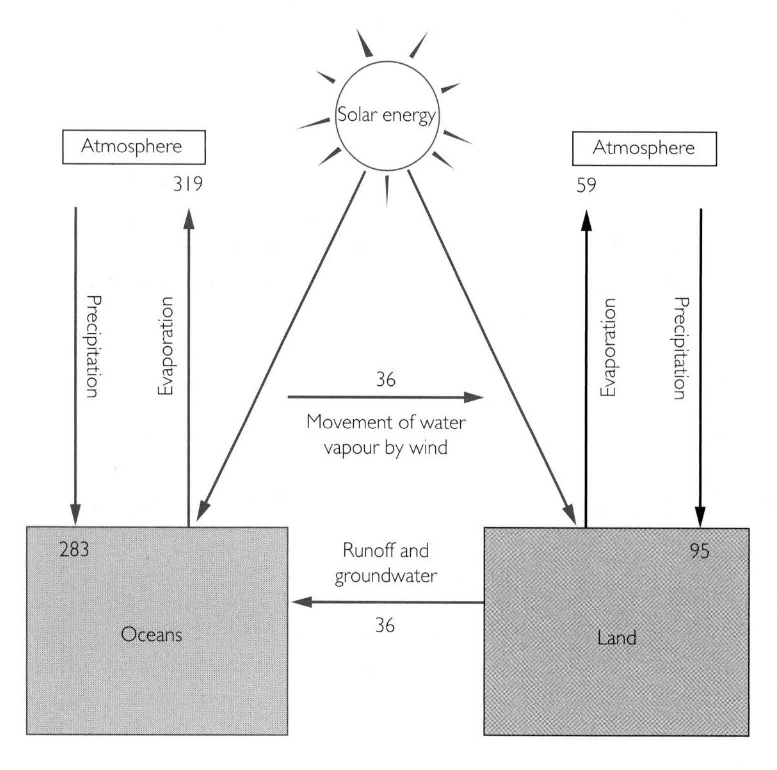

Fig. 12.7 The hydrological cycle. Numbers in the diagram represent 10^{18} g of water per year. The excess of evaporation over the oceans results in movement of water vapour from the ocean to the land. Excess precipitation over land results in the flow of surface and groundwater systems back to the oceans, completing the cycle.

taken up by plants through their roots, and lost from their leaves through transpiration. This soil water contains dissolved nutrients such as nitrogen and phosphorus. If nutrients dissolved in soil solution are not taken up by plants, they can be leached from the system. Thus, transpiration is a nutrient conservation pathway, it keeps nutrients in the ecosystem by taking them from the soil and allowing them to be incorporated into plants.

12.9.1.1 *Human alteration of the hydrological cycle*

Humans now use more than half of the world's runoff of freshwater. Most of the world's river systems have been altered through diversions and dams. In the United States, only 2% of the rivers run unimpeded, and within a few years, the flow of two-thirds of all the Earth's rivers will be regulated. At present, about 6% of Earth's river runoff is evaporated as a consequence of human impoundment and diversion to irrigation.

Humans also have large effects on local hydrological cycles by changing land cover. Converting forests to parking lots or houses will increase runoff dramatically. Logging of forests reduces transpiration and increases runoff, erosion of soil and leaching of nutrients.

Human alteration of the hydrological cycle can affect climate. Irrigation increases humidity, increasing precipitation and thunderstorm frequency. Land transformation to agriculture or pasture increases temperatures and decreases precipitation. Human drainage or canalization of water in large wetlands, such as the Everglades in Florida, can reduce evaporation, leading to decreases in regional precipitation.

12.9.2 The carbon cycle

Major storage pools of carbon on Earth occur in the atmosphere, oceans, sedimentary rocks and living matter. Despite its importance to global climates and plant life, only a small proportion of Earth's carbon is in the atmosphere. Most is stored as non-gaseous dissolved carbon in the oceans or as carbonate materials in sedimentary rocks (Fig. 12.8).

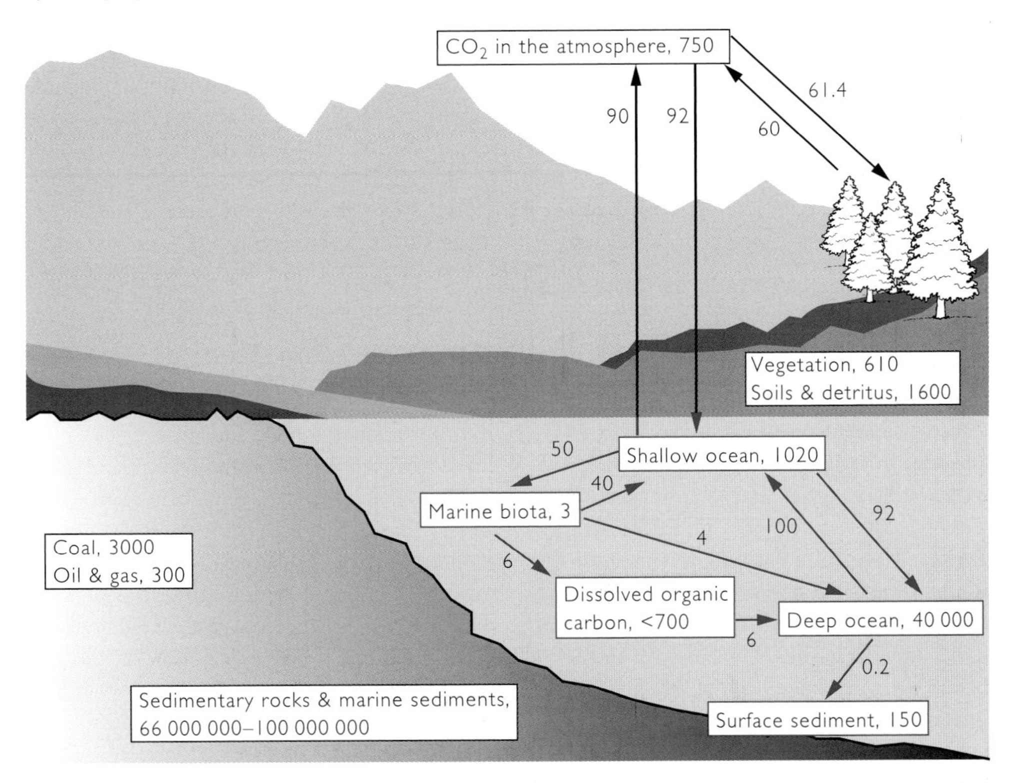

Fig. 12.8 The carbon cycle. The processes of photosynthesis and respiration dominate the major transformations and movements of carbon. Numbers are in billion tonnes of carbon per year. Numbers next to arrows represent yearly fluxes of carbon and numbers in boxes represent storage pools of carbon.

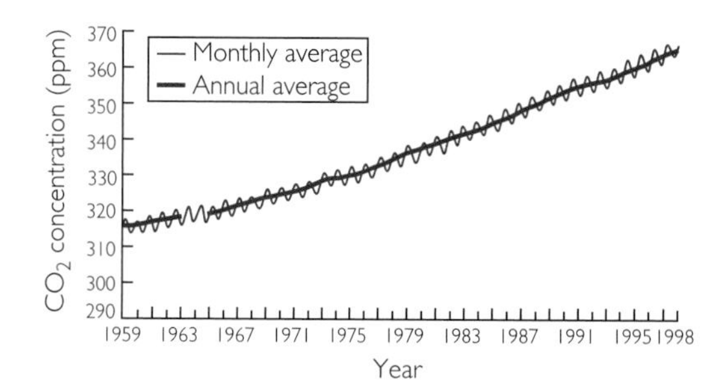

Fig. 12.9 Historic atmospheric carbon dioxide concentration in Mauna Loa, Hawaii. (Courtesy of Scripps Institution of Oceanography, University of California, 1998; with permission of UNEP GRID-Arendal.)

Although marine organisms make up a small part of the overall carbon budget of the Earth, they have a large influence on the distribution of carbon. This is because they convert soluble carbonate ions from seawater into insoluble ocean sediments, by depositing carbon in shells and skeletons that eventually sink to the bottom of the ocean. Biological processes are also important to the carbon cycle on land. Photosynthesis removes carbon from the atmosphere and places it into terrestrial storage. Respiration returns terrestrial carbon to the atmosphere pool. The great forests of Earth, especially in the tropics, are major pools of carbon.

12.9.2.1 Human alteration of the carbon cycle

The increase in carbon dioxide concentration in the atmosphere over the last 130 years represents the best-documented signal of human alteration of the Earth's system. Since measurements began in Hawaii in 1957, carbon dioxide concentrations have risen from 315 to 365 parts per million (ppm) (Fig. 12.9). Records of carbon dioxide concentration from air bubbles trapped in Antarctic and Greenland ice show that until the Industrial Revolution in the 1800s, carbon dioxide concentrations were relatively stable at about 280 ppm for thousands of years. Thus, humans have increased atmospheric carbon dioxide by about 30%, relative to preindustrial times. Humans add carbon dioxide to the atmosphere by burning fossil fuels and converting forests and grassland to agricultural and urban systems.

The carbon cycle includes the fluxes of carbon between the four main reservoirs: fossil carbon, atmosphere, oceans and terrestrial ecosystems. Human activities have altered carbon storage and fluxes between all four of these reservoirs (Fig. 12.10). Fossil fuel combustion and cement production adds approximately 5.5 billion tonnes of carbon dioxide to the atmosphere each year. Land-use changes, primarily deforestation, add another 1.6 billion tonnes, to give a total of 7.1 (± 1.1) billion tonnes of carbon per year added to the Earth's systems through human activities.

Of the carbon released, about 3.2 billion tonnes accumulates in the atmosphere, resulting in the observed increase in carbon dioxide concentration. Oceans are estimated to take up approximately 2.0 billion tonnes per year. Forest sinks and sources are highly variable. Current estimates include a yearly uptake of around 0.5 billion tonnes of carbon through the regrowth of northern temperate forests, an additional 1.0 billion tonnes taken up due to the stimulation of plant growth from increased atmospheric carbon dioxide, and perhaps 0.6 billion tonnes taken up due to a nitrogen fertilization effect (Section 12.9.3.1.).

However, land-use changes are a tremendous source of uncertainty. Urban and agricultural lands are less strong carbon sinks than forests. Changes in land use from 1850 to 1990 are thought to have released 122 (± 40) billion tonnes of carbon into the atmosphere. Future scenarios of land-use changes can result in carbon storage or release ranging from 100 to 300 billion tonnes over the next century. Obviously, with the size of these numbers, there is great variation in projections of future changes in the global carbon cycle.

12.9.3 The nitrogen cycle

Nitrogen ranks fourth, behind oxygen, carbon and hydrogen, among the commonest elements in living systems. It is a necessary component of proteins, nucleic acids, chlorophyll and other common organic compounds. However, despite its presence as the major gas in Earth's atmosphere, nitrogen is often the most limiting nutrient to ocean and terrestrial productivity. This is because the chemical form of nitrogen in the atmosphere, N_2, cannot be used by the vast majority of organisms. N_2 must first be converted into biologically useful forms such as nitrate and ammonia. This process, called **nitrogen fixation**, is performed by only a few species of bacteria and cyanobacteria. Nitrogen fixation also occurs during lightning strikes and volcanic activ-

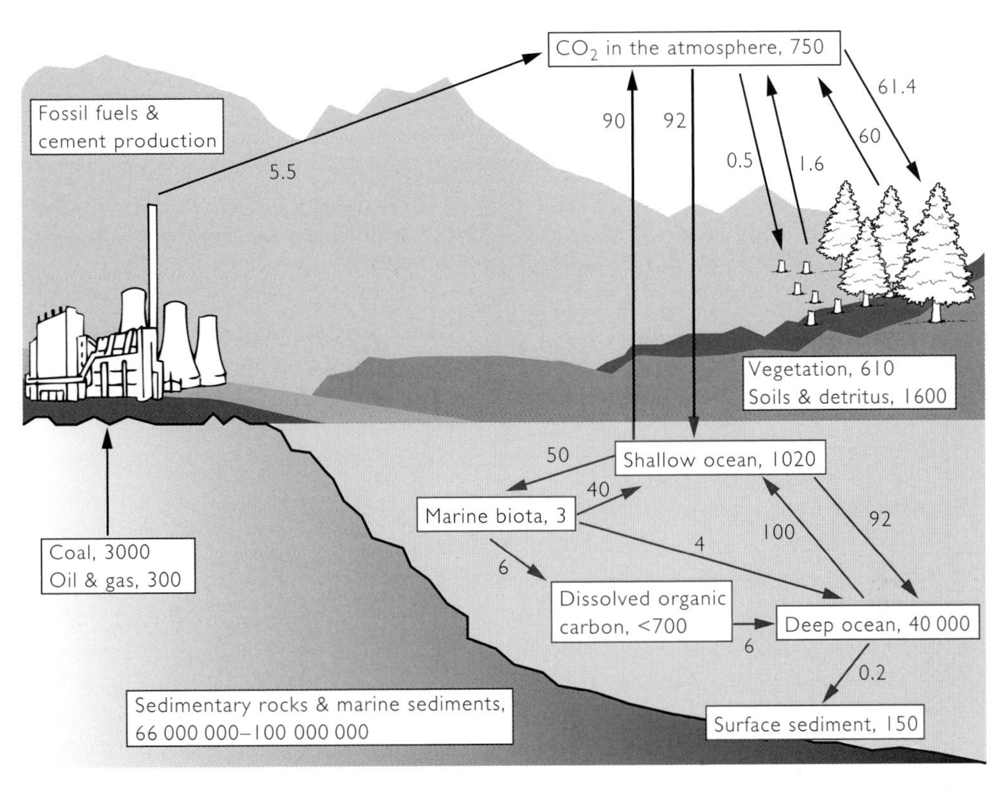

Fig. 12.10 The carbon cycle redrawn to indicate the estimates of human contributions between 1980 and 1989. Numbers are in billion tonnes of carbon per year. Numbers next to arrows represent yearly fluxes of carbon and numbers in boxes represent storage pools of carbon. (Adapted from Schimel 1995.)

ity. Once nitrogen is fixed, it is available for uptake by plants and enters the trophic structure of ecosystems (Fig. 12.11).

The amount of nitrogen being fixed naturally at any one time is small in relation to the pool of previously fixed nitrogen that is in living and dead matter. Most of the previously fixed nitrogen is locked up in soil organic matter, and must be decomposed by microbes before it becomes available again to living organisms. Nitrate and ammonia are soluble in water and are relatively mobile in the soil and in aquatic systems. They are easily leached from soils if they are not taken up by plant roots. Nitrogen, unlike carbon dioxide, is not respired directly back into the atmosphere. It must be converted to inorganic molecules, from amino acids or urea, in several stages.

Micro-organisms play major roles in the nitrogen cycle in addition to nitrogen fixation. They are involved in several chemical transformations. **Ammonification** is the process by which bacteria and fungi convert urea and amino acids to ammonia. Ammonia can be lost to the atmosphere, taken up by plants or converted to other forms of nitrogen. **Nitrification** is the conversion of ammonia to nitrate. This process is performed by bacteria that gain energy from the reaction. Nitrate can be transported in water or taken up by plants. **Denitrification** is the process where bacteria convert nitrate to nitrite, and then to N_2, where it can be lost back to the atmosphere. Movement of nitrogen among different organisms accounts for about 95% of all nitrogen fluxes on Earth.

12.9.3.1 Human alteration of the nitrogen cycle

In 1997, the first publication of a new series on issues in ecology was published by the Ecological Society of America. This paper addressed the human alteration of the global nitrogen cycle (Vitousek *et al.* 1997a). Nitrogen is unique in that there is a huge atmospheric reservoir that must be combined with carbon, hydrogen

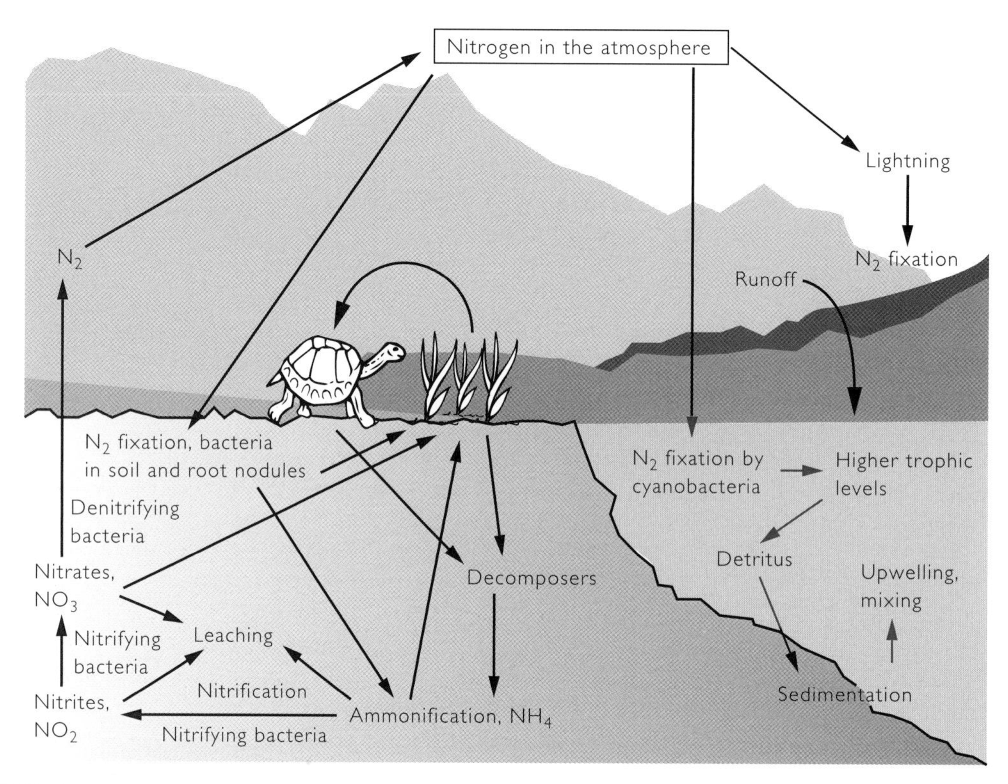

Fig. 12.11 The nitrogen cycle. The arrows in this diagram represent the various pathways and transitions of nitrogen in an ecosystem. Most of the nitrogen cycling occurs through food webs. Transformations between different forms of nitrogen by micro-organisms are essential to this cycle.

or oxygen before it can be used. Before human industrialization, lightning and micro-organisms were the primary mechanisms for nitrogen fixation. The supply of fixed nitrogen is a major factor in the type and diversity of plants in an ecosystem, the population dynamics of grazing animals and their predators, and the primary productivity and nutrient cycles of many ecosystems.

Prior to industrialization, about 90–130 million tonnes of nitrogen was biologically fixed each year on land. Fertilizer manufacture by humans in 1990 was about 80 million tonnes per year. In addition to fertilizers, agricultural land devoted to leguminous crops capable of nitrogen fixation, and the cultivation of rice (which is associated with nitrogen-fixing micro-organisms) has added another 32–53 million tonnes of nitrogen to Earth's ecosystems. The burning of coal, oil and gasoline in automobiles, factories and power plants adds another 20 million tonnes of available nitrogen per year to the atmosphere. Thus, humans have more than doubled the amount of nitrogen input into terrestrial and marine ecosystems.

In addition to increasing the amount of fixed nitrogen available worldwide, humans have also affected the turnover of nitrogen from long-term storage pools. Although estimates of mobilization of nitrogen have a great deal of uncertainty in them, these human activities could release half as much nitrogen again to the Earth's biosphere as is released in fertilizer manufacture, fossil fuel burning and cultivation of legumes. The burning of forests, wood fuels and grasslands emits more than 40 million tonnes of nitrogen per year. The draining of wetlands allows organic material that is decomposing slowly in anoxic conditions to be exposed to oxygen, which greatly increases the rate of decomposition. Estimates of release of nitrogen through this pathway are approximately 10 million tonnes per year. Land clearing for crops mobilizes another 20 million tonnes.

One of the consequences of human activity is the increase in nitrous oxide (N_2O), nitric oxide (NO) and ammonia (NH_3) in the atmosphere (Fig. 12.12). Nitrous oxide is a long-lived **greenhouse gas** that

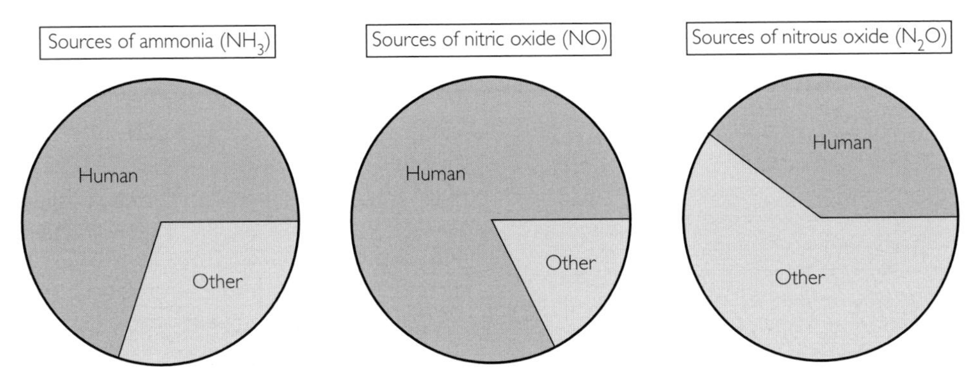

Fig. 12.12 Estimates of the percentage of nitrogen-containing trace gases that are released by human activities. (Data from Vitousek *et al.* 1997a.)

contributes to global climate change. Nitrous oxide absorbs infrared radiation in wavelengths that are not captured by carbon dioxide. Thus, its increase contributes significantly to predictions of global warming. Additionally, as nitrous oxide rises into the stratosphere, it catalyses reactions that destroy ozone, which absorbs ultraviolet radiation. The concentration of nitrous oxide is currently increasing at about 0.2–0.3% per year.

Nitric oxide is shorter lived than nitrous oxide because it is more reactive in the lower atmosphere. Nitric oxide catalyses the formation of photochemical smog, and in the presence of sunlight combines with oxygen and hydrocarbons to form ozone. Ground-level ozone negatively affects human health, as well as the productivity of forests and crops. Nitric oxide can also be transformed into nitric acid in the atmosphere, forming a major part of acid rain.

Human addition of nitrogen to the biosphere may also increase primary productivity and carbon storage in nitrogen-limited ecosystems. At first glance, this may seem to be a desirable outcome of human activities. Increases in carbon uptake and storage in forest, wetland and tundra ecosystems may be a direct consequence of extra sources of nitrogen. However, addition of nitrogen to natural ecosystems eventually leads to limitation by other factors such as phosphorus, calcium or water. This is called **nitrogen saturation**.

Nitrogen saturation leads to increased losses of nitrogen and elements such as calcium and potassium that are essential for long-term soil fertility. For example, as ammonium builds up in soils, more and more nitrate is formed from bacterial transformation. This acidifies the soil, which enhances emissions of nitrous oxide and the leaching of nitrate into streams or groundwater. As nitrate is leached, it carries with it calcium, magnesium and potassium. As calcium is depleted in soils and soil pH drops, aluminium ions are mobilized, reaching toxic concentrations that damage tree roots and kill fish in streams. The more nitrate and ammonium that build up in a soil, the more imbalanced the nutrients become. This may lead to stunted growth and eventual death of plants.

Nitrogen saturation also accelerates losses in species biodiversity, affecting most strongly plants adapted to low nitrogen conditions, and the animals and microbes that are associated with these plants. A long-term experiment where nitrogen was added to grasslands in England showed a marked decrease in species diversity in fertilized plots (Silvertown 1980). This was attributed to increased competitive dominance by a few nitrogen-responsive grass species, and the loss of other species adapted to lower nitrogen conditions. In the Netherlands, where some of the highest rates of nitrogen deposition occur, species-rich heathlands have been converted into grasslands and forests with significantly lower species diversity.

In extreme cases of nitrogen saturation, little nitrogen uptake may occur in soils. This results in dramatically increased nitrogen inputs into groundwater, lakes, ponds, streams, rivers and eventually oceans. Estimates of increases of nitrogen into rivers draining into the North Atlantic Ocean basin range between two and 20 times preindustrial levels. In rivers of the northeastern United States, nitrate concentrations have increased by a factor of between three and 10 since 1900. Increases in nitrate concentration of groundwater in agricultural areas have also been documented. High levels of nitrate in drinking water cause human

Table 12.2 Adverse effects of excess nitrogen on lakes, reservoirs, rivers and coastal oceans. (Adapted from Carpenter *et al.* 1998.)

Increased phytoplankton biomass
Shifts in phytoplankton species to bloom-forming species that may be toxic or inedible
Increases in biomass of gelatinous zooplankton
Increased biomass of benthic and epiphytic algae
Changes in macrophyte species composition and biomass
Death of coral reefs and loss of coral reef communities
Decreases in water transparency
Taste, odour and water treatment problems
Oxygen depletion
Increased deaths of fish
Loss of desirable fish species
Reductions in harvestable fish and shellfish
Decreases in aesthetic value

health problems, especially for infants. Nitrates are converted into nitrites by intestinal microbes. Nitrite in the bloodstream converts haemoglobin into methaemoglobin, which does not carry the levels of oxygen needed by tissues in the body. This can lead to anaemia, brain damage or death.

Increased nitrogen deposition into streams, lakes and rivers can also lead to **eutrophication**, blooms of nuisance and toxic algae, and reduced productivity of fisheries (Table 12.2). Additionally, nitric acid deposition can acidify lakes and streams, killing pH-sensitive plants and animals, thereby reducing productivity and biodiversity. Spring runoff of snowmelt containing high levels of nitric acid can lead to an acid pulse exceeding the pH tolerance of many species and especially affecting young fish.

12.9.4 The phosphorus cycle

Unlike carbon and nitrogen, the phosphorus cycle is a sedimentary cycle. New phosphorus becomes available through the weathering of rock or the uplift and exposure of marine sediments (Fig. 12.13). Phosphorus also has more limited mobility in soil solutions than does nitrate or ammonium. Phosphate dissolves readily in acidic or anoxic water, but with the presence of oxygen and neutral pH, it forms complexes with calcium or iron and becomes immobile. Any phosphorus that

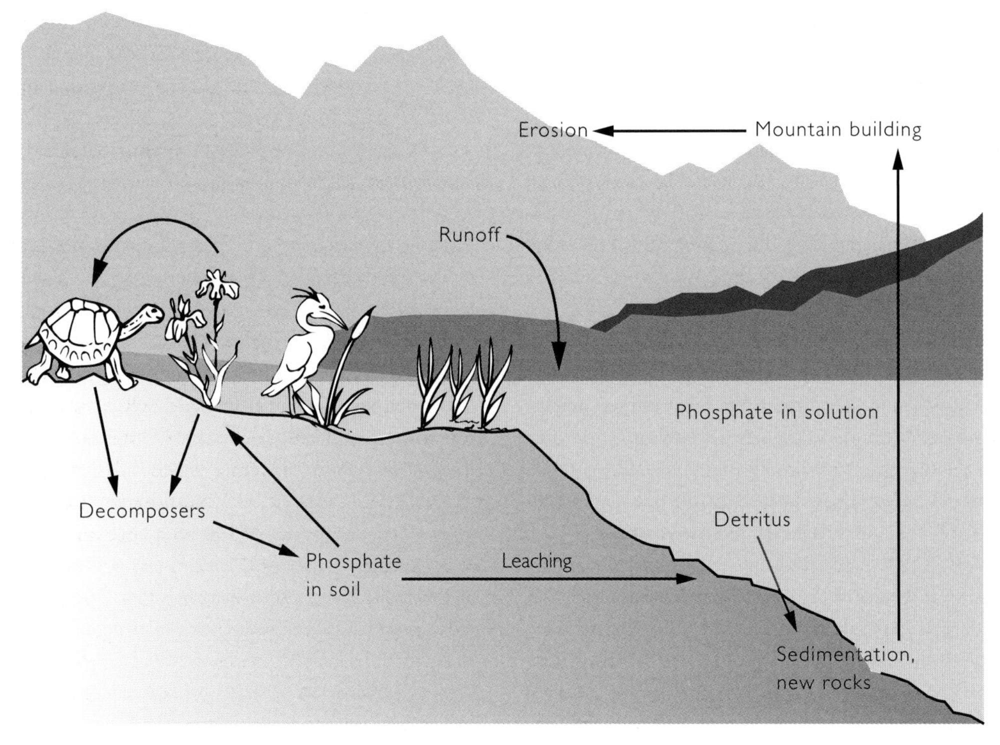

Fig. 12.13 The phosphorus cycle. Most of the cycling of phosphorus occurs locally, due to the absence of an atmospheric component.

becomes available is rapidly taken up by plants. Once incorporated into animals or plants, phosphorus is recycled through decomposition of excretory products and detritus.

12.9.4.1 *Human alteration of the phosphorus cycle*

Humans have had direct effects on the phosphorus cycle through mining phosphate-rich rock for fertilizers, and then applying this fertilizer to agricultural land. Much of the added phosphorus runs off into lakes, rivers and oceans. On a global scale, phosphate mining has more than doubled the supply of phosphorus to coastal seas compared to preindustrial levels. In some urban and agricultural areas, phosphorus availability is 10–100 times greater than preindustrial levels. Additionally, in urban areas, discharge of incompletely treated human waste into streams or rivers can be a significant source of phosphorus.

Phosphorus supply often limits the primary productivity of aquatic systems. For example, a typical freshwater lake may have concentrations of available phosphorus that are 0.0001 times that of living cells. Therefore, increased levels of phosphorus caused by human activities can lead to increased productivity of terrestrial or aquatic systems. It can also result in eutrophication of aquatic and marine systems, with many of the same effects as increased nitrogen supply (Table 12.2).

If productivity is increased, why are phosphorus addition and eutrophication cause for concern? The answer lies in the fact that productivity increases are short-lived, while ecosystem changes and effects on the biodiversity of aquatic systems are long-lasting. Eutrophication brings with it the growth of undesirable algal species, followed by massive die-offs and decomposition. Decomposition uses oxygen, depleting the water column of this necessary gas; most sensitive to oxygen decreases are desirable gamefish and other organisms high on the trophic pyramid.

Chesapeake Bay, located on the Atlantic coast of North America, is one of the largest estuarine systems in the world. However, there are dense human populations around this estuary. The current situation in Chesapeake Bay is the result of excess nutrient loading from sewage discharge, farm land runoff, deforestation, and overharvesting and disease mortality of oysters. Oysters feed by filtering large amounts of water for small food particles. In the past, high populations of

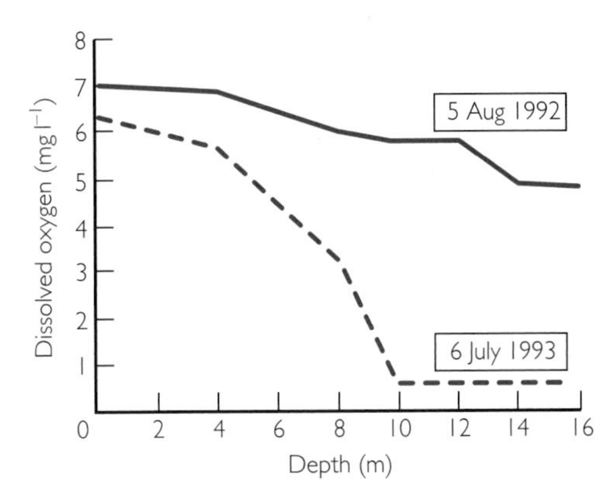

Fig. 12.14 Dissolved oxygen profiles for the Patuxent River, a tributary of Chesapeake Bay, at different times in the summer. (From Breitburg *et al.* 1997.)

oysters filtered so much water that they had a major effect on water quality and clarity. Additions of nitrogen and phosphorus have led to significant eutrophication and changes in water quality in Chesapeake Bay. In certain areas, oxygen concentrations in lower water layers often decline to stressful levels in the summer (Fig. 12.14). This is especially marked if temperature differences between the top and bottom water layers develop, thereby reducing mixing between layers. This allows microbial decomposition of a large phytoplankton biomass in the bottom layers to deplete oxygen.

Breitburg and co-workers (1997) were interested in how these low oxygen levels might affect aquatic food webs. The food web they examined involved juvenile striped bass, sea nettles (jellyfish), fish larvae and zooplankton (small animals such as copepods that eat phytoplankton) (Fig. 12.15).

Naked goby (*Gobiosoma bosc*) and bay anchovy (*Anchoa mitchilli*) are the two most abundant fish larvae in Chesapeake Bay. They are important prey of juvenile striped bass (*Morone saxatilis*), which is an important commercial and sport fishing species. Sea nettle (*Chrysaora quinquecirrha*) is a jellyfish that is an important predator in the bay. The copepod *Acartia tonsa* is an important component of zooplankton. All species overlap in the salinities and oxygen levels that they are naturally found in. However, species differ in their tolerances and optima for salinity and oxygen.

The effects of low oxygen were different for different organisms. Low oxygen levels had the effect of increasing predation on fish larvae by sea nettles, but decreasing predation on fish larvae by juvenile striped

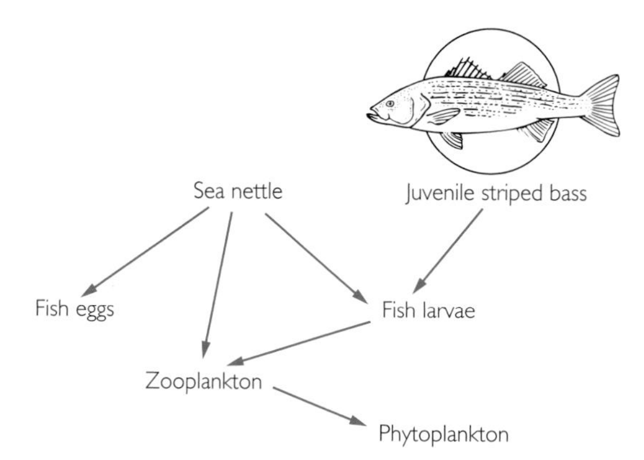

Fig. 12.15 A Chesapeake Bay food web involving the sea nettle (*Chrysaora quinquecirrha*), fish larvae—primarily bay anchovy (*Anchoa mitchilli*) and naked goby (*Gobiosoma bosc*)—and juvenile striped bass (*Morone saxatilis*). (Data from Breitburg *et al.* 1997.)

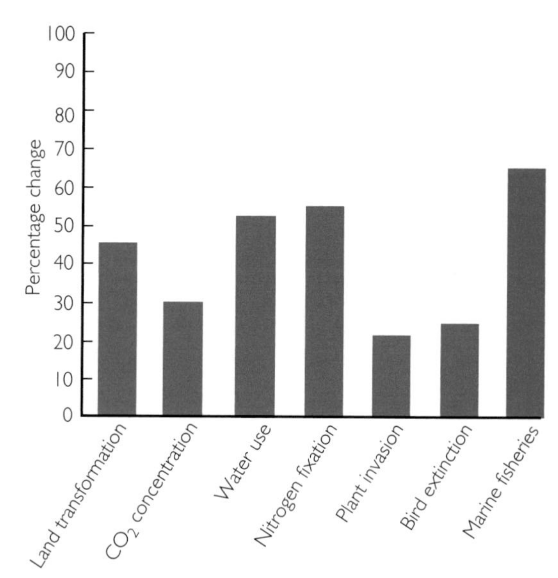

Fig. 12.16 Human-caused changes in the major components of Earth's systems, expressed as a percentage of the total resource. (Data from Vitousek *et al.* 1997b.)

bass. The increased predation by sea nettles on fish larvae was due primarily to reduced escape behaviour by the larvae. Fish larvae are more sensitive to oxygen depletion than jellyfish. At low oxygen concentrations, they swim slower and cannot escape as well. Juvenile striped bass are even more sensitive to oxygen than fish larvae, thus their ability to catch fish larvae was reduced at lower oxygen.

In contrast to the effect on fish larvae predation, the low oxygen reduced fish egg predation by sea nettles. Fish eggs float passively in water currents. However, because the nettles swam and contracted their bells more slowly, the capture rate for floating or passive objects was decreased.

All organisms in the food web migrated vertically to avoid low oxygen. Fish migrated first and most actively. However, in areas where low oxygen occurs throughout the depth profile, fish cannot avoid low oxygen. Thus, the more widespread the low oxygen levels, the greater the potential for low oxygen to cause major alterations in the food webs of Chesapeake Bay and its tributaries. Because of the differential sensitivity of organisms to oxygen, low oxygen also has the potential to affect the relative abundance of species that are important components of this estuarine ecosystem.

12.10 Human domination of Earth's ecosystems

With the explosive growth of the human population, humans are appropriating a growing percentage of total biosphere production. Current estimates are that humans use over 35–40% of terrestrial primary production for food, fibre, food for livestock and fuels. In addition, global productivity is indirectly modified through the conversion of forests and grasslands to farms, cities or grazing lands.

About 8% of Earth's aquatic primary production is used to support human fisheries. However, since fishing is concentrated in coastal areas, about 25–35% of the primary production of these areas is used by humans. In addition, humans use about 26% of global evapotranspiration to grow crops, and about 54% of available freshwater runoff for agriculture, industry and cities. As we have seen in looking at ecosystem engineers, most organisms alter their environment. However, as the human population has grown and technology has developed, the scope and nature of this modification has changed dramatically (Fig. 12.16). Most aspects of the structure and function of Earth's present-day ecosystems cannot be understood without looking at the influence of humans.

The growth of the human population, and use of Earth's resources, is maintained by agriculture, industry, international commerce, and recreational and commercial fishing and hunting (Fig. 12.17). These activities transform land, altering biogeochemical cycles and

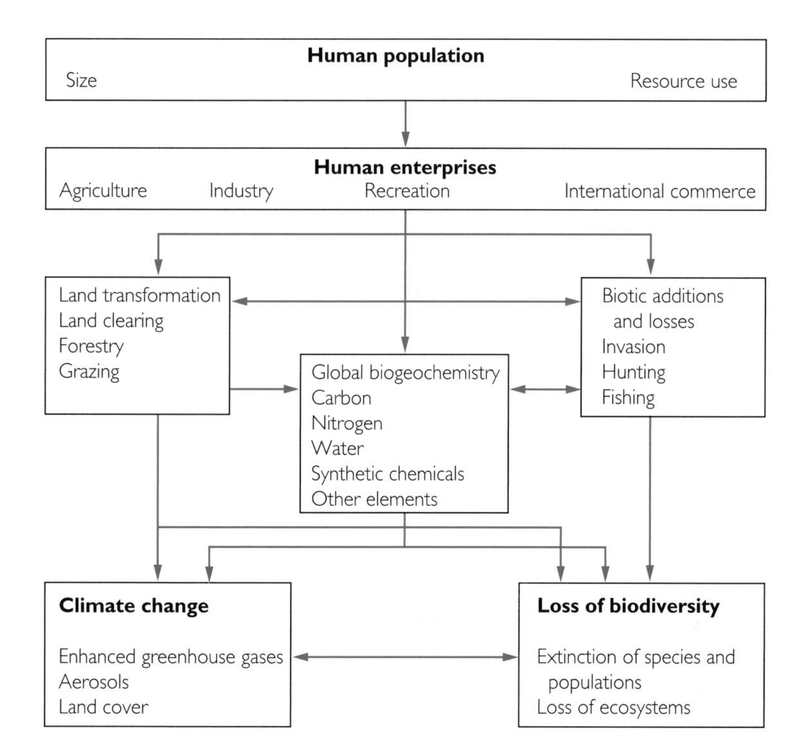

Fig. 12.17 A schematic of the direct and indirect effects of humans on Earth's ecosystems. (Reprinted with permission from Vitousek *et al.* (1997) Human domination of earth's ecosystems. *Science* **277**: 494–499. Copyright 1997, American Association for the Advancement of Science.)

changing biodiversity. Estimates of the fraction of land surface transformed or degraded by humans range from 39% to 50% (Fig. 12.16). Land transformation includes placing 10–15% of the Earth's surface into row-crop agriculture or urban areas, 6–8% into pastureland, and affecting almost all of it by hunting or other activities that involve extracting resources.

Estimates of land transformation are conservative, because much of the land that has not been transformed has been fragmented. **Fragmentation** alters species composition, nutrient fluxes and microclimate in the remaining patches. Land transformation represents one of the major driving forces in the worldwide loss of biodiversity. It can also affect climate at local and regional scales. It contributes about 20% to current human-related carbon dioxide emissions, and more to the increasing concentration of methane and nitrous oxide in the Earth's atmosphere. Finally, land transformation is associated with increased runoff of sediments and nutrients, and with resultant changes in stream, lake, estuarine and coral reef ecology.

Humans also have drastic effects on marine ecosystems. Sixty per cent of the world's population is located within 100 km of ocean coasts. Coastal wetlands have been drained and cleared for urban and agricultural uses. About 50% of the mangrove ecosystems in the world have been altered or destroyed by human activity. Humans use approximately 25% of the productivity in upwelling areas and 35% of the productivity in temperate continental shelves. In 1995, 22% of the world's fisheries were classified as overexploited or depleted, and 44% were at the limit of exploitation (Fig. 12.18).

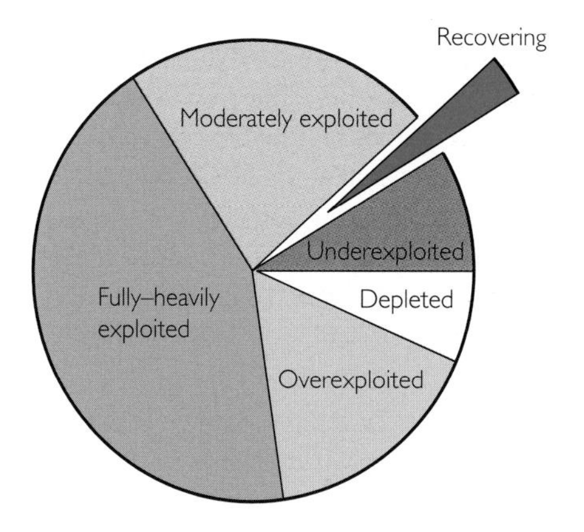

Fig. 12.18 The degree of exploitation of the world's fisheries. (Data from Botsford *et al.* 1997.)

Complicating the overall use of resources in these fisheries is the damage caused by fishing methods. Nontarget catches that are discarded average 27 million tonnes annually, about one-third of the total catch. Dredging and trawling damage seafloor habitats and may reduce their long-term productivity and biodiversity.

12.11 Human interactions with ecosystem function

Ecosystems provide many services to humans, including harvestable production, nutrient regeneration, medicines, breakdown and storage of pollutants, crop pollination, pest control and maintenance of clean water supplies. Two of the most dramatic ecological trends of the past century have been losses in biotic diversity and alterations to the structure and function of ecosystems. Ecosystem processes such as productivity, nitrogen mineralization and nitrate leaching respond directly to human activities. These processes also respond to changes in atmospheric composition and climate. Human-induced changes in biodiversity result from habitat conversion, land-use change and the introduction of exotic species.

12.11.1 Human modification of Earth's biological resources

Extinction is a natural process, but the current rate of loss of genetic variability, of populations and of species is far higher than the background rate that existed in the past. Recent calculations suggest that rates of species extinction are 100–1000 what they were before humanity's dominance. Land transformation is the single most important cause of extinction, but hunting and commercial uses have also played important roles. As many as 25% of Earth's bird species have gone extinct as a result of human activity over the last 2000 years, especially on oceanic islands. It is estimated that 11% of the remaining bird species, 18% of the mammals, 5% of the fish and 8% of the plants are currently threatened by extinction.

A disproportionate number of large mammals either have gone extinct or are threatened. These large animals are often ecosystem engineers or keystone species and have large impacts on ecosystem function and structure. Their loss has a domino effect that can affect the survival of other species.

Table 12.3 Types of effects caused by exotic, invasive plant species. (Adapted from Gordon 1998.)

Effects
Ecosystem level effects
Altered geomorphological processes
Erosion rate
Sedimentation rate
Elevation
Water channels
Altered hydrological cycling
Water-holding capacity
Water-table depth
Surface-flow patterns
Altered biogeochemical cycling
Nutrient mineralization rate
Nutrient immobilization rate
Soil or water chemistry
Altered disturbance regime
Type
Frequency
Intensity
Duration
Community/population level effects
Altered stand structure
New life form
Vertical structure
Altered recruitment of natives
Allelopathy
Microclimate shift
Physical barrier
Altered resource competition
Light absorption
Water uptake
Occupying growing sites
Nutrient uptake

The high rates of species loss do not reflect the fact that even greater losses have occurred in genetic diversity. As ranges decrease and locally adapted populations are lost, genetic variability is reduced in the species as a whole. This reduces the evolutionary potential of the species and its resilience in response to future changes.

Humans are homogenizing the world's biota by transporting species to new habitats at an unprecedented rate. Invading, non-native species are a problem throughout the world, and they have dramatic effects on native ecosystems (Table 12.3). In an analysis of the effects of introduced plant species on Florida ecosystems, Gordon (1998) found that 6 of 31 invasive plant species altered geomorphology, such as increasing rates of sedimentation and erosion (Fig. 12.19). Six species altered the hydrological cycle through processes such as

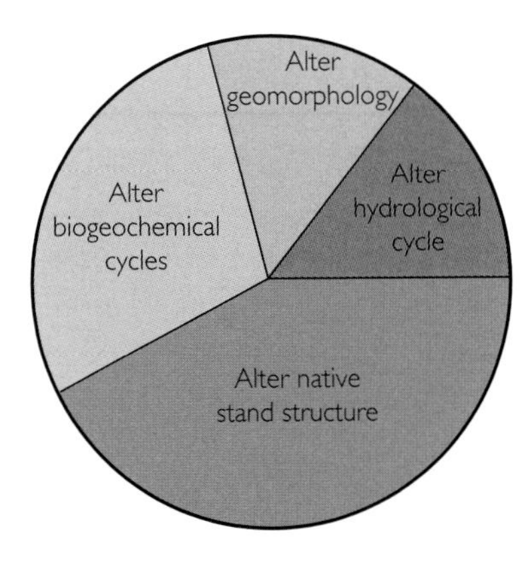

Fig. 12.19 Percentage of introduced species causing ecosystem perturbation of various types in Florida. Data are derived from a study of 31 introduced species. (Data from Gordon 1998.)

altering water-table depth, or surface-flow patterns. Ten to twelve species altered biogeochemical cycles by being nitrogen fixers, altering water chemistry or changing litter accumulation. Up to 18, or 58%, of the exotic species altered native stand structure by adding vine life forms, introducing trees to grasslands, or dominating competition for light in native stands of vegetation.

On many islands, more than 50% of the current plant species are non-native, and in many continental areas over 20% of the flora are non-native. Harbours and estuaries that incur heavy shipping traffic are also prone to introduction of exotic species. Studies in San Francisco Bay have shown that an average of one new species has been established every 36 weeks since 1850. This rate is increasing steadily, with one new species becoming established every 24 weeks since 1970, and one new species established every 12 weeks over the last 10 years. Many of these invasions are irreversible and can cause drastic changes in ecosystem function (Box 12.2). As a case in point, the introduction of exotic grasses into the western United States has altered the fire regime, reduced grazing quality and threatened many native species with extinction. Other introduced species can adversely affect human health and cause economic losses through loss of crops and the need to repair damage caused by exotic species.

12.11.2 Biodiversity and ecosystem function

Species differ in the rates and ways in which they use resources, in their effects on the physical environment and in interactions with other species. Thus, changes in species presence, absence or composition may alter ecosystem processes. For example, the presence of ecosystem engineers such as elephants will affect the proportion of grasses, trees and shrubs in an ecosystem. This, in turn, will affect interception of rainfall, evapotranspiration and regional temperatures and rainfall. Therefore, biotic changes can influence ecosystem processes sufficiently to alter the future state of the world's ecosystems and the services they provide to humans. Ecologists are only now beginning to establish the theoretical, empirical and experimental frameworks to understand and predict how changes in species composition affect ecosystem processes (Fig. 12.20).

There is growing recognition in the ecological community that biodiversity may have important consequences for ecosystem processes. First, the number of species in a community is a substitute measure of the probability of the presence of species with important traits, for example nitrogen-fixing plants. Second,

Box 12.2 Exotic species are a major cause of the loss in biodiversity

It is estimated that the introduction of exotic species is second only to land transformation in causing species extinctions and loss of biodiversity. As an example, the introduction of the brown treesnake (*Boiga irregularis*) into Guam (the largest island of Micronesia) around 1950 led directly or indirectly to the extinction of 12 out of 17 native bird species and 5 out of 12 native reptile species. In addition, the Mariana fruit bat (*Pteropus mariannus* ssp. *mariannus*) has been reduced to one surviving colony on the island. This snake was able to eliminate so many different species because: (i) it was a superior climber, able to access nests and roosting sites; (ii) it was nocturnal, attacking many animals while they slept; and (iii) it was a generalist. As a generalist, it was able to eat many different prey, allowing it to maintain its own population when a particular prey species became rare.

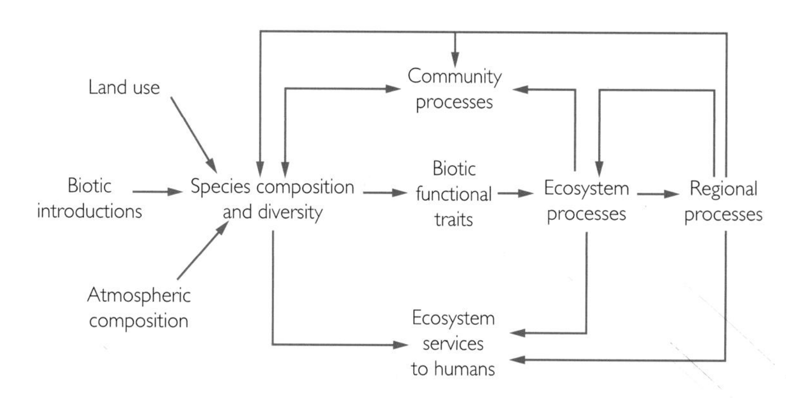

Fig. 12.20 The linkages between species composition, diversity and ecosystem processes such as productivity and nutrient cycling. Regional processes include trace gas fluxes to the atmosphere and nutrient fluxes between terrestrial and aquatic systems. Community processes include competition, mutualism, pathogenicity and predation. Ecosystem services are benefits to humans from ecological processes. (Reprinted with permission from Chapin *et al.* (1997) Biotic control over the functioning of ecosystems. *Science* **277**: 500–504. Copyright 1997, American Association for the Advancement of Science.)

greater diversity allows a greater range of traits to be represented in the ecosystem, increasing the chance for more efficient resource use in variable environments. Third, higher species diversity may increase ecosystem stability by the following:

1 By leading to greater diversity in trophic interactions, alternative pathways for energy flow are provided and energy flow between trophic levels is stabilized.

2 By reducing the susceptibility of the ecosystem to invasion by species with novel traits.

3 By slowing the spread of pathogens through increasing the average distance between susceptible hosts.

There are different hypotheses about the mechanism whereby biodiversity enhances ecosystem function. One hypothesis states that biodiversity is analogous to rivets on an airplane. Several rivets may be lost without detrimental effects. However, eventually the loss of one rivet will lead to catastrophic consequences.

The redundancy hypothesis states that some ecosystem processes can be carried out by a relatively small number of species. Other species are redundant or less necessary for maintaining function. Thus, ecosystems may lose a certain amount of biodiversity without compromising their ability to carry out their functions.

Finally, there is the insurance hypothesis. In this scenario, species replace each other in functions when stress or environmental change occur, so species diversity is especially important when conditions change. Thus, the loss of species may have unforeseen consequences when extreme events such as drought or storms occur.

These disparate views may be reconciled by grouping organisms into **functional groups**, and examining the effects of functional group diversity on ecosystem structure and function. As described in Section 11.3, a functional group is similar to a guild—it is a group of species that serve similar functional roles in an ecosystem. For example, there is the functional group of primary producers, or plants, that form the base of the trophic structure of an ecosystem. This functional group could be subdivided into functional groups of grasses or trees, each of which would access and use resources differently. Differences in the representation of different functional groups can lead to large-scale changes in ecosystems in response to the environment. For example, rising carbon dioxide concentrations can reduce plant transpiration, increasing the level of soil moisture. This may shift the competitive balance in the community from grasses to shrubs, causing replacement of grasslands by shrublands or forests. These changes can in turn feed back into climate change by altering regional temperature and precipitation. An ecosystem may be especially susceptible to the loss of a species if it is the sole member of a functional group.

Species differences within the same functional group can provide stability or increase resilience and resistance to change. This is because a decrease in one species may be compensated for by an increase in another species. In the grasslands of Minnesota, David Tilman found that plots with higher plant diversity maintained higher productivity during drought than plots with lower diversity (Tilman *et al.* 1996). He hypothesized that this was due to the presence of drought-tolerant species in higher diversity plots, which allowed productivity to be maintained.

In Western Australia several species of *Acacia* (a small tree in the pea family that forms mutualisms with bacteria in its roots that fix atmospheric nitrogen) differ in their temperature threshold for germination. Different species will germinate after fires of different intensity. These differences in germination ensure the replacement of soil nitrogen after fire across a broad range of burn conditions.

As timescales increase, an ecosystem will experience a greater range of conditions. This increases the importance of diversity among functionally similar species. Thus, genetic, population and species diversity are important to the long-term maintenance of ecosystem structure and processes.

Despite theoretical expectations, many studies have failed to show a clear relationship between ecosystem function and species diversity. This may be due to the fact that other factors beside the number of species are also important. Those species that modify the availability, capture and use of soil resources often have a dominant influence on ecosystem processes. In other words, the presence or absence of ecosystem engineers can make a large difference in ecosystem function, irrespective of the total number of species present. For example, the introduction of mycorrhizal fungi to mine tailings will greatly enhance phosphorus uptake and plant growth. The introduction of the exotic tree *Tamarix* (salt cedar) to the deserts of the southwestern United States alters water runoff patterns, salinization of reservoirs and recharge of aquifers. This is because the deep-rooted *Tamarix* taps into previously inaccessible deep-water sources, lowering water tables. The introduction of *Eucalyptus* trees into Mediterranean ecosystems has had similar effects. Termites are able to change forest ecosystems to grasslands by bringing clay particles from the subsoil to the surface. This increases water and nutrient retention, allowing shallower rooted grasses to outcompete trees. Due to their open canopy and shallow roots, dominance of grasses changes the fire regime and can cause a general drying and warming of regional climate.

12.12 Investigations into global change and ecosystem function

12.12.1 Tundra plants and climate change

Global circulation models all indicate that global warming in response to increased greenhouse gases will occur first and with greatest intensity at high latitudes. The long-term consequences for tundra ecosystems are projected to include melting of the permafrost, a deepening of the active soil layers, increased release of carbon dioxide and methane from buried frozen organic deposits, reorganization of tundra ecosystems by changing the competitive relationships of existing species, and invasion of the low arctic tundra by subarctic species. The International Tundra Experiment (ITEX) was established in the 1990s to monitor phenology, growth and reproduction of major vascular plant species in tundra ecosystems in response to climate variations and environmental manipulations.

Many different experiments were established, including manipulations of temperature and snow depth. The initial results of these experiments showed that virtually all species responded to temperature increases in some way. However, the responses were highly species-specific, and no general patterns in type or magnitude of response were observed. Some experiments showed that early snowmelt increased the carbon/nitrogen ratios (C : N) in plant tissues. This was an important result, since sustained growth and reproductive responses to global warming will depend on nutrient supply. Increased C : N ratios in litter slow down decomposition and could slow nutrient cycling and plant growth.

12.12.2 Response of stream invertebrates to global warming

Global warming will increase the temperature of running water systems such as streams. Increased temperatures could alter respiratory rates, growth, fecundity, adult size and emergence of stream organisms. A large-scale experiment on an intact system designed to provide information on the potential effects of temperature changes was conducted on a small, first-order stream near Toronto, Ontario (Hogg & Williams 1996). The manipulated stream was 1 m wide by 60 m long by 3.5 cm deep. The stream was divided into two by a metal barrier. Temperatures were intentionally increased by 2°C in spring and summer, and by 3.5°C in the winter in one branch of the stream.

The temperature changes did not cause immediate alterations in species richness or community biomass. However, there were strong effects on some individual species and considerable variation in the effects on different species. There were decreased total animal densities, particularly in fly larvae in the order Diptera. There were also earlier onset of adult insect emergence, increased growth rates, precocious breeding, reduced size at maturity, and altered sex ratios in some species. The scientists conducting the experiment concluded that changes in life history parameters are likely to be more sensitive indicators of shifts in environmental temperature than are changes in species composition, richness, biomass or density. Dispersal of individuals with differing thermal tolerances may be an important mechanism of response to changes in thermal regime. In many ways, these results are similar to manipulations of carbon dioxide and nutrients on plants, where responses are species-specific and often unpredictable.

12.13 Responses to human effects on the biosphere

The recognition of the many and varied effects of humans on the world's ecosystems suggests three possible responses. First, we can work to reduce the rate at which we alter the Earth's systems. This may involve reducing human population growth and increasing the efficiency of resource use. Second, we can accelerate our efforts to understand Earth's ecosystems and how they interact with human-caused global changes. By increasing our understanding of world systems, we may be able to mitigate or manage future changes. Finally, we can increase human involvement in the maintenance of as much natural biodiversity and ecosystem function as possible. Because of the pervasive human influence already present and that which is certain to occur in the future, it is already too late to take a hands-off approach. No matter what combination of these approaches are used, ecological theory and knowledge should drive much of this future involvement.

12.14 Chapter summary

An ecosystem includes all of the organisms as well as all of the abiotic components of a defined area. Ecosystem ecologists examine the flow of energy and matter between organisms and the environment. An early framework for these studies is called the trophic–dynamic concept of ecosystem structure. It involves dividing organisms into different trophic levels, and examining the exchange of energy and matter between these levels. Because of inefficient transfer of energy from one trophic level to another, there is a rapid decline in total energy and biomass with increasing trophic levels. These relationships are often portrayed in trophic pyramids. Decomposer trophic levels are important in the recycling of nutrients within ecosystems. Food webs, detailing the feeding relationships among organisms, have been valuable tools in the study of ecosystem structure and function.

There is evidence to support both bottom-up and top-down control of trophic levels. Aquatic ecosystems often have strong top-down control due to the high efficiency of predation and herbivory. Fewer examples of top-down control are available in terrestrial systems, but there is evidence that this occurs.

Ecosystem engineers are organisms that control resource availability for other organisms through physical changes in the environment. As such, these organisms play major roles in determining the structure of an ecosystem.

Biogeochemical cycles are essential to the flow of nutrients within and between ecosystems. These cycles include hydrological, nitrogen, carbon, phosphorus, calcium, sulphur and any of several other essential nutrients. The growing human population has altered many of these cycles, both on regional and global scales. Alterations in the nitrogen, carbon and hydrological cycles have been especially marked. These have altered global climate and ecosystem structure.

Human population growth has also resulted in a homogenization of the Earth's biota through the introduction of non-native species to new areas throughout the world. These introductions, combined with habitat loss and land transformation, have resulted in a major decrease in worldwide biodiversity. Biodiversity and ecosystem function are linked, but not always in a straightforward way. Various models have been developed to explain this linkage, including an airplane-rivet analogy, the redundancy hypothesis and the insurance hypothesis.

Recommended reading

Harrington, R., Woiwod, I. & Sparks, T. (1999) Climate change and trophic interactions. *Trends in Ecology and Evolution* **14**: 146–149.

Malhi, Y. & Grace, J. (2000) Tropical forests and atmospheric carbon dioxide. *Trends in Ecology and Evolution* **15**: 332–337.

Marcogliese, D.J. & Cone, D.K. (1997) Food webs: a plea for parasites. *Trends in Ecology and Evolution* **12**: 320–324.

Polis, G.A., Sears, A.L.W., Huxel, G.R., Strong, D.R. & Maron, J. (2000) When is a trophic cascade a trophic cascade? *Trends in Ecology and Evolution* **15**: 473–475.

Rapport, D.J., Costanza, R. & McMichael, A.J. (1998) Assessing ecosystem health. *Trends in Ecology and Evolution* **13**: 397–402.

Wilhelm, S.W. & Suttle, C.A. (1999) Viruses and nutrient cycles in the sea. *BioScience* **49**: 781–788.

Chapter 13

The structure and composition of ecological communities

13.1 Patterns in the coexistence of species

When scientists began to explore the islands of the Indian and Pacific Oceans in the eighteenth and nineteenth centuries, they noticed that although neighbouring islands were generally similar in climate and appearance, they tended to be home to different sets of species. Often the differences were only small, and the species on nearby islands, although not identical, were very similar to one another.

In the Indian Ocean, the island of Mauritius was the home of the flightless pigeon known as the dodo (*Raphus cucullatus*), but the species was absent from the nearby island of La Réunion, which had its own similar but unique species, the Réunion dodo, or solitaire (*R. solitarius*). The large-flowered parasitic plant *Rafflesia* lived on some South East Asian islands, while the closely related *Sapria* lived on others.

Modern American birdwatchers are familiar with the same phenomenon on the mainland; the eastern wood-pewee (*Contopus virens*) lives only in the area to the east of a line drawn from Texas to North Dakota, while the very similar western wood-pewee (*C. sordidulus*) is found only to the west of this line (Fig. 13.1a). There are many other such patterns in the distributions of organisms: in Australia, the red-necked wallaby (*Macropus rufogriseus*) lives in a strip along the southeast of the country, but is not found north of the Tropic of Capricorn, while the closely related agile wallaby (*M. agilis*) is found only in a similar strip in the north and east, but is not found south of the Tropic of Capricorn.

The large-scale, regional, processes that determine the overall distributions of species may not be precisely the same effects that control the composition of a local community. Although the eastern wood-pewee has a regional distribution that covers most of eastern America,

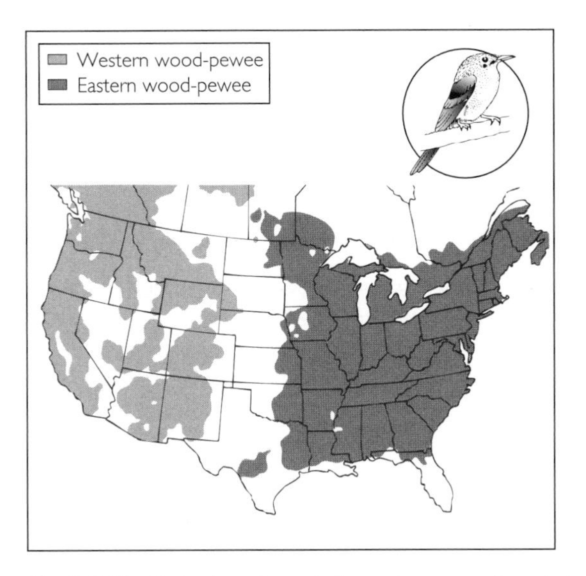

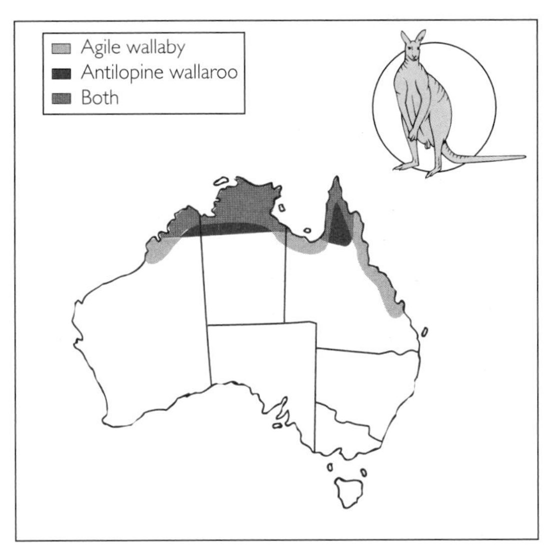

Fig. 13.1 The distributions of some similar species are mutually exclusive, while in others there is considerable overlap: (a) non-overlapping distributions of the eastern wood-pewee (*Contopus virens*) and western wood-pewee (*C. sordidulus*) in North America, and (b) the overlapping distributions of the agile wallaby (*Macropus agilis*) and antilopine wallaroo (*M. antilopinus*) in northern Australia.

it does not occur in every community, because some communities have the wrong habitat, or contain other species—competitors, predators or parasites—that exclude the wood-pewee.

Negatively associated distributions is the term sometimes given to the phenomenon of similar species having distributions that never overlap. It is one of the weaker ideas in ecology, however, mostly because patterns of distribution are rarely characterized by a strict separation of species with sharp boundaries.

Some Asian islands are home to both *Rafflesia* and *Sapria*, and there are plenty of families of North American birds or Australian mammals in which closely related species occupy identical habitats at the same sites. Nevertheless, the ranges over which these relatives live are rarely identical. The antilopine wallaroo (*Macropus antilopinus*) and agile wallaby (*M. agilis*) both live in the open forests of northernmost Australia, but the agile wallaby has a distribution that extends slightly further south into Western Australia than its relative, while the antilopine wallaroo is the only one of the two to inhabit the central part of Cape York Peninsula in Queensland (Fig. 13.1b).

Trying to explain why one species lives in one place, while a similar but slightly different species lives in another place is not a simple exercise. Distributions may be governed partly by historical accidents, which are no longer evident to a modern observer. But, as Chapter 1 described, the distribution of organisms is one of the two fundamental aspects of ecology, along with their abundance.

In some cases, explanations of individual patterns may be very simple. It is possible to postulate that the dodo evolved on Mauritius, and since it was flightless, it could never reach any other part of the world. Likewise, the seeds of *Rafflesia* plants are not easily transported between islands. Such explanations are a powerful reminder that in the search for a general understanding of the distribution and abundance of organisms, care must be taken not to ignore the specific details of individual organisms. Flightlessness is (or was) a feature of an individual dodo, not of a population or entire species.

But this sort of explanation is not universally applicable and will not provide a general understanding; it will not, for example, explain the distribution of wood-pewees. There is no obvious or convincing reason why eastern wood-pewees could not travel a few hundred miles to colonize Montana or New Mexico. At one level, there is a simple explanation for this pattern—a population of eastern wood-pewees may not be able to become established in Montana because of competition from western wood-pewees. Such an explanation does not explain why the two species evolved in their current habitats, and the relevant processes, called biogeography, will be examined in Chapter 14. Nevertheless, the concept of competitive exclusion, which was examined in Section 8.6.1, offers a single, comprehensible cause for the observed distribution of these birds.

Unfortunately, such simple explanations will not always suffice in the face of the complex systems that exist in nature. Most observed patterns are not as simple either to observe or to describe as the sharp boundary between the distributions of two virtually identical bird species, or the absence of a plant species from one island despite its presence on a neighbouring land mass. A glance through field guides with the distribution maps of various groups of organisms in a region—such as European fungi, Australian grasses or Caribbean fishes—reveals a complicated pattern of presence and absence among the various species. Moreover, the regional distributions displayed by such maps describe only the overall geographical ranges of species. The maps say nothing about the pattern of local communities that the different species actually inhabit.

Such patterns do not lend themselves to simple explanations that rely on a single cause, such as the ease with which organisms can travel between different sites, or the suggestion that two species might have mutually exclusive distributions, with one replacing the other in some geographical locations.

13.2 Ecological succession and stability

To explain why an ecological community contains a particular complement of species, it is necessary to start by realizing that communities are not fixed entities —they change as individual organisms are born or immigrate from elsewhere, or as they die or emigrate. Examining changes in communities may reveal some of the explanations for the various patterns of species coexistence that exist in nature.

If a camera were placed in any habitat anywhere in the world, and set so that it would automatically take a photograph at the same time every day, the resulting snapshots would give a great deal of information about the stability of ecological communities. We would not expect to see identical photographs as the weeks, months and years passed, because individual organisms would grow, die and move around. If the photographs were shuffled, it would probably be difficult to reconstruct the exact order in which they were taken just by looking at any differences among the various images. This would be true if communities were stable, in the sense that they tended to stay largely the same over long periods of time, but if a community was constantly undergoing significant changes, it would probably prove easier to sort the shuffled photographs back into the order in which they were taken. Without some information about the starting or finishing conditions, someone might still put them in the reverse order to the sequence in which they were originally taken.

Of course, any major physical changes would affect the community in obvious and predictable ways, and the effects of such disturbances will be investigated in Chapter 14. But even ignoring the effects of fires, volcanic eruptions and global climate change, there remains an important question: How stable are ecological communities?

It is clear that some communities are not stable, and Section 1.10.2 described one type of process by which communities change—ecological succession—which occurs in a newly created patch of habitat, such as a piece of abandoned farmland or a sandbank deposited from the sea. In the course of this succession, the first species to colonize the habitat are eventually replaced by other species, which are then replaced by another set of species, and so on. The weeds that spring up on a disused field are ultimately substituted by scrubland, which is, in time, replaced by mature woodland.

Ecological succession is a complex process. It can be differentiated into **allogenic** succession, in which changes in the physical environment drive changes in the ecological community, and **autogenic** succession, in which features of the biological environment cause successional changes in the populations of organisms in a community. Ecologists also distinguish between **primary succession**, which occurs on bare rock that has never been colonized before, and **secondary succession**, which takes place on existing soil from which an existing community has been removed by fire, by human activity or by natural disasters.

Sections 13.2.1–13.2.4 will continue to examine the process of ecological succession, before Sections 13.3–13.7 go on to investigate other aspects of stability in communities.

13.2.1 The facilitation concept of succession

Imagine that you are standing on the surface of the Earth 4 billion years ago. It does not matter whether you are standing on the piece of land that is now Beijing, or Melbourne, or Nairobi. There would be no life in any form; no animals, no plants, no fungi, no micro-organisms. If your imagination shows anything other than bare rock, you have probably accidentally included something organic. Although some of the rock may have been smashed into sand by the wind and rain, there would be no soil, and not even any real dust, because soil and dust contain large quantities of organic matter—debris from the dead cells of organisms.

When life first evolved, it consisted of tiny single-celled organisms not unlike modern bacteria. Leaving aside the difficulties of temperature and humidity, animals could not have existed because there was nothing for them to eat, and even plants would not have survived, partly because there was no soil in which their roots could grow.

Now imagine that you are standing in Melbourne, or Nairobi, or Beijing today, looking at a piece of bare rock. Of course, there is little bare rock in these cities, but there are huge areas of reconstructed rock in the form of roads, pavements, garage forecourts and car parks. Just as the bare rock of 4 billion years ago could not support plant life, so a city parking lot is not a suitable home for a grass plant or a pine tree.

Yet it is clear that abandoned towns are rapidly taken over by nature. Green plants very quickly become established on disused stretches of road or abandoned forecourts. Part of the difference between an abandoned car park and a piece of real bare rock is that the car park is built on top of what was once a field or a stretch of woodland, so there is organic material underneath the concrete. But as the expanse of concrete reverts to nature, many of the changes that come about are brought about by the organisms themselves. If the car park is not cared for, the winter frost will crack the surface of the concrete, allowing small plants to take root. If the pot holes are not filled in, they will also be home to lichens and mosses, which will eventually die, and rot away under the influence of fungi, bacteria and springtails or other tiny animals; the result will be a mixture of organic material and powdered concrete; in other words—soil.

This soil can be used by other, larger plants and when there is a large enough pocket of soil, an acorn or a chestnut or a seed from a pine cone will germinate, and its roots will hasten the cracking process. As time passes, the plant community will develop from mosses and liverworts, through grasses and herbaceous plants, to a woodland of trees and undergrowth.

This is just one description of the way that ecological succession occurs, but it is easy to see that it is a plausible hypothesis, especially when the succession process starts with a barren area of bare rock.

The hypothesis assumes that each stage in the succession prepares the habitat for the next stage. This process is known as facilitation, because the mosses, by creating some soil, facilitate the habitat for the grasses and herbs; then those grasses facilitate the habitat for oak trees. The principal feature of this hypothesis is that the living (and dead) organisms create changes in the physical environment, and these changes have some effect on other organisms.

It is extremely difficult to prove that the different successional species actually cause changes in the physical environment, but there is considerable support for the hypothesis. Mitchell *et al.* (1997) asked whether the plant species on heathland in southern England might cause changes in the soil nutrients. The original heathland can change in a number of ways, with the resultant habitats dominated by one of five different species: the birch (*Betula* spp.), the pine (*Pinus sylvestris*), the bracken fern (*Pteridium aquilinum*), the rhododendron (*Rhododendron ponticum*) or gorse (*Ulex europaeus*). Mitchell and colleagues studied in detail the chemical composition of the soil in various areas of the habitat that had reached different stages in the succession.

On those patches where the pine dominates, the soil nutrients were similar in composition to those on the original heathland. On the patches characterized by rhododendron, the concentration of sodium increased several times. Concentrations of various nitrogen-rich compounds (such as ammonium, nitrates and nitrites) increased during the succession of bracken or gorse. The greatest effect on the chemistry of the soil appeared to be caused by the birch succession, with decreased acidity, and increased phosphorus and calcium (Fig. 13.2).

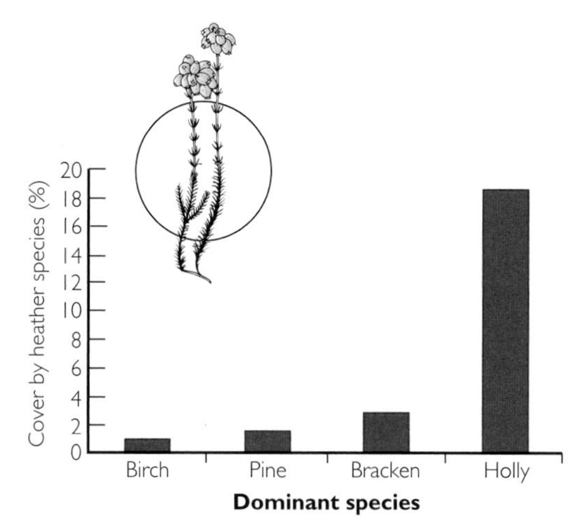

Fig. 13.2 The percentage cover of acid-tolerant heather species (*Calluna* and *Erica* spp.) in areas of heathland that have undergone succession in different ways to be dominated by birch, pine, bracken or holly. Heather species are particularly scarce in areas dominated by birch trees, where the physical environment has changed—the chemistry of the soil is less acidic and more rich in calcium. (Data from Mitchell *et al.* 1997.)

However, Mitchell and his colleagues were unable to be completely certain that various plant species had actually been the direct cause of the changes in soil nutrients. An alternative hypothesis was that some unmeasured variation in the original composition of the soil among the different areas caused the areas to be susceptible to invasion by different species. On balance, however, it seems that it was more likely that facilitation was taking place: birch trees caused soil to become less acidic, which potentially prepared it for invasion by acid-hating plants at a later stage of the succession, and made it unsuitable for acid-loving plants like the rhododendron.

13.2.2 Succession does not only happen on bare rock

Ecological succession can take place in any habitat, not only on bare rock. A ploughed field can also be viewed as an empty habitat, and if it is left alone, a succession of different species will colonize the habitat until the land reverts to a 'natural' woodland. Of course, the empty field is not really empty; the soil contains many micro-organisms, small animals and fungi, as well as the seeds and spores of plants. Section 13.2 above

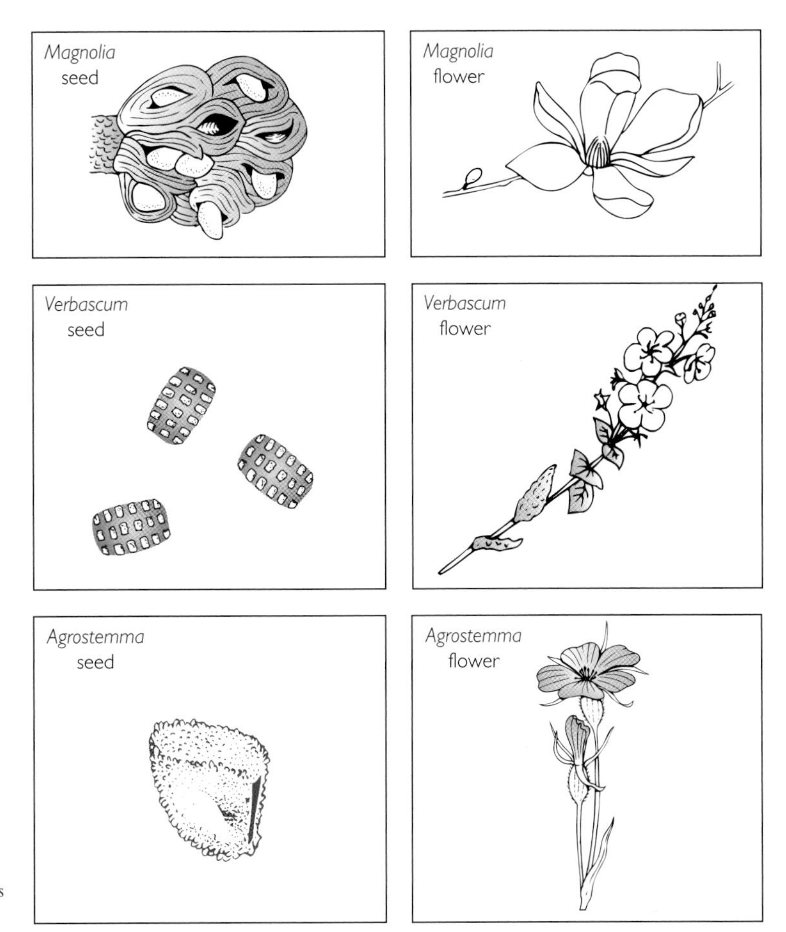

Fig. 13.3 Variation in the longevity of seeds: (a) a *Magnolia* seed germinated after remaining dormant for 2000 years, (b) the seeds of the moth mullein (*Verbascum blattaria*) can survive in the soil for about a century, and (c) seeds of the corn cockle (*Agrostemma githago*) die after less than 5 years of dormancy.

noted that succession on existing soil is called secondary succession.

Ecologists who have studied secondary succession have tended to take a different view of the process from those people who have studied primary succession on bare rock or other barren habitats, such as newly formed sand banks.

A simple but crucially important difference between bare rock and abandoned soil is that the soil contains the seeds of plants that lived on the site in the past. The seeds may not simply be those of the crops and weeds that grew on the farmland in the recent past. There may well be seeds of the plants that grew in the original, natural community. Many seeds can survive for long periods; one *Magnolia* seed germinated and grew into a healthy plant after lying in the archaeological ruins of a Japanese village for over 2000 years (Fig. 13.3). The seeds of the moth mullein (*Verbascum blattaria*) have a longevity of about 100 years. Not all seeds last for long periods—the corn cockle (*Agrostemma githago*) has seeds that die in less than 5 years—and the survivability of the various seeds in the soil strongly affects the nature of the community that develops (Bewley & Black 1994).

13.2.3 Other ways of explaining succession

Because succession can occur in such a variety of conditions, it would be foolish to imagine that facilitation is a universal process that is always an accurate description of ecological succession.

There are various other theories about how succession might progress, but the one that is perhaps most different from the concept of facilitation is that of **inhibition**. This theory is based as much on mathematical certainties as it is on ecology. Suppose that once a plant is established at a precise point in the ground, it cannot be replaced until it dies. The principal driver of succession in this situation is that the presence of one plant inhibits all other plants from occupying the site. In these circumstances, the only factors governing botanical succession will be to do with the identity of the seeds that happen to arrive in the habitat, and the order in which they arrive. Species with seeds that arrive first (or are already in the soil) will appear first in the succession, and these species may eventually be replaced by the species with seeds that arrive later.

In general, the seeds that arrive first will be those of plants that produce many small, easily dispersed seeds. These tend to be the short-lived, weedy plants that are typical of highly disturbed places, the kinds of species that were described in Chapter 7 as being '*r*-selected'. That is why the rosebay willowherb (*Chamaenerion angustifolium*), which produces many seeds with silky plumes that drift through the air, is often found in bare ground that has been disturbed or burned, and why it is also known as the fireweed. As time goes by, the seeds of longer lived, larger (more *K*-selected) plant species will arrive and establish populations. That is one reason why woodlands of oak (*Quercus* spp.), beech (*Fagus* spp.) or pine (*Pinus* spp.) were typical of large areas of the northern hemisphere before human clearance of extensive areas.

Plants of one kind will give way to those of another kind, with the appearance of a reasonably ordered succession.

Something like the process of inhibition appears to be taking place in the succession of plants on newly exposed substrate at the forefront of the Lyman Glacier in Washington State in the United States (Fig. 13.4). Jumpponen *et al.* (1998) noticed that two species of willow (*Salix commutata* and *S. phylicifolia*) tended to become established reasonably early in the succession, and the botanists performed observations to ask whether the willows were responsible for inhibiting the establishment of populations of other species. Although the willow shrubs did not appear to inhibit the establishment of all the other invaders, they undoubtedly inhibited the germination of some species, and some other plant species were generally found only in areas at some distance from the established willows. The rush *Juncus mertensianus* and two willowherbs (*Epilobium alpinum* and *E. latifolium*) were both inhibited by the willows, but another rush (*J. drummondi*) and a speedwell (*Veronica wormskjoldii*) were not.

13.2.4 How does succession actually occur?

Given that succession appears to be a key process in the understanding of the composition of ecological communities, it may seem that it is important for ecologists to understand exactly how it occurs. But Sections 13.2.1 and 13.2.3 have illustrated that there is evidence for two quite different procedures by which it might occur. Facilitation and inhibition are completely separate explanations for the way in which succession occurs, and there are others that this chapter has not investigated.

In fact, the truth is that different processes apply in different circumstances, and that more than one process may be occurring in the same community either at the same time, or at different stages during the successional sequence. At the Lyman Glacier, the two species of willow inhibited only some of the other potential invaders, and some other process must also have been involved in the succession.

Houle (1997) studied the succession of the sandwort (*Honckenya peploides*), the lyme grass (*Elymus mollis*) and the sea pea (*Lathyrus japonicus*) on sand dunes in Quebec in Canada, and found that different processes were occurring at different stages of the succession. The early stages were under the control of strong abiotic conditions, but as those conditions changed, partly due to facilitation, interspecific interactions became more important, and inhibition began to play a role.

The most important lesson to learn from the somewhat confusing evidence about succession is that the composition of ecological communities is determined by a complex mixture of all the physical and biological interactions that were investigated in Chapters 4–10. Facilitation is partly to do with the physical, abiotic conditions in which the physiology of particular organisms is able to operate; that is why Olff *et al.* (1997) found the salt-tolerant sea spurrey (*Spergularia maritima*) early in the saltmarsh succession on the Dutch Frisian Islands, described in Section 1.10.2. But inhibition has more to do with biological interactions, namely interspecific competition. The example of the willow inhibiting germination of other species in the Lyman Glacier succession is remarkably similar to the example described in Section 8.2.1, in which the annual worm-

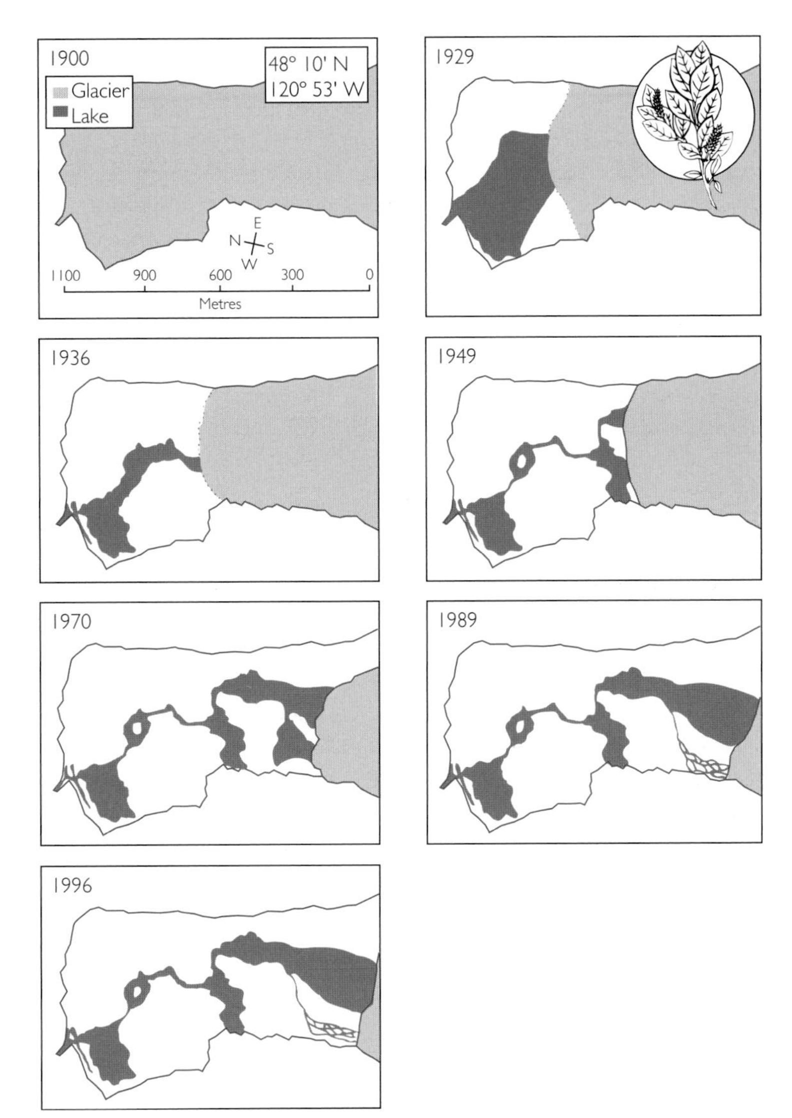

Fig. 13.4 Retreat of the Lyman Glacier between 1900 and 1996. As new substrate was exposed, ecological succession took place and a complex plant community was built up. At least part of the succession was governed by inhibition. The presence of the willow *Salix commutata* inhibited the germination of some other species of plants. (Adapted from Jumpponen *et al.* 1998.)

wood (*Artemisia annua*) was observed to prevent the germination of other plant species.

13.3 Stability in climax communities

It is clear that succession is a process by which a community is continuously changing, whatever the exact mechanism that governs the changes. However, it also appears that, if left undisturbed for long enough, a community will reach a climax, after which its basic structure changes relatively little, unless it is disturbed in some way. That is why the imaginary automatic camera, described in Section 13.2, left in the same place, will continue to take photographs that are essentially indistinguishable from one another.

The nature of climax communities is much less clear than the progression of stages through which an ecological succession passes. A climax community is not a completely fixed entity, because individual organisms will grow, reproduce and die. However similar the basic structure remains, the collection of individual organisms will be in constant flux.

One set of fundamental questions that can be asked about communities is concerned with the stability of the climax community. The precise questions that ecologists ask are not always the same and, unfortunately, it is sometimes difficult to know the extent to which different studies can be compared.

The main problem is that stability is difficult to define. As a single concept, it can measure a number of different things. One of the things it measures is usually known as the '**equilibrium**' of a community; the idea of ecological equilibria is investigated more fully in Section 13.4.

The other two aspects of stability can be distinguished as **resilience** and **resistance**. Resilience is a measure of a community's capacity to return to its original state if it is perturbed, and resistance is the degree to which the community can remain unchanged in the face of such a perturbation. The difference between resilience and resistance will be investigated more fully in Section 13.7.

13.4 Equilibrium

One of the most important questions that ecologists ask about communities is whether or not they exist in some sort of equilibrium. The precise meaning of this question is not always clear, but in essence it asks whether ecological communities contain elements of randomness, or whether there are factors that strictly determine the populations that are present, and at what abundance. For example, does ecological succession lead to an inevitable climax community, or do chance events play an important role in structuring the final result? Is the concept of a climax community a good description of reality, or does succession never come to a conclusion? Do long-standing communities contain empty niches or will the processes of invasion and succession eventually fill all the niches that are potentially available?

There is an endless list of questions that can be asked about the equilibrium of ecological communities. But before beginning to ask such questions, ecologists must ensure that they do not confuse the scientific concept of a stable equilibrium with the arbitrary concept of a 'balance of nature'. The 'balance of nature' is an idea that tends to be used by people who oppose human-induced changes to the environment, and it is an unhelpful, idealized view of the world that we might wish to exist, or of some mythical Paradise or Garden of Eden. Environmental protesters are often quite right to point out that one ecological change may lead to other, unforeseeable, alterations in a particular habitat, and they are certainly right to be concerned with the sustainability of any changes. But in studying the science of ecology, it is essential to remember that there is no one 'correct' community that should inhabit a given location.

13.4.1 Empty niches

One of the ways that scientists approach the issue of equilibria within communities is to ask whether existing natural communities contain empty niches. In other words, are there sets of resources that remain unused, but which could sustain populations of suitable organisms if those organisms were to invade or evolve.

One relatively simple solution to the problem of empty niches is to observe what happens when alien species are introduced into habitats from which they would otherwise be absent.

All over the world there are many introduced species, some of which were deliberately introduced by people, and others that were moved accidentally. To know whether communities tend to contain empty niches, two factors must be examined. First, do introduced species tend to succeed in their new habitats? Second, if a population of an alien species is successfully established, does it tend to cause extinctions among the existing species? If an alien species can be successfully introduced into an existing community without causing extinctions among the resident populations, then the original community must have contained an empty niche.

In fact, various species have been introduced successfully in all kinds of habitats in many countries around the world. Many of the most obvious are European bird species that were taken around the world by eighteenth and nineteenth century colonists in North America, Australia, India, the Pacific Islands and Africa. The house sparrow (*Passer domesticus*), for example, was such a familiar and popular bird that individuals were transported from their native range in Europe and India to Australia in 1863, to Argentina in 1872, to Brazil in 1906, and at other times to Zanzibar, Jamaica and Brooklyn in New York, from where the species has spread to much of North America (Fig. 13.5).

But although town-dwelling songbirds are among the most noticeable successful introductions, there are many insects, plants and micro-organisms that are equally successful. Among the most successful are the

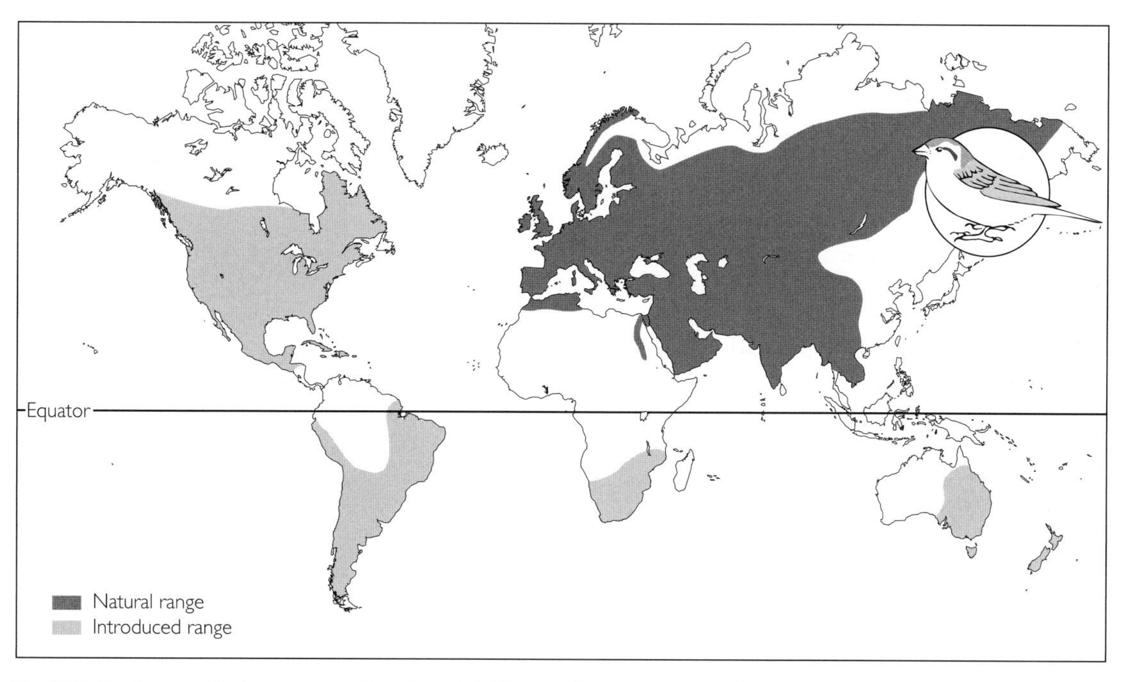

Fig. 13.5 Distribution of the house sparrow (*Passer domesticus*). The natural range covers most of Europe and large parts of Asia, but the species has also been deliberately introduced to southern Africa, Australia and North and South America.

bacteria and viruses that cause disease in humans. Syphilis, which is caused by a bacterium called *Treponema pallidum*, was unheard of in Europeans until it was brought to their continent by members of the crew of Christopher Columbus's American expedition in the fifteenth century. Similarly, brown rats (*Rattus norvegicus*) and their fleas (*Xenopsylla cheopis*) were spread by hitching rides on ships during the Middle Ages, thereby transmitting the bubonic plague bacterium *Yersinia pestis*. Every year, new strains of the influenza virus evolve and are spread around the globe by international travellers.

Not all introductions of alien species succeed. In New Zealand, for example, most introductions of bird species have failed, even those that have been given appropriate preparation (Veltman *et al.* 1996, see Fig 3.4). However, there have been many successful introductions of alien species of birds, plants, insects, fungi, mammals and other organisms, all over the world.

13.4.1.1 Do successful introductions cause extinctions?

If communities do not contain empty niches, a species that is successfully introduced can only do so by replacing an existing species and taking over its niche. Alternatively, the alien species could upset the nature of other interactions and cause extinctions in ways other than filling an existing niche.

There is no doubt that many populations of introduced species do indeed cause the extinction of other populations. A dramatic example is seen in feral populations of the American mink (*Mustela vison*) that originally escaped from fur farms in England. The presence or absence of a population of mink is now the main factor in determining the distribution of the water vole (*Arvicola terrestris*). Voles are absent from sites that previously harboured healthy populations (Barreto *et al.* 1998). Similar situations were described in Section 9.2, in which the Pacific rat (*Rattus exulans*) is in the process of exterminating the tuatara (*Sphenodon punctatus*) on some islands in New Zealand, and the introduced Nile perch (*Lates nilotica*) is responsible for the extinction of a number of fish species in Lake Victoria.

These examples both involve predation by the alien species. Introduced species may also cause extinctions by outcompeting native species. It is widely believed that the American grey squirrel (*Sciurus carolinensis*), which was introduced into England in the nineteenth century, is partly responsible for the disappearance of the native red squirrel (*S. vulgaris*) over most of its range (Corbet & Harris 1991). The mechanism by which this might have occurred is unknown, but either interference competition, or exploitation competition, or both, almost certainly played a strong part.

Tilman (1997) provided evidence of empty niches in the grassland community of the Cedar Creek area of Idaho, by causing the experimental invasion of plant species that did not occur there naturally. He placed the seeds of 57 species of plants onto small grassland plots, including seeds from species that were not formerly present. Other plots were left as controls, with no added seeds. The community of plants on plots that had received the seeds became significantly more diverse than those on control plots, and the effect lasted for several seasons. Tilman interpreted these results as a demonstration that for some of the species, their niches had previously been vacant, and that their absence from the original community was due merely to their inability, for whatever reasons, to disperse into the area.

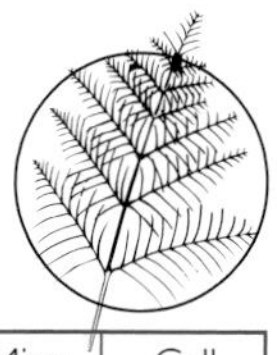

(a) **England**

	Chew	Suck	Mine	Gall
Main stem				
Leaflets				
Leaf stems and veins				

(b) **Papua New Guinea**

	Chew	Suck	Mine	Gall
Main stem				
Leaflets				
Leaf stems and veins				

Fig. 13.6 Feeding sites and feeding methods of insects eating bracken (*Pteridium aquilinum*) in: (a) England, and (b) Papua New Guinea. Shaded squares represent niches that are occupied and open squares represent niches that are not occupied. Although the evidence from England shows that it is possible for insects to feed by sucking the main stem of bracken, no insect in New Guinea does so, suggesting that the community in New Guinea contains a genuinely 'empty' niche. (Adapted from Lawton 1984.)

13.4.1.2 Insects that eat bracken

Previous sections have made a number of inferences about empty niches by studying what happens when exotic species are introduced into habitats from which they are naturally absent. The problem with this approach is that it is never possible for ecologists to be certain that they have observed all the effects of the introduction. In addition, they cannot be certain that the changes that they see were not caused by some other process that happened at around the same time as the introduction of the alien species. For example, the habitat of the water vole underwent significant changes at the same time as the spread of feral mink, but it is difficult to disentangle the effects of the mink from those of the habitat alteration.

Lawton (1984) made a very clever set of observations that allowed him to be as certain as anyone could be that he had demonstrated the existence of empty niches. He observed that the same species of fern, known as bracken (*Pteridium aquilinum*), occurs over a wide geographical area. In fact, there are probably a number of very closely related species, but they are so similar that even experienced botanists have difficulty distinguishing one from another. Lawton studied the plant in areas as far apart as Britain, Hawaii, Papua New Guinea and New Mexico. A wide range of insects occurs on bracken, and there are four different ways for them to feed: they can chew, suck, burrow or create galls. There are also three different parts of the plant: the main stem, the leaf stems and veins, and the leaflets themselves. With four feeding methods and three parts of the plant to eat, there are 12 different potential feeding niches (Fig. 13.6). Of these, one is not occupied by any insects anywhere: no insects chew the leaf stems and veins.

It is clear that all 11 of the other potential feeding niches can be occupied, because at least one species of insect fills each niche in at least one of the four widely dispersed locations. Lawton now asked whether any of these 11 niches remained unoccupied in any of the countries. Such vacant niches do indeed exist. In Britain, no insect species chews the main stem of the bracken even though this niche is occupied by insect populations in Papua New Guinea. It is extremely unlikely that any insect species have been overlooked in Britain because it has a particularly well-studied fauna.

13.4.2 Do empty niches exist?

Overall, the evidence appears to be quite clear that many ecological communities do indeed contain empty niches. It is true that many introduced species do not become established, or that after they have become established, they cause the extinction of populations of other species. But that implies only that one particular niche was not vacant in the original community; it does not mean that there were no vacant niches in the community.

But the evidence of Lawton's bracken insects, and of those established alien species that have not caused extinctions, strongly suggests that all sorts of communities contain vacant niches of various kinds.

Insofar as an ecological equilibrium implies that communities do not contain vacant niches, observational evidence shows that many communities are not in equilibrium. Chance, or apparent randomness, plays a part in structuring them.

13.5 Communities are not random

Although the various strands of evidence point to the fact that communities are not in equilibrium, it would be wrong to assume that their structure is completely random. This is easy to see by studying the distribution of ecological biomes in Fig. 2.1.

Rainforests, deserts and prairies are not scattered randomly across the Earth's surface, and Chapters 4 and 5 described how distribution is largely determined by the climate, which is governed by certain physical processes. If an area of rainforest is cleared and then abandoned, the habitat that developed would look something like the original forest; it would not look like the Arctic tundra or the Australian desert.

Occasionally, humans bring together collections of species that are almost random, such as exotic gardens with plants from all over the world. But when these areas are neglected, many of the local species tend to take over again, and the ultimate community is very similar to the original habitat that existed before the garden was created, although some of the exotic garden species may persist.

Although ecologists cannot predict exactly the structure of the local community that would inhabit the area, they can be reasonably sure of many of its basic features because of the relatively deterministic nature of the processes that control the regional distribution of species. In other words, randomness and chance are important at a small-scale, local level of distribution, but at broad regional scales, deterministic factors are more important.

13.6 What happens when communities are perturbed?

Another way of observing that communities have neither a completely random nor a completely predictable structure is to follow what happens when communities are perturbed from their current structure. In fact, Section 13.4.1.1 has already examined what happens when exotic species are introduced into communities, such as the introduction of the American mink into the British countryside. There are, however, many other kinds of perturbation, such as the removal or extinction of a species, or the effects of a period of extreme weather or disturbance caused by humans.

Section 13.2.2 looked at the process of secondary succession, in which organisms recolonize an area such as an abandoned farm, where a natural community has been highly disturbed by human activity, and the area is then left alone. If such communities recover their original structure precisely, it is possible to argue that this structure is highly determined, that it contains only a very small element of randomness, and that the communities have reached some kind of equilibrium. If on the other hand, the communities that develop on land abandoned by humans are unlike the communities that originally inhabited the same areas, then it is possible to conclude that there is a large element of chance in their structure. It would be clear that the original community was only one of many possible sets of species that might live in the same place, depending on the circumstances.

It is possible to be sure that some areas abandoned by humans can recover to something very similar to the original community, because explorers have discovered habitats that appeared at first sight to be pristine, but on closer examination were found to contain ruined buildings and other signs of large-scale human habitation. For example, the rainforests of Central and South America contain large structures, such as pyramids, built by the Maya people over 1000 years ago. At Tikal in Guatemala are six pyramids, one of which is more than 70 m high. It is inconceivable that the number of people needed to build these structures could have existed without large-scale destruction of the rainforest, but since the sites were abandoned, the ecological community has recovered to something that is very similar to our assumptions about the original structure, and the ruins are surrounded by thick rainforest.

However, it is also clear that communities that regenerate after human disturbance are not precisely identical with the original communities that inhabited the same sites before they were perturbed. This is obvious from the fact that many species survive long after the effects of human interference have ceased or have been greatly diminished. *Rhododendron ponticum* is a shrub that is native to Spain and Portugal, but which was introduced into the gardens of castles and manor houses in Wales in the nineteenth century. Many parts of these enormous gardens are now left untended, and have largely reverted to the natural oak woodland that inhabited the Welsh hillsides for 10 000 years since the last ice age. *Rhododendron* is now a naturalized part of the local community, and clearly does not rely on human interference for its continued survival. In fact, conservationists have put a great deal of effort into trying to eliminate the species in some areas, in order to promote the regeneration of the native shrubs and bushes. *Rhododendron* is such an established part of the ecological community that these conservationists find the task extremely difficult, and their success rate can be dispiritingly low.

The ecological community in this area looks superficially as it did before nineteenth-century landlords employed teams of gardeners to create the kind of woodland park that was fashionable at the time. But it is not identical; it contains species like the *Rhododendron* that were not previously present, and moreover, the abundances of other species have changed substantially. The red kite (*Milvus migrans*) remains much rarer in the same areas of mid Wales than it was 150 years ago, despite current human intervention to protect it.

Section 14.4 will deal in more detail with the effects of perturbation and disturbance on ecological communities.

13.7 Resilience and resistance

There have been many problems with the study of the effects of perturbation on community stability, partly because of difficulties with defining the terms. To improve on a rudimentary understanding, it is essential to distinguish between resilience and resistance in ecological communities. Resistance is also called **inertia**, and is the capacity of a community to remain largely unchanged even when faced with a potentially damaging disturbance. Resilience is the capacity of a community to recover its original state having been altered by a disturbance or perturbation.

The volcanic eruption of Mount Saint Helens in the United States in 1980 provided a useful natural experiment to study resistance and resilience in plant communities (Zobel & Antos 1997). Much of the area of the Cascade Range of mountains to the north of the volcano was covered in a layer of ash and lava, which had a pronounced effect on the individual plants in the area. The plant community consists of three elements, namely: (i) larger trees and bushes, such as two hemlock species (*Tsuga heterophylla* and *T. mertensiana*) and a fir (*Abies amabilis*); (ii) herbs, including several willowherbs (*Epilobium* spp.) and a hawkweed (*Hieraceum albiflorum*); and (iii) mosses (e.g. *Ceratodon purpureus* and *Pohlia annotina*) and other bryophytes.

The area was not in the immediate blast zone of the eruption, and the amount of volcanic debris that covered the ground after the eruption varied across the Cascade Range, from a relatively thin layer of 4.5 cm to a thick carpet of 15 cm. The three different elements of the community suffered differential damage as a result of the variable thickness of the layer of ash.

For each of the three elements of the plant community, Zobel and Antos defined resistance to the perturbation as the degree to which the element maintained its abundance (in terms of the percentage of the ground that it covered) immediately following the disturbance. In other words, they used a measure of how limited were the effects on the plants caused by the initial disturbance of the eruption of Mount Saint Helens.

The ecologists also defined resilience in a variety of different ways, but the simplest was the degree to which the element of the community had recovered its abundance 10 years after the eruption. Again, abundance in the community was measured in terms of the proportion of the ground that was covered by each of the three elements (Fig. 13.7).

The results of the study showed that the abundance of herbs and mosses was reduced dramatically by the volcanic activity and, at least in some areas, the species had failed to recover even 50% of their initial

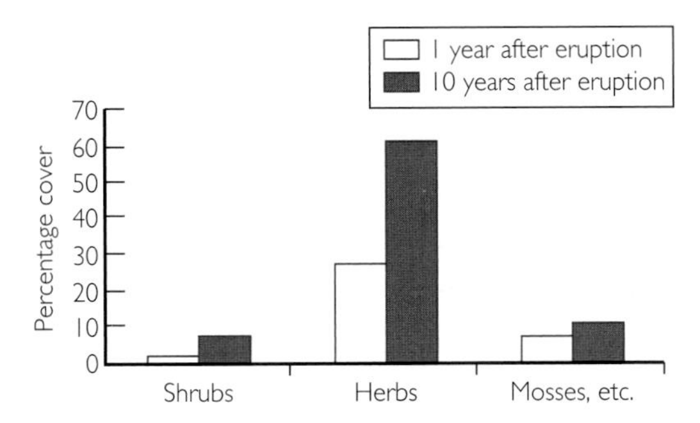

Fig. 13.7 Recovery of the plant community on a group of sites that were covered with a deep layer of ash and debris after the eruption of Mount Saint Helens in 1980. Most types of plant have increased in abundance over the course of a decade but many species have not recovered their original abundances.

importance after a decade. In other words, they were neither resistant nor especially resilient to this particular kind of perturbation. The shrubs were also affected significantly by the disturbance but recovered more quickly. In some parts of the study area, the shrubs had recovered at least half of their original cover within 3 years.

When the effects on the community were studied in more detail, using a set of plots on different kinds of soil and with different amounts of initial damage, Zobel and Antos found that, overall, there was a strong correlation between resistance and resilience. Elements of the plant community that were most strongly affected by the initial disturbance were also those that appeared unable to recover from the effects of the perturbation.

This study of resilience and resistance in a community suffering from a serious and rapid disturbance suggests that the two measures are very similar and that they both encapsulate aspects of the same feature of the community, which is often called stability. Stability is another aspect of equilibrium, which means that there are four concepts—equilibrium, stability, resilience and resistance—all of which are intended to express something about the **persistence** of a community through time.

All of these terms are defined in slightly different ways by different ecologists, with the result that many people find the subject of community stability to be very confusing. To try to understand the issue more fully, ecologists might recall some of the basic principles of ecology that were established in Chapter 1. Changes in ecological communities can only be examined by making **comparisons** of the same community at different times, or of different communities at different stages of a process such as ecological succession.

But evolution by **natural selection** is an important process for all organisms, so ecologists need to take account of the timescale over which the comparisons are made. A community that is broadly stable—or in equilibrium—over a period of a century may change dramatically over a period of a millennium.

This reveals a problem with general conclusions that are drawn from studies such as the 10-year investigation of the effects of the eruption of Mount Saint Helens. Ten years may be long enough for a population of shrubs to recover from being covered by a thick layer of hot ash, but it is not long enough for any noticeable evolution in a population where each shrub might live for 50 years.

13.8 Equilibrium over evolutionary and ecological timescales

On the one hand, it is clear that communities have empty niches and appear to be structured partly by chance events; they are not in a precise equilibrium. On the other hand, it is easy to predict the main features of the community that inhabits an area; the basic structure of a community is not governed by chance.

At one level, communities do not appear to be very persistent and are somewhat unstable, but at another, they appear to be highly persistent and stable. To resolve this apparent disparity between conflicting observations about equilibrium within communities, it is necessary to consider the different timescales over which different processes operate.

The broad-scale pattern—the fact that the distribution of biomes is fairly fixed—is built over an evolutionary timescale. Desert plants, like the welwitschia (*Welwitschia mirabilis*) or Namaqua daisy (*Dimorphotheca sinuata*), live in deserts because natural selection has governed their evolution over millions of years. A million years from now, species may have evolved that are even better fitted to the desert environment. But over a hundred generations or so, it is reasonably certain that desert communities will not change very much.

The pattern at the finer scale—which particular species happen to live in which precise locations—is governed by processes that act over shorter, ecological timescales. Some of these processes, such as competition, predation and mutualism, were dealt with in some detail in Chapters 6–10. Other processes, collectively known as biogeography, will be examined in Chapter 14, and some (but not all) of these will also be important in governing the fine-scale pattern of the structure of ecological communities. The important point about these processes is that they are concerned with events that occur within the lives of individual organisms.

Over long periods of time, the effect of many smaller chance events may be averaged out and there may be an appearance of continuity. But on shorter, ecological timescales chance events can be important. In other words, the precise composition of ecological communities is subject to a degree of randomness, but the general nature of the community may be in broad equilibrium. The habitats known as the mallee of Australia and the fynbos of South Africa are essentially the same now as they were a millennium ago.

This appreciation of how short-term ecological changes in community structure can be considered separately from longer term evolutionary processes brings a deeper understanding of complex patterns of distribution among populations. It begins to draw together the separate elements of each of many individual explanations. Section 13.1 contrasted the distribution of dodos and solitaires on islands in the Pacific and the distribution of wood-pewees in North America. The limited distribution of the dodo was determined by evolutionary processes—the species evolved on the island of Mauritius, and because individual dodos were flightless they were unable to invade any other communities. But the distributions of wood-pewees are probably determined more by the ecological process of competition—the eastern wood-pewee competitively excludes the western wood-pewee in some areas, while in other places the circumstances favour the western species.

However, there are many examples throughout this book showing that ecology and evolution cannot be divorced, and in this example, it is essential to recognize the importance of ecological processes to the dodo, and of evolution in the interaction between the wood-pewees. In fact, it was the process of evolution by natural selection that generated the differences between the wood-pewees, and it is the same process that maintains those differences. The eastern species is better adapted than the other to the physical and biological environment in Georgia because it has evolved some adaptive differences that have not been quantified. Likewise, the western species is better adapted to the conditions in Utah.

Ecological factors may also have been important in the evolutionary history of the dodo. Its flightlessness may have evolved because natural selection favoured birds that did not waste resources and energy on developing wings that were useless in an environment that lacked predators, or due to some other biological or physical feature of the environment.

Although it is convenient for ecologists to make this distinction between long-term evolutionary processes and short-term ecological processes, there is no strict boundary. As Section 1.4 examined, evolution by natural selection is what shapes ecological communities, and it is the ecology of an individual organism that determines whether or not its genes contribute to the next generation in the evolutionary process. It is the interaction between the individual organism and its biological and physical environment that determines whether or not it survives, and if it does, how many offspring it produces.

Once ecologists appreciate this tight link between short-term processes and those that take place over many generations, they can see that the two, apparently contradictory views of equilibrium in ecological communities, are in fact, simply different manifestations of the process by which natural selection shapes the make-up of the collections of species that inhabit particular areas.

The difference between evolutionary events that take place in an instant (such as the death of an individual wood-pewee) and those that operate over millions of years (such as the evolution of flightlessness in a population of the dodo) explains the difference between an emphasis on the randomness of community structure and a view that stresses the equilibrium nature of communities.

When a community develops on an abandoned field, it resembles the community that originally inhabited the same area because populations of organisms have evolved to be suited to the region's climatic and geological conditions. But the new community is not identical because of unpredictable variations in the death, immigration, growth and emigration of the individual organisms. Those apparently random variations are caused by all manner of interactions between individual organisms and their physical and biological environments.

13.9 Chapter summary

It is difficult to explain or predict the precise composition of ecological communities, but relatively easy to explain their gross structure. The very existence of large-scale patterns suggests that community structure is defined by deterministic rules, and that many communities are in, or are close to, an equilibrium, which is not the same thing as a mythical 'ecological balance'.

However, the fact that communities appear to contain empty niches, and that their precise structure changes, sometimes significantly, even within a human lifespan, points to a lack of equilibirum.

In fact, these seemingly contradictory observations can be united by the fact that the links between ecology and evolution exist over a variety of timescales. Long-term stability and equilibrium is promoted by natural selection driving evolution in **populations of organisms**. Unpredictable, shorter term variation and deviations from equilibrium are caused by the randomness of environmental effects on **individual organisms**.

Recommended reading

Baskin, Y. (1999) Yellowstone fires: a decade later. *BioScience* **49**: 93–97.

Finegan, B. (1996) Pattern and process in neotropical secondary rain forests: the first 100 years of succession. *Trends in Ecology and Evolution* **11**: 119–123.

Holway, D.A. & Suarez, A.V. (1999) Animal behavior: an essential component of invasion biology. *Trends in Ecology and Evolution* **14**: 328–330.

Nyström, M., Folke, C. & Moberg, F. (2000) Coral reef disturbance and resilience in a human-dominated environment. *Trends in Ecology and Evolution* **15**: 413–417.

Chapter 14

Species richness, abundance and diversity

14.1 Seeing the bigger picture

Throughout the first 13 chapters of this book, a wide variety of ecological concepts have been discussed, including biodiversity, population interactions, equilibrium, distribution, individuals, genes and populations. Because it is necessary for ecologists to understand each of these different ideas fully, it sometimes seems that a detailed investigation of each one is completely separate from studies of the others. In fact, one of the greatest difficulties in studying ecology is that ecologists must always remember to keep each of these fundamental concepts in their minds. Even the cleverest person cannot concentrate on too many ideas at once, and in many situations ecologists can only make progress in their studies by focusing on one small section of the subject at any given time.

This is a general feature of all scientific study: scientists must define specific, limited questions to investigate. Once investigators have the answers to narrowly defined questions, they can try to put them together into a larger, integrated understanding of the complete subject.

Thus, by studying the dynamics and evolutionary change of individual populations, and the interactions between them, and by studying the carbon cycle, the phosphorus cycle and other aspects of the flow of energy and matter, ecologists can begin to build up a bigger picture of how communities are constructed and why each community has the characteristics that it does. Why, for example, do some communities contain many species while others contain very few species? Why are some communities, like the pine forests of northern Europe, dominated by a few common species, while the species in others, such as coral reefs, have more even distributions of abundance?

This understanding of community ecology allows us to understand and appreciate how communities change through time. This is particularly important in the modern context because human activity forces the pace of change in ecological communities. We use vastly more resources than any other species, and consequently have a greater effect on the ecology of the planet. Section 4.7 explained how human activity is rapidly increasing the amount of carbon dioxide in the atmosphere, with potentially major effects on the structure of ecological communities and ecosystems.

So, in this chapter, it is necessary to take a step back from the intricacies of competition, ecological niches, and so on. This approach helps ecologists to see the bigger picture of how all these factors fit together to create the living ecological communities that we can see around us. We might see how those factors combined to change the world of the dinosaurs in such a way that we ended up with the world we now inhabit, 65 million years later, or how they changed the world of the last ice age (12 000 years ago) into the modern world. The network of processes that brought about these changes was complex, and once it is understood, ecologists are in a better position to predict and understand the possible developments that could happen in the future.

This understanding of the place of humans in the ecological world requires a return to the concepts that were introduced in Chapter 1. It is necessary to understand long timescales of millions of years and to think about changes that might happen in the shorter timescale of the next 100 years. It is also essential to understand how populations evolve to adapt to changing environments, so that predictions can be made about how crops and forests may adapt to a world in which global warming may be constantly changing the environment. It is also necessary to appreciate the interactions between individual behaviour and the status of whole populations.

14.2 The concept of species richness

Chapters 4–10 examined the ways in which the physical and biological environments are important to populations of organisms, and dealt with populations of one or two species at a time. Several times, we noted that more realistic situations would involve more species. Chapters 11–13 started to investigate how to integrate a knowledge of the distribution and abundance of individual populations into an understanding of more complex communities, ecosystems and landscapes. It was accepted that ecologists could group together many different interactions between individual organisms and their physical and biological environments. These sets of interactions are called communities and ecosystems.

So far, however, no attempt has been made to interpret an understanding and set of observations in such a way that it is possible to know **how many** different populations are likely to occur in any community or ecosystem. Many people are impressed by the enormous diversity of life forms that live in tropical rainforests, and are sometimes mesmerized by the beautiful simplicity of ecological systems like the Arctic tundra, where a handful of species appear to interact in a limited number of ways.

The truth is that ecologists have not yet developed a convincing theory that thoroughly explains the variation in species diversity among different habitats around the world. To do so would be a major leap forward in our understanding of the natural world. It would be at the heart of ecological awareness, for the simple reason that it would unite the various strands of thinking inherent in ecology. To know why a given number of species exists in a given place requires a complete comprehension of the distribution of all the populations involved, which depends in turn on an appreciation of the ecology of the individual organisms in the population.

14.3 A simple pattern, with many theories and no real explanation

One of the patterns that dawned on early ecologists as they travelled the world is obvious to everyone, now that television can show us the details of ecological communities from all over the Earth's surface.

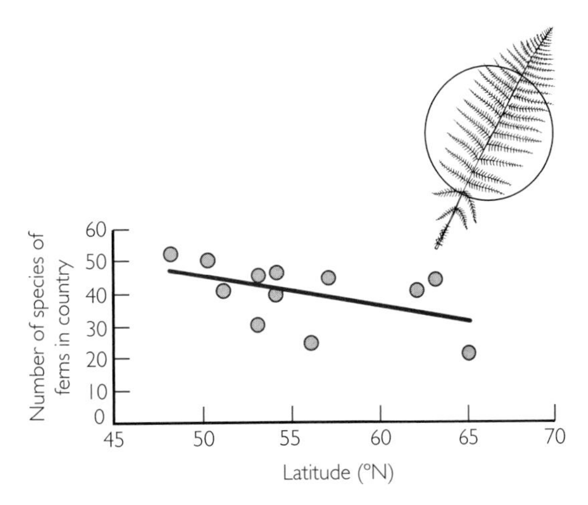

Fig. 14.1 The number of species of fern living in each country in northeastern Europe tends to be lowest in those countries that are furthest from the equator.

Communities in the tropics contain many species, but those communities that are further away from the equator, nearer the North or South Pole, tend to be less rich in species. There are countless examples of this pattern. For example, there are only two species of true crabs found in the seas around Svalbard in Scandinavia, but 54 species are known from the English Channel; in the area around the Atlantic coast of Spain and Portugal, there are about 100 species of true crabs (Lindley 1998).

Figure 14.1 shows how the number of fern species in the most northerly European countries, such as Iceland and Norway, is lower than the number in countries slightly nearer the equator, such as France and Germany.

In fact, proving beyond all doubt that there is a real ecological pattern to explain is often difficult because of complexities in the data. How do ecologists know, for instance, that 'the seas around Svalbard' is a comparable habitat to 'the English Channel', or that European countries are comparable (Sweden is more than 10 times the size of Denmark)?

In fact, even the most careful studies generally find the same pattern. Kaufman and Willig (1998) asked whether there was good evidence of a systematic pattern that an area of tropical America was home to more mammal species than a similar area that was more distant from the equator. They considered different sampling measures and controlled for other statistical artefacts, and found that in both North America and South America, there really was a strong latitudinal gradient in mammal species richness.

Ecologists can appreciate the paucity of their knowledge of the subject by examining the theories that have been proposed for the simple observation that communities near the equator contain more species than those near the poles. In all, there are at least a dozen suggested explanations of the pattern, some of which are purely ecological, while others are evolutionary in nature. Some are mutually compatible, and others are mutually exclusive. What all the theories have in common is that none of them is supported by unequivocal evidence.

14.3.1 How ecological productivity affects species diversity

Perhaps the simplest explanation for the high species diversity in tropical communities is based on the primary productivity of the ecosystems. It is undeniably true that a square kilometre of the tropics receives more of the sun's energy in a year than does a square kilometre of the Arctic. Argentina and Japan receive about 71% of the amount of solar radiation received by a point on the equator.

To many people, this is in itself a good enough explanation for the latitudinal gradient in species diversity. It seems obvious that when more energy is available for photosynthesis, there will be more plants to use the energy, and consequently, there will be a greater number of plant species. The greater the diversity of plants, the greater the diversity of fungi and herbivorous animals that eat those plants, and the greater the diversity of fungivores and carnivores that eat those fungi and herbivores.

The truth is that such a 'simple' explanation relies on a series of assumptions. It assumes that the extra energy falling on the Earth's surface at the equator is necessarily available for plants to use in photosynthesis. It also assumes that because there is a greater mass of plant material, there must necessarily be a greater species diversity, rather than just more of the same. It is possible to see that this is not necessarily true by looking at a huge field of wheat (*Triticum aestivum*); as farmers put fertilizer on the soil to increase the productivity, they produce more wheat, not a higher diversity of plant species. They may even reduce the diversity, as the stronger wheat is better able to outcompete weeds that rely on more open patches of ground.

Despite this observation, there is strong evidence that differences in productivity, caused by the climatic determinants explored in Chapters 4 and 5, are indeed important factors governing differences in biodiversity among communities.

The easiest way of relating primary productivity to species richness is to look at individual ecosystems. Observing those that are very productive, such as forests and coral reefs, it is easy to see that they also tend to contain many different species. Habitats where the climate does not allow high primary productivity, such as the Arctic tundra or the hot deserts of Africa, Asia and Australia, generally contain fewer species.

A more general, less anecdotal, test of the suggestion that species richness is driven by primary productivity requires a more systematic set of comparisons. Currie (1991) achieved this by examining the trees and vertebrates of North America (excluding the fish). He divided the continent into 336 squares and counted the number of trees, mammals, birds, reptiles and amphibians in each square. Using data from other sources, he was able to make a good estimate of the potential evapotranspiration in each square. Evapotranspiration is the loss of water from the leaves of plants, and is a good measure of primary productivity. Currie found that of all the measures he included in his analysis, his estimate of productivity was most strongly correlated to the species diversity of his 336 squares (Fig. 14.2).

Thus, the weight of the evidence points to the fact that different communities have different numbers of species largely because their physical environments allow different levels of primary productivity.

14.3.2 Other causes of variation in species diversity

Although there is no doubt that variation in primary productivity is a major cause of variation in species diversity, it does not explain all of the patterns that are observed in nature, and ecologists cannot dismiss other potential explanations.

Of the other 12 or so suggested theories, some are based on strictly ecological processes that operate within the lifespan of individual organisms, while others are based on evolution by natural selection, and take place over millions of years.

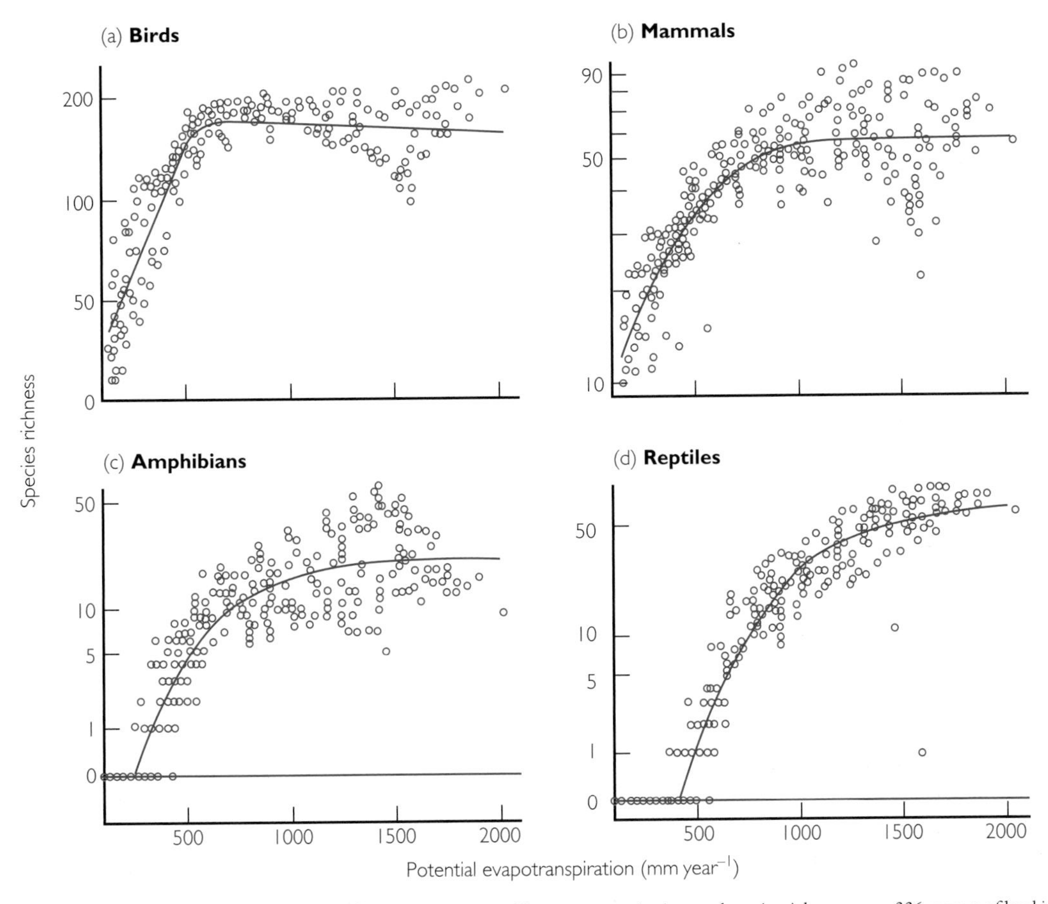

Fig. 14.2 The relationship between primary productivity, as measured by evapotranspiration, and species richness across 336 squares of land in North America for: (a) birds, (b) mammals, (c) amphibians, and (d) reptiles. Although the precise form of the relationship varies in each group, high species richness is associated with high primary productivity. (From Currie 1991.)

14.3.2.1 Features of the physical environment

Some of the ecological theories to explain why some communities contain many species (especially those near the equator) are based entirely on inferences about features of the physical environment. This is exactly the basis for examining the effects of primary productivity, described in detail in Section 14.3.1. As Chapters 4 and 5 described, variation in primary productivity is based on inescapable facts of the physical universe, such as the Earth's movements in relation to the sun.

Most of the theories about the physical environment relate to the differences in climate experienced by different ecological communities. Those that deal with simple descriptions—Bangladesh is hot and wet but Mongolia is dry and cold—are mostly variations of the theme that high primary productivity leads to a high diversity of species.

However, there is more to a climate than an average description of how hot and wet a certain place might be. Other theories are more concerned with the variability in the climate of particular sites. Someone visiting Malaysia in summer could predict that it would be hot and wet, but if someone were to visit Germany in autumn, it might be hot and sunny, or it might be warm and raining, or if winter had started early, it might be frozen and snowy, or frozen and dry.

In fact, there are two slightly different factors at work in such a comparison. The German climate is both more variable and less predictable that the Malaysian climate. German weather is more variable in that, over the course of a year, the temperature could fluctuate

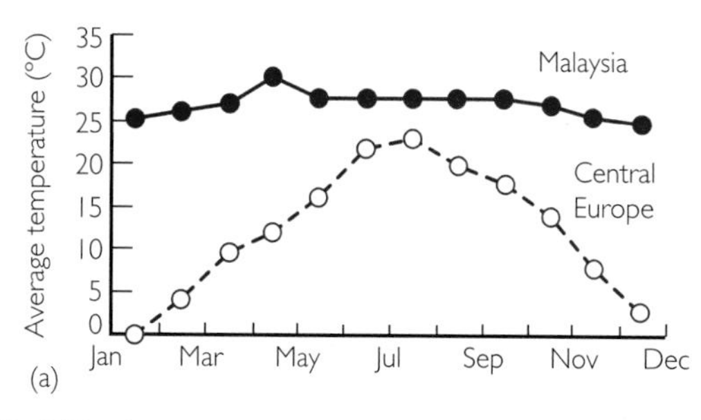

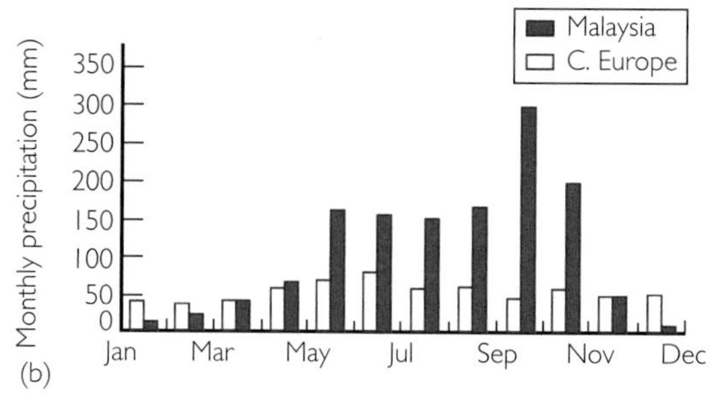

Fig. 14.3 Climatic variation in Central Europe and the Malaysian Peninsula: (a) monthly average temperature, and (b) monthly precipitation. Through the year, both temperature and precipitation are less predictable in Central Europe than in Malaysia.

between extreme cold (many degrees below freezing) and extreme heat (more than 30°C) (Fig. 14.3) (see Section 4.2). The German climate is also less predictable, in that although it can be very hot in summer and exceptionally cold in winter, there is no guarantee that either will happen. In some years, the summer is disappointingly cold and grey and the winter is not as severe as expected.

Some ecologists have argued that variable, unpredictable environments cannot support as many species as constant, predictable ones. Natural selection can lead to the evolution of many highly specialized organisms in constant climates, with no danger of the population being wiped out by a freak climatic event.

Thus, some species of hummingbirds are so highly specialized that they have bills precisely the right shape to eat nectar from flowers of a particular form, and those plant species may have evolved to depend almost entirely on the hummingbirds to pollinate their flowers (see Section 11.2.1.3). For example, in Brazil, a number of species of bright-flowered heliconias (*Heliconia* spp.) are said to have bracts that are shaped exactly like the bill of a single species of hummingbird, known as the planalto hermit (*Phaethornis pretrei*).

Another example of highly specialized tropical organisms with narrow niches appears to come from the palms of the Yasuní National Park in Ecuador. Svenning (1999) studied more than 30 species of palms, including several in the genera *Aiphanes*, *Geonoma*, *Bactris* and *Attalea*, and also included six 'palm-like' species including a bromeliad (*Achmea* spp.) and two tree ferns (*Alsophila cuspidata* and *Cyathea laesiosora*). He asked whether each of the species had very particular niche requirements, and found that the distribution of each species within the park was strongly related to the precise altitude, slope, drainage and topographical position of individual sites.

However, the evidence is not universally in favour of the theory that tropical diversity might be promoted by high degrees of specialization. By no means all hummingbirds are sufficiently specialized that they rely on small groups of plant species for their nectar. In Trinidad and Tobago, for example, the white-necked jacobin (*Florisuga mellivora*) commonly feeds on nectar from the flowers of at least six different genera of plants, including vines in the family Marcgraviaceae (such as the canopy vine *Norantea guianensis*) and shrubs and herbs in the family Leguminosae, such as the pigeon peas (*Cajanus* spp.) and the coxcombs (*Erythrina* spp.) (Fig. 14.4). Moreover, these plant species do not rely entirely on jacobins for pollination. Trinidadian *Erythrina* flowers are also pollinated by the green-throated mango (*Anthracothorax viridigula*), the blue-chinned sapphire (*Chlorestes notatus*) and the white-chested emerald (*Amazilia chionopectus*) (ffrench 1991).

Furthermore, highly specialized species, with precise resource requirements, also occur outside the tropics. In England, the large blue butterfly (*Maculinea arion*) has two different specialist requirements. During the first three stages of their development, the caterpillars eat nothing except the leaves of wild thyme (*Thymus praecox*). Once the fourth stage is reached, ants (almost always red ants of the genus *Myrmica*) carry the larvae into their nests. Here the butterfly larvae feed on ant larvae; in return the larvae secrete a sugary solution called honeydew, which provides energy for the ants. Populations of large blue butterflies cannot survive in habitats that do not have both wild thyme and red ants.

Varying degrees of specialism can also be observed within a single species that has a wide geographical

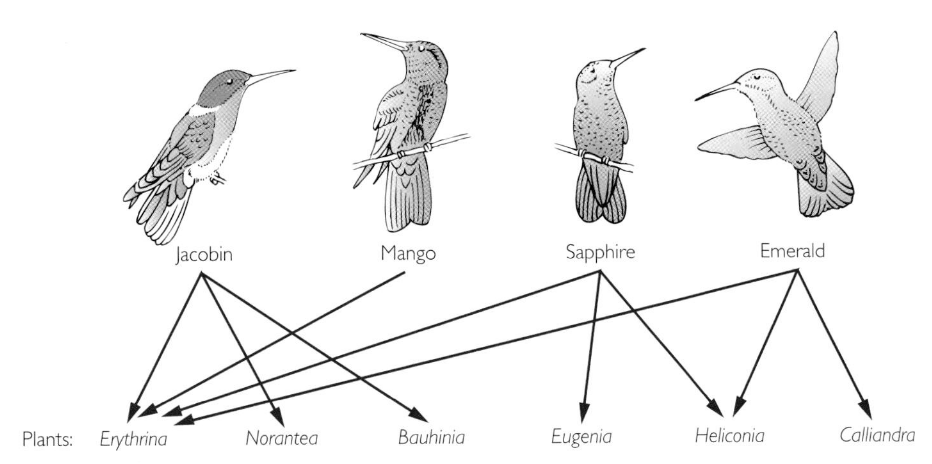

Fig. 14.4 Some feeding relationships among a group of Trinidadian hummingbirds and flowers. The hummingbirds are not specialized, but generally feed on a number of different kinds of flowers, and the flowers do not necessarily rely on a single species of bird to pollinate their flowers.

distribution, and it is not always true that tropical populations are more specialized than temperate ones. Domestic cats (*Felis catus*), including feral populations, live in many areas throughout the world, and generally eat rodents and rabbits, together with small birds and some insects. Pearre and Maass (1998) studied the variety of food types eaten by several populations of cats and found that at lower latitudes, cats tend to eat a wider range of species. In other words, cats near the equator are *less* specialized than cats living further from the tropics.

14.3.2.2 Features of the biological environment

As well as theories based on explaining how features of the physical environment can promote high levels of species diversity, there are potential explanations that rely more on understanding biological interactions such as competition, predation or parasitism (see Chapters 8 and 9).

Theories involving ecological competition One example of a theory based on ecological interactions that may explain why tropical communities are especially diverse is the possibility that competition tends to be more intense in the tropics than it is in the temperate and polar regions. As a result, the process of niche differentiation becomes important. Recall from Section 8.6.2.1 that niche differentiation occurs where a population of one species excludes a population of a second species from part of its fundamental niche. The presence of the long-eared myotis bat (*Myotis evotis*) reduces the realized niche of a second species, the southwestern myotis (*M. auriculus*), such that individuals of the second species are forced to concentrate on eating moths rather than a wider diet of insects, including beetles.

Such intensified competition could lead to many species occupying a tropical community, with each species utilizing only a very narrow set of resources. The observable effects of such a process might be much the same as the theory that climatic variability or stability was driving tropical diversity. The result could be many, specialized species in the tropics, with narrow ecological niches, as appears to be the case in the palm trees of Ecuador (see Section 14.3.2.1).

The problem with this explanation for the high diversity of species in the tropics is that it offers no independent reason for the suggestion that competition should be more intense near the equator than it is in the temperate zone.

An alternative theory is directly at odds with the idea of intense tropical competition. This explanation argues that, because conditions become harsher as organisms travel away from the equator and towards one of the poles, life becomes harsher for individual organisms, which exaggerates the experience of competition. By this line of reasoning, competition is less intense near the equator than it is in the temperate and polar zones. To explain the relatively high species diversity of the tropical zones, we must now postulate that the intense competition in the temperate and polar regions leads to competitive exclusion, a process that was described in Section 8.6.1.

In principle, it should be easy to test whether either of the theories based on the varying intensity of competition has any validity. If ecologists could simply measure whether or not competitive interactions are stronger or weaker in the equatorial region than they are in the temperate regions, they would know whether or not one of the theories could potentially be valid.

In practice, measuring the intensity of competition between various sets of populations in different places is not easy, and attempts to answer the question in relation to latitude have given disparate and inconsistent results.

For example, James and Partridge (1998) found no latitudinal variation in the competitive ability in populations of the fruitfly *Drosophila melanogaster* from Queensland and Tasmania, while Cotgreave and Stockley (1994) found that in small insectivorous mammals there was some evidence that there may be greater competition at lower latitudes than nearer the poles. Miller and Hay (1996) found the opposite result when they studied the coral *Oculina arbuscula* in North Carolina. They examined competition for space between the coral and various species of seaweeds, and concluded that competition might be strongest in the temperate zone, and that it might explain why few corals thrive outside the tropics.

The fact that no consistent latitudinal pattern can be found does not preclude the possibility that competition is important in causing differences in species diversity. It may even be that the two conflicting theories are both true, in different places. There is no reason in principle why it cannot be the case that competitive exclusion could cause a relatively low diversity of insects in northern China, while niche differentiation leads to a high diversity of bacteria in Mozambique.

Theories involving other ecological interactions There are other theories about how biological interactions might lead to high species diversity in Indonesian rainforests and Caribbean coral reefs, and low biodiversity in the Falkland Islands and Siberian grasslands. There might be more predation near the equator, or more intense mutualisms near the poles, or fewer parasites in the temperate zone.

There are three problems with all such explanations. First, they are difficult to test because they require detailed knowledge, collated in a comparable manner, from large numbers of interactions in large numbers of communities, over huge geographical areas. Second, the evidence that does exist is often difficult to interpret and sometimes inconsistent for theories involving competition (this was evident above). The third and most important problem with such explanations is that, even if they are true, they give only a partial understanding of why the world's communities vary so greatly in the number of species they contain. If competition is intense near the equator, ecologists must still ask why it is so. If mutualisms are weak near the equator, we need a theory about how this situation has come about.

14.3.2.3 Theories of tropical diversity based on ecological interactions have a major problem

In fact, any theory about the distribution of biodiversity over the planet cannot rely entirely on explanations relating to the current ecology of biological interactions. Any variation in the intensity or frequency of particular types of interactions must have a root cause. The causes fall into two kinds. They may be due to factors of the physical environment of the kind that were examined in Section 14.3.2.1, which looked at climatic effects on productivity, or at the predictability of the climate. Alternatively, the variation in intensity of biological interactions has been caused during the evolution of the populations involved. Evolutionary aspects of the distribution of biodiversity will be examined in Section 14.9.3.1.

Evolution is a process of change, and before investigating how it might affect the diversity of communities, it is necessary to examine the ways in which change occurs in communities.

14.4 Changes in communities

Chapter 11 observed that ecological communities are difficult to define precisely, but that they consist of populations living together in the same place and at the same time. But communities, like everything in ecology, are dynamic, and at different times a community can change for a variety of reasons.

There are many ways in which such changes may occur, depending on alterations in the biological and physical environments of the populations within the community. Different types of change are sometimes given different names; changes to the physical environment are often perceived as disturbance, while alterations in the biological environment are known by names like invasion, colonization or extinction, depending on the precise dynamics of the populations involved.

Ecologists actually define the word **disturbance** in a variety of ways, but the definitions usually include any change that is relatively discrete and which either results in the death of organisms or changes the physical or biological environment in such a way as to alter the availability of resources.

In fact, if a community has reached the sort of equilibrium that was described in Section 13.4, then any change is a perturbation from the equilibrium. If no such equilibrium exists, it is not sensible to think of a change as a perturbation; it is just an alteration in the physical or biological environment.

Physical disturbances, such as a fire or a tornado, affect individual organisms, and hence change the biological environment. Equally, a biological change can alter the physical environment. This was observed in Section 13.2, which examined the process of facilitation in ecological succession. The invasion of birch trees (*Betula* spp.) into the heathland in England increased the amount of calcium and phosphorus available in the soil. All over the world (except in the Arctic and Antarctic), members of the plant family Leguminosae have a mutualistic interaction with bacteria of the genus *Rhizobium* that allows them to use nitrogen in the atmosphere and to convert it into useful nitrates and other compounds. This 'nitrogen-fixing' is a widespread example of biological change altering the physical environment for other organisms.

To understand the dynamic nature of communities, it is sensible to start by examining some features of physical disturbances, and then to concentrate on changes in the biotic environment. Integrating the two will give information about the processes that link biology and geography. These were mentioned briefly in Section 1.6.1, which also introduced the name **biogeography** for the topic (Section 14.9).

14.4.1 Physical disturbances

For an organism, disturbance can come in many forms. The wind may buffet a flying bird. A rotten tree may fall over and crush the herbs below it. A colony of army ants may march through a section of rainforest, disturbing everything in their path, including insects and even mammals. And, of course, humans may plough up, walk over or build on a piece of habitat, disturbing whatever organisms were living there.

When we look at things in this way, we tend to see disturbance as an unquestionably bad thing. In doing so, we are concerning ourselves with the effects of disturbance on the individual organism. The buffeted bird, crushed herb or mouse running away from the approaching column of ants is disturbed. But that does not mean that the population is negatively affected, or that other features of the habitat and ecological community are necessarily influenced in ways that might be thought of as negative.

Many species of plants, for example, rely on the disturbance created by fire in the African savanna or Australian bush. The same is true of a variety of plants in many other places. The rosebay willowherb (*Chamaenerion angustifolium*) thrives in the open ground created by fires in Britain and became very common in parts of Europe during World War II, when it easily spread into the waste ground created by bombs. This feature of its biology is recognized in its American name—fireweed.

Many common animals and plants that live in association with humans are also dependent on the disturbance that our activity brings. The grey partridge (*Perdix perdix*) almost certainly evolved in central Asia, but the bird species spread throughout much of Asia and Europe, probably following the development of agriculture several thousand years ago. Farming practices that regularly disturb the soil, such as ploughing, create a habitat that is similar in some ways to the open, sparsely vegetated plains of Afghanistan, where the species first evolved. The partridge is now a familiar species in France, Austria, Italy and Sweden.

Because disturbance can have beneficial effects on some populations, but is inevitably disastrous for others, it is often seen as an important factor in structuring ecological communities.

As an example of the effects of disturbance in natural communities, Whitney (1994) provides a summary of known information about hurricanes and tornadoes in the northeastern quarter of the United States of America. Central New York State has the lowest average level of disturbance, with about 0.01% of the area

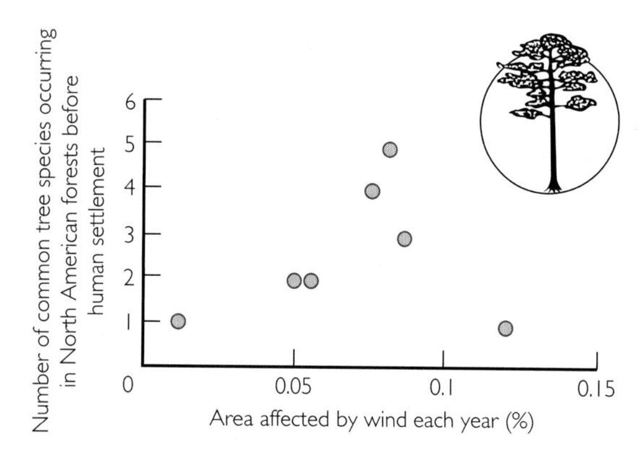

Fig. 14.5 Natural American forests are most diverse at intermediate levels of disturbance. (Data from Whitney 1994.)

being affected by severe wind each year; as a rough estimate this means that in an area of woodland, one tree in every 10 000 is likely to be affected on average each year. In Indiana, however, the comparable figure is one tree in 800 being affected each year (about 0.12% of the area). Figure 14.5, shows that the diversity of common tree species in New York and Indiana is low compared with their diversity in Michigan, which experiences, on average, intermediate levels of natural disturbance from hurricanes and tornadoes (about 0.05% of the area is affected).

It is not unusual to find that the most diverse ecological communities seem to be those that experience a level of disturbance that is somewhere in the middle of the range of possibilities. The probable explanation for this appears to be fairly obvious. Communities that suffer intense levels of disturbance are unlikely to become very diverse. Think of a pathway in a park, which is trampled by thousands of human feet each year. With such high levels of disturbance, it is impossible for many individual plants to survive, so the diversity simply cannot be very high. But in communities that never experience any disturbance, there is no environmental change. In such circumstances, those species that are well adapted to the very specific conditions may come to dominate the community, leaving little room for new species to invade.

Thus, the most diverse communities are likely to be those in which disturbance is not so great that individual organisms are so constantly disturbed that they cannot persist, but where there is sufficient disturbance to create a subtly changing set of conditions, favouring a changing set of species.

This **intermediate disturbance hypothesis** was proposed by Connell (1978), and is confirmed in many natural communities. For example, Stewart and Samways (1998) studied the dragonflies of the Kruger National Park in South Africa. They found 51 different species, including *Trithemis annulata, Pseudagrion sublacteum* and *Elattoneura glauca*, and they asked whether there tended to be more species in disturbed or undisturbed rivers. In general, there was a greater number of dragonfly species in rivers that were moderately disturbed, and the ecologists concluded that a small amount of human interference promoted species diversity in their study area.

Yoshimura *et al.* (1997) studied the algae of the Langtang region of the Himalayas in Nepal. These 'snow algae' include species from at least eight different genera, although Yoshimura and colleagues were not able to give positive identifications to all of them. The lower zone of their study area, up to about 5200 m above sea level, was characterized by a stable ice environment, and had an assemblage of algae that did not have many species, but did have one dominant species, *Cylindrocystis brebissonii.* The upper zone (higher than 5300 m above sea level) was also poor in species, and was characterized by a stable environment of snow. The 100 m between the two zones was a much more environmentally unstable area, with continuous and somewhat unpredictable changes from snow to ice and back again. This middle zone was again dominated by a single species, *Mesotaenium berggrenii*, but it had more different species of algae than either the upper or the lower zone. It appears that species diversity was promoted by the moderate level of constant changes in the environment.

14.4.2 Problems with the intermediate disturbance hypothesis

It is common for scientists of all kinds to accept general principles for which there is a good amount of evidence. Ecologists are no exception, and the intermediate disturbance hypothesis is a widely accepted generalization. However, it is possible to learn a more important general lesson from this hypothesis: generalizations in ecology are rarely firm rules. In fact, many studies have failed to show any clear relationship correlating the highest levels of species diversity with intermediate levels of disturbance.

In the shrublands of South Africa, known as fynbos, disturbance is caused by natural and human-induced fires. Schwilk *et al.* (1997) studied the effect of fire on the species diversity within this community and showed that it was not true that there tended to be more species in places that suffered moderate or average levels of disturbance by fire. When they studied areas of about 1000 m^2, they found most species at the sites where fires occurred least frequently (about once every 40 years). When they studied much smaller areas (of 1 m^2), they found that the highest species diversity occurred at a wide range of levels of fire disturbance (from one fire every 26 years to one every 4 years).

The fact that there were different results at different scales is interesting in itself, and is an important reminder that, when ecologists are studying communities, they must be careful to think of a community in terms of the organisms that live within it. Community ecologists can define their community at whatever scale they choose, but if they choose a scale that is inappropriate for the individual organisms that they are studying, their results will not necessarily answer the questions they really wanted to ask (see Box 11.2).

14.5 Patch dynamics

The plausible arguments in Section 14.4.1 seemed to explain the observation that moderate levels of disturbance appear to promote species diversity. But they should not be accepted as a proven fact. They rely on certain assumptions that may not be true in all situations. It seems unarguable that a very high level of disturbance will prevent the existence of high levels of diversity because of its negative effects on individual organisms. But it is not possible be so sure that very low levels of disturbance lead to lower levels of diversity.

However, there is a significant line of evidence supporting the theory that moderate disturbance promotes species diversity, relative to a total lack of disturbance. To understand it, it is necessary to recall the process of ecological succession (Section 13.2), by which different species invade an area as time goes by, eventually resulting in a climax community that is difficult for new species to invade. Theory concerning disturbance assumes that when there is no disturbance, this climax community results, and that the number of species in the area is then set by the nature of the climax community. Section 13.5 investigated the possibility that some communities do indeed reach this kind of equilibrium.

Moderate levels of disturbance would therefore promote diversity by temporarily arresting the development of the climax community in particular parts of an area. Those parts would then revert to an earlier stage of the successional sequence, and would be occupied by species that are typical of the earlier stage. If different parts of the habitat were disturbed at different times, a patchwork might result, with each piece of the habitat at a different stage of the succession, and each harbouring a different subset of the species that form part of the succession.

One element of this patchwork of habitats can be seen on the rocky shore of Santa Catalina Island in California. Here, it is sensible to define the climax community as the carpets of red algae that dominate the surface of the rocks in all of the sheltered areas, and most of the less sheltered ones. The algal seaweeds are of many species, including members of the widespread genera *Corallina*, *Gigartina*, *Rhodoglossum* and *Laurencia*, all of which are typical of temperate rocky shores in the northern hemisphere. In part, this community depends on the existence of spiny lobsters (*Panulirus interruptus*), which clear the rocks of mussels (*Mytilus* spp.) and allow the algae to colonize the bare rocks.

However, some of the less sheltered patches of the habitat are occasionally vigorously disturbed by the sea, which appears to prevent the lobsters from inhabiting the patch. In these conditions, larvae of the Californian mussel (*Mytilus californianus*) are able to take hold on the rock (Robles 1997). In other words, the effect of periodic disturbance in various patches of the habitat is to allow the mussels to maintain a population that might otherwise be unable to persist.

Similar patch dynamics occur in the conifer forests in the Lake Duparquet area of Quebec in Canada, where gaps in the forest created by an insect herbivore, the spruce budworm (*Choristoneura fumiferana*), increase the species diversity of plants by allowing the coexistence of plant species such as the rosebay willowherb (*Chamaenerion angustifolium*), which is very intolerant of shade, and the shade-loving Canadian yew (*Taxus canadensis*) (de Grandré & Bergeron 1997).

In Victoria Mayaro reserve in Trinidad, Wood and Gillman (1998) were interested in conserving the butterfly species, and came to the conclusion that the most effective way to ensure the survival of the maximum number of species would be to maintain a mosaic of habitat patches that included undisturbed and disturbed areas. This strategy would allow populations of each species to persist, while leaving the reserve entirely undisturbed may have excluded some species that were more typical of the earlier stages of succession.

14.6 Problems of definition

One of the problems with trying to understand the effect of disturbance on ecological communities is that ecologists lack an agreed definition of what they mean by disturbance. Everyone would agree that ploughing a field, or burning a mountainside, or flooding a forest are all forms of disturbance. But it is not easy to compare them in terms of their severity, and as Section 1.7.1 described, it is only by making comparisons that ecologists can decide what features of a situation make it interesting. The comparisons that have been made in Sections 14.4.1 and 14.4.2 strongly suggest that physical disturbance can play an important role in promoting species diversity, but that ecologists must be careful to understand this role in the wider context of the biological environment of the communities that they are studying.

14.7 More gradual changes

Many of the changes that have been considered so far have been the kind of events that happen rather suddenly. Hurricanes, floods and invasions are all discrete events that happen at a particular, distinct moment in time. The day after a hurricane, the populations in an ecological community will begin to evolve according to the new set of evolutionary pressures.

But many changes in the physical and living environments occur gradually and continuously over a period of time (see, for example, Box 14.1). When this happens, evolutionary pressures will also change in a continuous way. In a strict sense, extinctions are abrupt—extinction occurs at the precise moment when the last individual in a population dies. In reality, however, extinction can be viewed more as a gradual process. It took 100 years of hunting by humans to bring about the extinction of the passenger pigeon (*Ectopistes migratorius*); this represented four or five generations of the predators and tens of generations of the prey.

The most obvious gradual alteration in physical conditions is the global climate change that most scientists believe is currently underway. For simplicity, this is generally presented as a continuous increase in the average temperature around the world, although it also includes changes in rainfall patterns and other climatic features (see Sections 4.7 and 12.9).

The current intense interest in this area is focused on climate changes brought about by human activity, such as warming caused by so-called greenhouse gases, including carbon dioxide, the rising level of which was mentioned in Chapters 11 and 12. But in fact, weather is constantly changing all over the world. Between about 1650 and 1715, the whole of Europe experienced the 'Little Ice Age', a period of uninterrupted cold weather compared with the years before or after this period. In general, summer temperatures were 1°C colder than usual. This may not seem very much, but it reduced the growing season for wheat (*Triticum aestivum*) by at least 3 weeks each year, and lowered the maximum altitude at which cereals could grow by almost 200 m.

14.8 Biological upsets of equilibria

As well as physical changes, communities can be perturbed by biological events. New species can invade existing communities, and in doing so can alter the balance of interactions among the existing populations. Alternatively, the extinction of a population might alter the details of interactions among the remaining populations.

In many cases, species invade communities by natural processes. An extreme example of such an invasion comes from the black-browed albatross (*Diomedea melanophris*), which nests on islands in the southern oceans, such as South Georgia and the remote islands of Chile. It winters throughout the cold seas of the southern hemisphere and is observed around New Zealand, South Africa and southern Australia. However, because they are such good fliers, individual albatrosses sometimes turn up in similar habitats in the northern hemisphere. They have been recorded in Scotland, Norway, Iceland and Spitzbergen, and one lived in a seabird colony on the Faeroe Islands for 34 years; some of these European visitors even prepare nests every year.

Although the black-browed albatross sometimes invades coastal habitats in the North Atlantic, populations

Box 14.1 Plate tectonics

One of the long-term processes that affects the distribution and abundance of species within communities is continental drift—the constant, gradual movement of the Earth's great plates. As recently as 50 million years ago, the Indian subcontinent was as close to Africa as it was to Asia, which is why India has many organisms that are more closely related to African species than to those of China; elephants, mongooses and hornbills are good examples in the animal kingdom, while among the plants, the families Musaceae (which includes the banana) and Sonneratiaceae show similar patterns.

The movement of the plates plays an important role in determining the distribution of different habitats. The highlands of the Himalayas were formed by a collision between two such plates. Of course, climatic conditions on a particular patch of land can change dramatically as the continental plate drifts. When the dinosaurs first evolved, Antarctica was a temperate continent with its own diverse fauna and flora, and even 10 million years ago, there was tropical forest over much of what is now the central Australian desert.

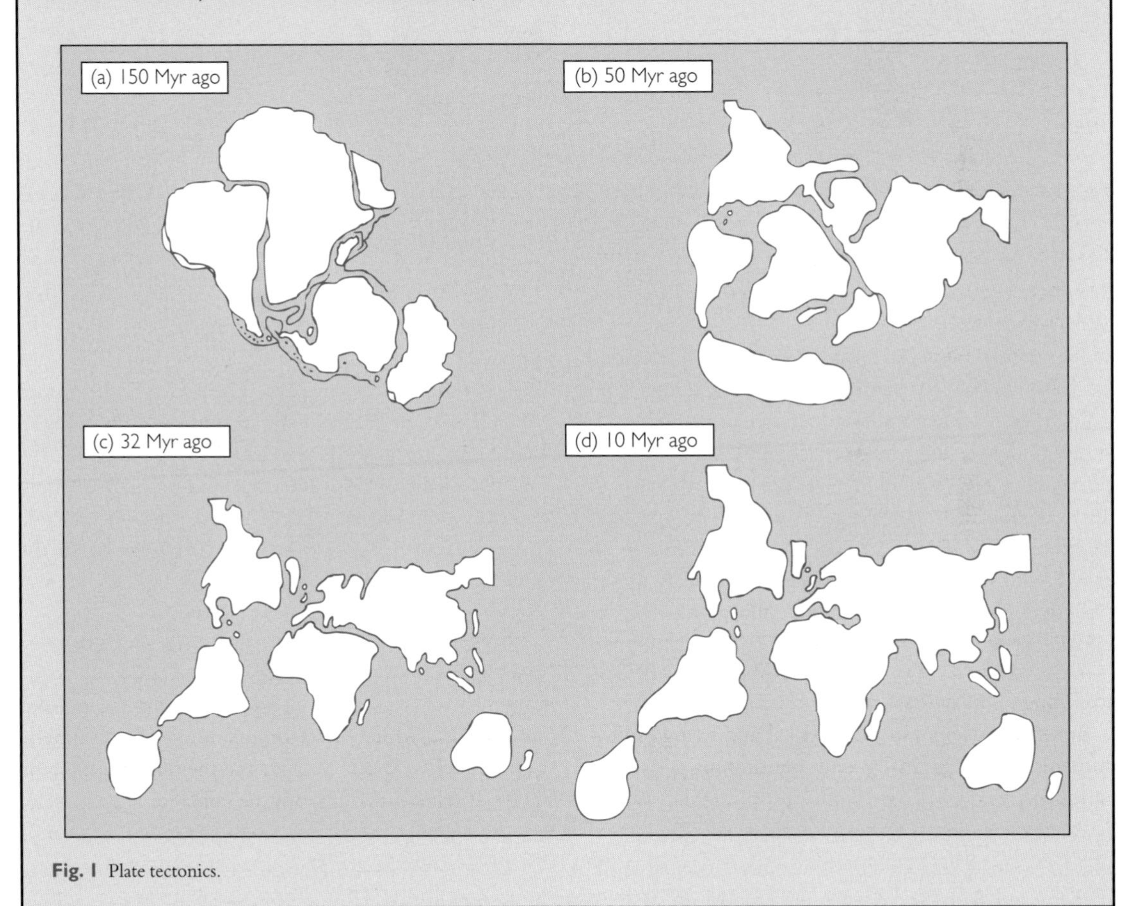

Fig. 1 Plate tectonics.

never become established, partly because individuals occur so infrequently that a male and female have never turned up in the same place at the same time.

But many populations do become established in communities from which they were previously absent. When a 40 ha piece of farmland in Illinois was abandoned in the 1940s, the red-headed woodpecker (*Melanerpes erythrocephalus*) was unknown in the area. But in 1960, the species invaded and bred. Ten years later, the woodpecker was a regular breeder, with a

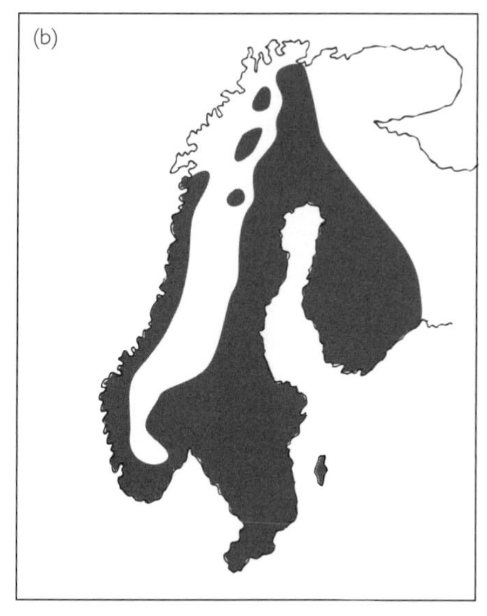

Fig. 14.6 (a) Distribution of the speckled wood butterfly (*Pararge aegeria*) in Scandinavia at the end of the 1990s. (b) Simulated future distribution for the period 2070–2099, based on probable climatic changes. (From Hill *et al.* 1999.)

population size of about four or five pairs breeding each year (Kendeigh 1982).

The natural ranges of many species are not bounded by absolute, discrete barriers, and shifts ranging from a few metres to many kilometres can occur. There is some evidence, for example, that bird and butterfly populations are expanding their range northwards in areas of North America and Europe. The speckled wood butterfly (*Pararge aegeria*) expanded its European range northwards throughout the second half of the twentieth century, partly as a result of changes in climate of the kind believed to be caused by human activity, and Fig. 14.6 shows how that northward movement is likely to continue.

However, a majority of the most dramatic invasions that are documented came about not through natural dispersal of organisms, but through human intervention. Some species have been deliberately introduced into new areas, and others have either escaped from gardens and farms, or have gone unnoticed as they hitched a ride on ships or planes. In North America, invertebrates such as the hessian fly (*Mayetiola destructor*), the honeybee (*Apis mellifera*) and the earthworm (*Lumbricus terrestris*) have all been introduced, as have a range of plants, such as henbane (*Hyoscyamus niger*) from Europe and Jimson weed (*Datura stramonium*) from Asia. The state of Massachusetts alone has 27 introduced species of fish (Whitney 1994).

The effects of invaders and alien species on communities give a great deal of information about the processes that normally operate in the populations that make up those communities. These effects depend on the interactions between the invaders and the populations that are already present. The presence of the invader changes the biological environment of those existing populations. The invasive population will interact with other populations, through one of the mechanisms described in Chapters 8 and 9, such as competition, parasitism or mutualism.

Section 13.4.1.1 demonstrated how the spread of alien species can provide information about the structure of ecological communities. The reason for this is that ecologists can compare communities before and after they have been invaded by an alien species, and learn from the differences. They can also see how naturally invasive species, as well as those introduced by humans, could have significant effects on the structure and composition of a community, just as a change in the physical environment might affect the ecological structure of a community.

14.9 Biogeography

The black-browed albatross, described in Section 14.8, is a rather dramatic example of how organisms can invade communities from which they were previously absent. Even though the albatrosses have not managed to become established in the northern hemisphere, it is not unknown for them to appear in Europe, several thousand kilometres away from their normal breeding grounds. But the Falkland steamerduck (*Tachyeres brachypterus*), which also nests on islands of the South Atlantic, has never turned up in Europe, or in Australia, Africa, North America, or anywhere else outside the tight limits of its normal geographical range. There is a simple reason for this difference: Falkland steamerducks cannot fly.

The difference between the albatross, which is a superb flier, and the flightless steamerduck is a very obvious and significant difference in dispersal ability. More subtle differences exist among other kinds of species, but in general, there is a huge variation in the ability of organisms to disperse from one community to another. The fruits of the coco de mer (*Lodoicea maldavica*) are heavy and look like large coconuts, but they are killed by seawater. So, when the fruits wash up on beaches a long way from their natural habitat on the Seychelles, they cannot germinate, and the plant species is restricted to the Seychelles. By contrast, cotton (*Gossypium* spp.) has seeds covered in fine filamentous fibres, which allow them to be carried on the wind over great distances. Cotton is native both to America and to the Old World. A large spider like a tarantula must walk everywhere it goes, but many smaller spider species, such as *Erigone dentipalpis* and *Savignya frontata*, are so light that they can waft on currents of air, using a thread of silk to catch the wind.

Ecological communities around the world bear the imprint of these differences in ability to disperse. Ireland, for example, has no snakes, because they failed to get there before the Irish Sea divided the island from Great Britain at the end of the last ice age. Although grass snakes (*Natrix natrix*) are good swimmers, they do not normally live on the seashore, and would have difficulty in making the 50 km crossing across the choppy Irish Sea. Similarly, lemurs are restricted to the island of Madagascar, the Celanese cow plant (*Gymnemum latiflorum*) is found only in Sri Lanka, and the long-toed pigeon (*Columba trocaz*) lives only on the Atlantic island of Madeira.

14.9.1 Different kinds of islands

The places listed in Section 14.9 above—Ireland, Madagascar, Sri Lanka and Madeira—are all islands. It is easy to appreciate that the Irish Sea, the Mozambique Channel, the Palk Strait and the Atlantic Ocean are real barriers to movement, which prevent the cow plant or the grass snake from dispersing between one location and another.

In studying biogeography, ecologists need to concern themselves with the interface between biology and all manner of geographical features. They cannot restrict themselves to oceanic islands. When ecologists choose to look at mountain peaks as if they were high-altitude islands in a sea of lowlands, they find that the mountain 'islands' have a number of restricted species —the white-tailed sabrewing hummingbird (*Campylopterus ensipennis*) lives only in the uplands of Tobago and Venezuela, the plant known as Elisha's tears (*Leycesteria formosa*) is restricted to the Himalayas in Nepal, and the fish *Lythrurus snelsoni* is found exclusively in the Little River of the Ouachita Mountains of Oklahoma.

The reason that oceanic islands make such clear examples of ecological barriers is because the land and the sea are so very different that the boundary between the two is clear and unmistakable. The boundary between mountain peaks and lowlands can also be defined so as to be unequivocal; mountains are traditionally defined as any peak over 1000 m above sea level.

Other boundaries are less obvious, but they exist nevertheless. For example, Section 8.6.1.1 described an equally abrupt divide between the ranges of the bush pig (*Potamochoerus larvatus*) and the red river hog (*P. porcus*).

If ecologists looked closely enough, they could probably find some kind of boundary just about anywhere they looked—one side of a field will always be wetter, or more rich in nitrogen, or more alkaline than the other. Because of this, it is possible to use a knowledge of islands as an analogy to help in understanding the biogeography of any habitat. Ecologists can simply say that a patch of bare earth surrounded by dense woodland is an island of soil in a sea of trees. A pond is an island of water in a sea of land, and the Sahara Desert is an island of sand in a sea of various other habitats.

14.9.2 The theory of island biogeography

Treating any patch of habitat as an island, it is feasible to compare different islands in an attempt to understand how geography affects the structure of ecological communities.

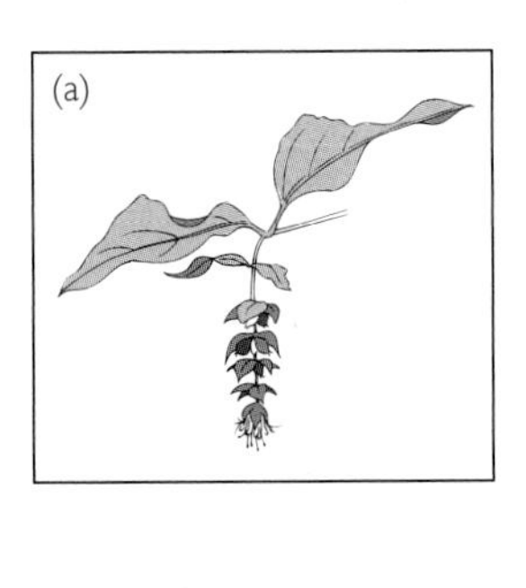

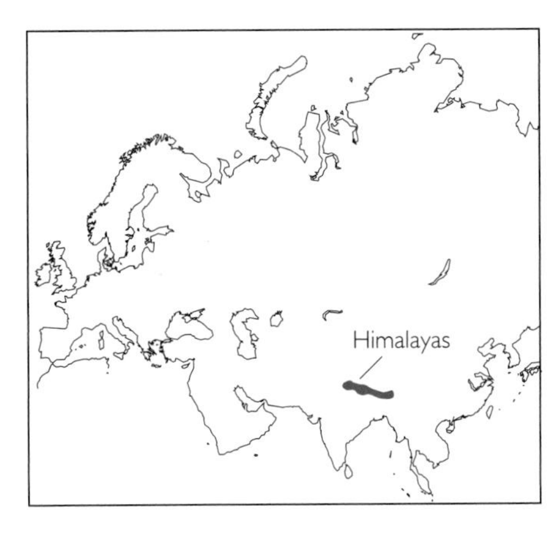

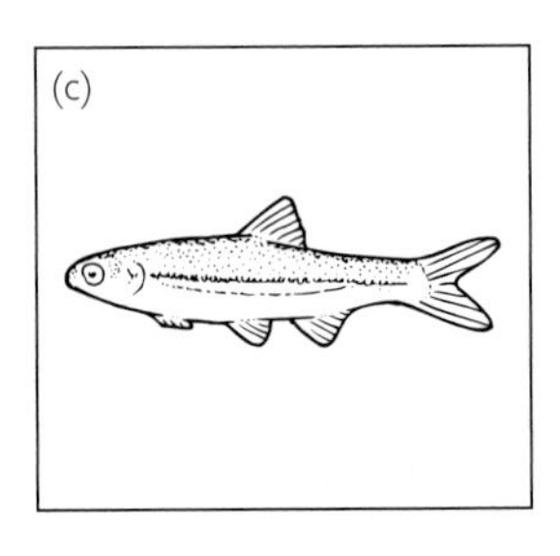

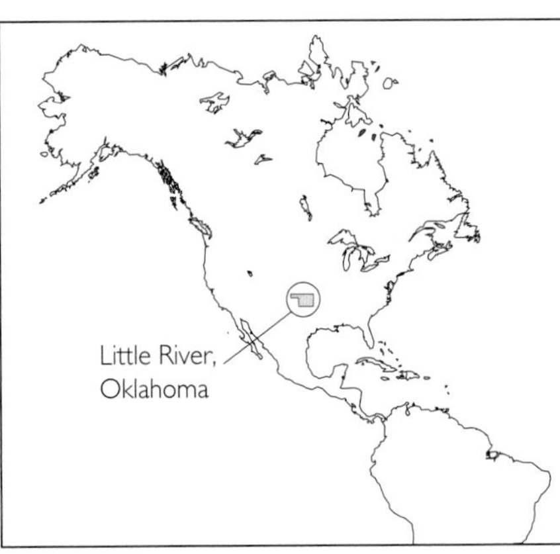

Fig. 14.7 Examples of species that are restricted to islands. In some cases, the islands are real islands of land in a sea of water, but in others, the 'island' refers to an island of mountains in a sea of lowland. (a) Elisha's tears (*Leycesteria formosa*) is restricted to the Himalayas in Nepal, (b) the long-toed pigeon (*Columba trocaz*) is restricted to the island of Madeira, and (c) the Ouachita Mountain shiner (*Lythrurus snelsoni*) is restricted to the Little River of the Ouachita Mountains of Oklahoma.

Section 14.9 above has already described how some real islands do not have species that exist on the mainland nearby—Ireland has no snakes, and it is also lacking the mole (*Talpa europea*) and weasel (*Mustela nivalis*), both of which are common just across the sea in Britain. It is also clear that differences in dispersal ability determine which species are absent. There are no bird species that are common in Britain but entirely lacking in Ireland, because, unlike snakes and mammals, most birds can easily make the short flight, even if they do so by accident in a storm.

A similar pattern is seen elsewhere. Few of the mammals of northern Australia are found in New Guinea, but many of the birds are.

But some islands do not have many of the bird species that are present on the nearest mainland. Very few of the birds of mainland North America occur in Hawaii, for example. Likewise, most of the native birds of New Zealand are endemic, and do not occur in Australia.

One of the most obvious differences between Ireland and Hawaii, or between New Guinea and New Zealand, is the distance from the mainland. Ireland is very close to Britain, and New Guinea is not far from Australia. But New Zealand and Hawaii are very remote, and even excellent fliers would find it difficult to reach them.

The longer the distance between patches of similar habitat, whether they are real islands or not, the less likely a colonist is to invade the habitat. Because of this, remote habitat patches tend to have fewer species than those that are near to a larger habitat island that can act as a source of colonists.

In the rainforests of the Ivory Coast in West Africa, Porembski *et al.* (1995) studied the vegetation on the tops of granite outcrops. The climate was characterized by prolonged drought interrupted by short periods of very heavy rain. In naturally occurring depressions on the outcrops, the vegetation was more similar to the grasslands typical of the savannas further east than to the surrounding rainforest. The botanists examined how the patches of grassland on the granite outcrops changed as they penetrated more deeply into the forest, becoming more isolated from large expanses of savanna. The hollows nearest the savanna contained many more species than the more remote hollows deeper inside the forest. There was no climatic reason to expect a reduction in the number of species in those areas furthest away from the savanna; indeed the surrounding rainforest showed no such fall. In other words, the isolation of the depressions in the outcrops was undoubtedly one of the causes of the lack of diversity of the plant communities within them.

The distance between habitat patches is only one of the factors that determine how likely it is that colonists will happen to travel between them. All sorts of other features of the landscape will be important. The features that turn out to be most important will depend on the organisms doing the dispersing. Section 14.9 above observed that seawater is a total barrier to the dispersal of the coco de mer (*Lodoicea maldavica*), while coconuts (*Cocos nucifera*) frequently germinate after drifting long

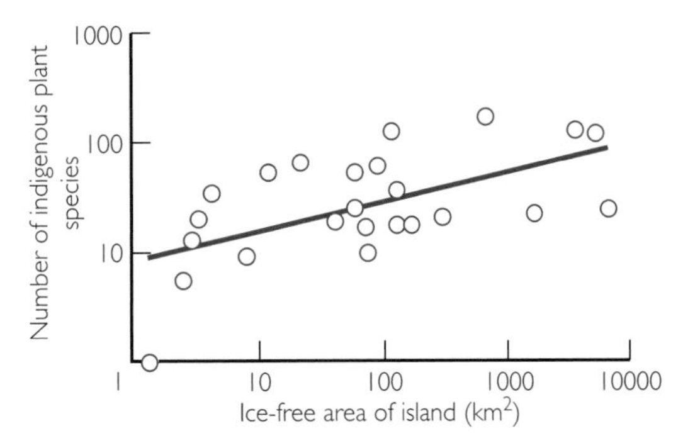

Fig. 14.8 Larger islands in the southern oceans generally have more indigenous higher plant species than smaller ones. (Data from Chown *et al.* 1998.)

distances by sea. Similarly, in some places, roads are known to be effective barriers to dispersal by some butterfly species, while even the narrowest strip of land would prevent an entirely aquatic organism from moving between two patches of water.

14.9.2.1 The species–area relationship

One of the strong points of the theory of island biogeography is that it goes some way to explaining the common observation that larger habitat patches tend to harbour more species than smaller ones.

Chown *et al.* (1998) investigated this relationship among the plants of several islands in the southern oceans. Their sample of islands varied in size from the tiny Bounty Island (less than 2 km^2) to Kerguelen Island (6450 km^2). Figure 14.8 shows how the number of indigenous higher plant species is greatest on the largest islands. There are no plant species on Bounty Island, while East Falkland (which covers 5000 km^2), has 149 species.

At first sight, it seems obvious that a large island, or a large habitat patch such as a large wood or a large lake, will be occupied by more species of organisms than a small wood or a pond. But on closer examination, the pattern of species diversity cannot be explained away so easily.

Island biogeography in its simplest form helps to explain the species–area relationship by postulating that a seed, or spore, or animal, in dispersing, is more likely to hit a large habitat island than to land on a small one. Thus, populations become established on habitat islands in proportion to the size of the island. This explanation relies entirely on chance events, and essentially assumes that dispersal is a random sequence for the pioneer seeds or animals from which populations are originated. It is unconcerned with the biology of the organisms concerned.

Another factor by which random events are postulated to affect the number of species in a habitat patch is concerned with extinction rates. Small habitat islands will tend to have populations with few individuals, which are more likely to become extinct if they experience a major density-independent event, such as a hurricane or extremely severe winter.

Taken together, the lower colonization rate and higher extinction rate in small habitat patches could explain the species–area relationship. But this explanation relies on the assumption that ecological communities consist of an 'equilibrium' set of populations that is constantly changing, with some species going extinct while others are fresh colonists. As Section 13.4 indicated, communities are not always in equilibrium.

14.9.2.2 Other explanations for the species–area relationship

The dynamic equilibrium outlined by the basic theory of island biogeography is not the only explanation for the observation that larger habitat patches tend to be occupied by more species than smaller ones.

Many people find it unconvincing that when ecologists treat habitat patches as islands, they pretend that similar resources are distributed in a similar way throughout habitat islands of different sizes. In fact, larger habitat patches may contain more variation in resources than smaller ones.

Imagine that ecologists were comparing the number of plant species in habitat patches of different sizes, and that, for convenience, they defined habitat patches as entire countries. In small countries like Bahrain, Brunei or Guadeloupe, there are few species relative to larger countries at similar latitudes, but this pattern might easily be caused by a smaller diversity of resources in the small nations. Annual rainfall is less than 200 mm throughout the entire area of Bahrain, and greater than 100 mm throughout Guadeloupe. But in the various regions of India, which is centred at very roughly the same latitude, annual rainfall varies between less than 200 mm and more than 800 mm. Such variation in resources must inevitably lead to a diversity of plant species, with vegetation ranging from rainforest around Calicut, through dry forest around Bangalore, to dry, sparse grasslands around Jaipur.

14.9.3 Ecological and evolutionary timescales

All the aspects of biogeography that have so far been investigated have taken place over an ecological timescale. Large habitat islands that are near to the source of immigrants harbour more species than small, remote ones because of processes that take years, decades or perhaps centuries. Even global warming is discussed in terms of changes taking place over a century. For an oak tree (*Quercus* spp.), this represents less than a lifetime, and it is no more than 100 generations of many animals, fungi and plants. Of course, some of these organisms, and most bacteria and other microorganisms, have more than one generation per year. But for many species, only a few hundred generations are represented during the biogeographical processes that have been described.

Habitat patches (and real islands) continue to exist, and change, over much longer periods of time. Likewise, the communities that inhabit these habitats exist in one form or another, constantly changing, for millions of years. This means that the populations in these communities experience evolutionary pressures as well as ecological processes like invasion, colonization and physical disruption.

There is, of course, no true distinction between evolutionary and ecological timescales. The genetic make-up of populations changes in every generation, so evolution is actually happening constantly; every time an animal is born, or a fungus dies, or a bacterium divides, or a seed germinates, evolution takes place as the gene pool of the population changes. However, the effects of evolution can only be perceived by humans when they think about changes in communities over very long time periods.

14.9.3.1 Species diversity may build up over time

Tropical rainforests, such as those in Brazil, Malaysia and Cameroon, are well known as places of extraordinary biological diversity. A variety of explanations has been proposed throughout this chapter for why such communities might be so diverse. Section 14.3.2 observed that most of these explanations are attributed to ecological processes, to do with the ways in which individual organisms and populations interact either with their physical surroundings or with the biological environment.

In fact, there are other proposed explanations for patterns in biodiversity suggesting that at least some of the vast diversity in the tropics exists because of the enormous periods of time during which the rainforest habitat has existed in an almost unchanging form.

The precise details of how long rainforests have persisted in their current form are debatable, but it seems likely that most of the Amazon basin has been home to moist, warm areas of large trees for several hundred-thousands of years. In the temperate zone, where ice ages have led to massive changes in the physical environment, it is reasonable to assume that populations have experienced constantly changing evolutionary pressures. Indeed, large parts of Europe were covered by glaciers until 10 000 years ago, and the plants that lived there were all killed. But if the tropics have been relatively constant for long periods in evolutionary time, it is possible that there has been time during periods of great stability for specialized evolution to have taken place.

There is some evidence that the Amazon forest has indeed been relatively stable for at least 50 000 years. Haberle (1999) was able to examine a column of soil from beneath the Earth's surface, which contained a record of the pollen from the region's plants over several tens of thousands of years. The evidence suggested that many of the same types of plants, such as *Cecropia* species, had been present in the Amazon basin for the past 50 000 years. There was little evidence that the habitat had changed during the last ice age, so it is entirely possible that many of the groups of species in the Amazon rainforest have existed for a very long time. They would have been wiped out by the harsh changes in the physical environment if they had evolved further away from the equator.

14.10 An important trap to avoid

Studying the effects of change on the ecology of communities is an effective way of reminding ecologists of an important lesson. Scientific ecologists should avoid falling into the trap of believing that there is some mythical 'balance of nature' that makes one particular set of species superior to another. Causing disturbance undoubtedly alters the ecological community present in a particular place, as the populations that remain adapt to the new conditions. But no one, unique set of species is inherently superior or necessarily more balanced than any other. This fact is not, of course, a permit for humans to cause any disturbance that they choose. Most people would probably think that in the absence of any other considerations, it is better not to introduce species from one part of the world into habitats on the other side of the Earth.

But populations of organisms will evolve, become extinct and invade according to the pressures of the biological and physical environment, according to the principles of natural selection. They do so without any consideration of balance in nature; the only thing that affects what occurs is which genes happen to be passed on to the next generation.

14.11 Abundance and diversity

There is a crucial question that has, so far, not been answered by any of the biogeographical or other ecological factors that have been considered in trying to explain why some communities have more species than others, or why a particular community happens to have a particular set of species. So far, no attempt has been made in this book to study in any detail the reasons why some species are common within communities, while others are present but very rare. The original definition of ecology, developed in Chapter 1, stressed the importance of both the distribution and the abundance of a population. Moreover, Section 11.1.1 observed that diversity was more than a simple measure of the number of species present in a community. A community that has 100 species, each of which is equally abundant, seems intuitively to be more diverse than a community of 100 species, in which one population is overwhelmingly dominant while the other 99 are extremely rare.

Within the boundaries of a community, the abundance of a population is directly proportional to the density of individual organisms within the population. Section 6.3.2 described how mathematical models could be used to understand either the total number of individuals in a population, or the population density.

In fact, these mathematical models, together with detailed investigations of biological interactions between populations, have already gone a long way towards helping to understand the commonness and rarity of populations within communities. Competing populations can reduce one another's density by a factor that depends on the intensity of the competition between them, and the effects of that competition may be asymmetrical, with one population suffering more severely than another. The effect of a herbivore population on a population of plants depends in part on the efficiency with which the physiology of individual herbivores can convert the plants' nutrients into meat and bones.

14.11.1 Community patterns

One of the questions that has long perplexed community ecologists is: What determines the abundance of each of the species within a particular community?

Searches for patterns in abundance in communities have tended to use the size of individual organisms as a broad surrogate measure that summarizes much of the information about the biology of an individual. For example, large animals like African elephants (*Loxodonta africana*), basking sharks (*Cetorhinus maximus*) and giant squids (*Architeuthis* spp.) share biological features that are not apparent in a marsupial mouse (*Antichinus* spp.), a guppy (*Poecilia reticulata*) or a dog-whelk (*Buccinum undatum*), and the large animals all live at lower population densities than the small ones. The same is true of large plants, such as oak trees (*Quercus* spp.), redwood pines (*Sequoia sempervirens*) and giant water lilies (*Victoria amazonica*).

It is obvious that large organisms cannot live at the same densities as small ones: if you packed elephants tightly, side by side, trunk-to-tail, you could fit 67 000 into 1 km^2; whelks and other molluscs live at densities hundreds of times greater than this.

But this is not the complete explanation for the pattern that large organisms live at lower densities than smaller ones, because almost no organism lives near its maximum possible density. In fact, elephants live on average at only 1% of their potential maximum density.

14.11.2 A general rule of species' abundances in communities

A general rule for animal populations is that the population density of larger animals is lower than that of smaller ones, roughly in line with the metabolic rates of the individual organisms. A large animal has a total mass of tissue that is greater than that of a small animal, so it

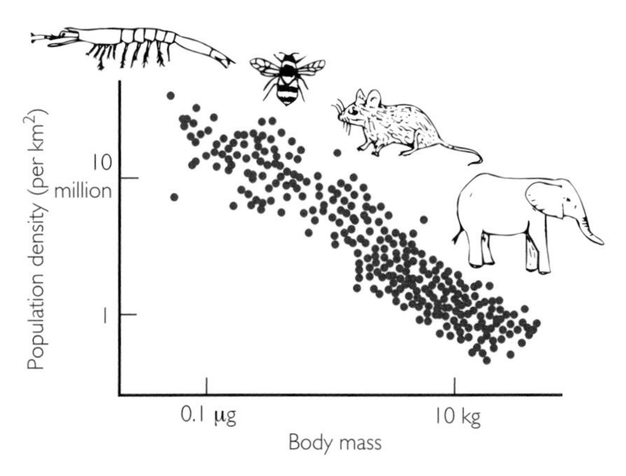

Fig. 14.9 The relationship between body size, or metabolic requirement, and population density across 700 species of animals. There is a strong pattern for larger animals to live at lower population densities than smaller ones. (Adapted from Damuth 1987.)

needs more resources; that is why larger animals must live at lower densities than smaller ones. But the total amount of active muscle tissue in a large animal is not as great as might be imagined merely by scaling up small ones. Large animals that live on land need a greater proportion of hard structural material—skeleton—than smaller ones. A lion (*Panthera leo*) is not simply a scaled up domestic cat (*Felis catus*); if it were, its legs would break when it walked. Think how thick an elephant's leg bones are in relation to the muscle wrapped around them, and compare them with a human leg.

Something similar is true of plants. Large mahogany trees (*Swietenia* spp.) are made up of a much greater proportion of structural materials than pampas grass (*Cortaderia selloana*), which is in turn tougher than a water lily (*Nymphaea* spp.). In essence, the only difference is that in animals the structural materials are calcium carbonate (in bones and seashells), chitin (in insect cuticles) or keratin (in fingernails or rhinoceros horns), but that in plants the structure is created by other materials, such as the complex polymer lignin, which is what makes plant cells 'woody'.

Figure 14.9 shows how the metabolic energy requirement of an animal is related to its body size. Comparing all sorts of animals, from the tiniest zooplankton to large mammals, the density of a population of animals is proportional to the metabolic requirement of the individual animals.

The reasons for this are not fully understood, and deeper investigation reveals more complex patterns. For example, some groups of birds, such as woodpeckers

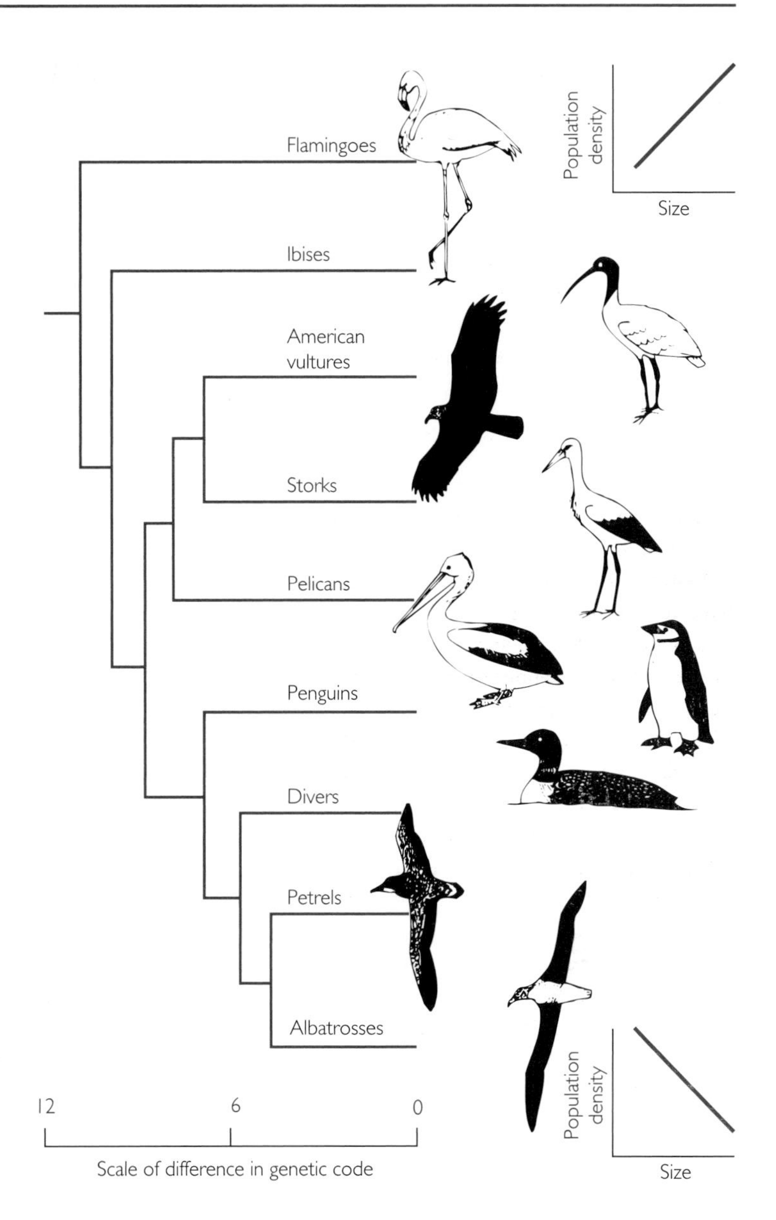

Fig. 14.10 Families of birds that have no close relatives, such as flamingoes, tend to show an unusual pattern of abundances, in which the larger species live at higher population densities than the smaller ones. Families on shorter branches of the evolutionary tree, such as petrels and albatrosses, tend to show the more usual pattern, in which larger species live at the lowest population densities.

and parrots, do not appear to fit the pattern—the larger species tend to live at higher population densities than the smaller ones. The groups that show the abundance patterns exhibited by woodpeckers and parrots tend to be those with a particular pattern of evolutionary history (Fig. 14.10).

14.11.3 Returning to basic concepts in ecology

No single clear theory can explain the intricate patterns of population abundances within ecological communities, but the promising lines of research are those that keep in mind the principles that were defined in Chapter 1. Progress is made by studying different levels of diversity (populations and communities, for example), by making comparisons between them (such as comparing woodpeckers with sparrows), by remembering that ecological theories must take account of natural selection (the pattern in Fig. 14.10 proves that tens of millions of years of evolution are important) and by understanding the ecological requirements—the niche—of the organisms involved, in terms of both the biological and the physical environment. It transpires that ecological competition might begin to explain the pattern in Fig. 14.10.

All of the complex and various patterns that have been investigated in this chapter—the high diversity of species in the tropics, the low diversity of remote small habitat patches, etc.—have required a return to those concepts with which the subject of ecology was introduced: evolution by natural selection, the niche of a population, the different levels of diversity and the practice of making comparisons between species, populations and habitats.

14.12 Chapter summary

Biological and physical processes combine to give each ecological community a unique composition. Communities near the equator tend to have more species than those at higher latitudes, but there is no clear reason that fully explains this and other broad patterns. However, there is no doubt that at least part of the explanation is to do with the amount of primary productivity in each community, which is largely controlled by aspects of the physical environment, particularly the climate. Other explanations divide into those based on ecological processes and those based on evolutionary processes.

Explanations based on ecological processes generally only provide partial answers, because latitudinal patterns in the intensity of competition, predation or other interactions require further explanation.

Within communities, larger organisms, which have greater energetic requirements, generally live at lower population densities than smaller ones, but this pattern is not rigid, and the exceptions appear to be correlated with evolutionary aspects of the species involved. This leads to the conclusion that a full appreciation of the structure of communities will depend on an understanding of both current ecology and past evolution.

Recommended reading

Adams, G.A. & Wall, D.H. (2000) Biodiversity above and below the surface of soils and sediments: linkages and implications for global change. *BioScience* **50**: 1043–1048.

Dukes, J.S. & Mooney, H.A. (1999) Does global change increase the success of biological invaders. *Trends in Ecology and Evolution* **14**: 135–139.

Reid, W.V. (1998) Biodiversity hotspots. *Trends in Ecology and Evolution* **13**: 275–279.

Smith, F.D.M., May, R.M., Pellew, R., Johnson, T.H. & Walter, K.R. (1993) How much do we know about the current extinction rate? *Trends in Ecology and Evolution* **10**: 375–378.

Snelgrove, P.V.R. (1999) Getting to the bottom of marine biodiversity: sedimentary habitats. *BioScience* **49**: 129–138.

Waide, R.B., Willig, M.R., Steiner, C.F. *et al.* (1999) The relationship between productivity and species richness. *Annual Review of Ecology and Systematics* **30**: 257–300.

Concluding remarks

Ecology is a vast subject. It is concerned with millions of species of organisms from bacteria and viruses to whales and redwood trees, and with the physical environment in which they live. In order to understand that physical environment, ecology is even concerned with the angle at which the sun's rays hit the Earth.

But at its heart, ecology has a simple aim—to explain the distribution and abundance of the world's animals, plants, fungi and micro-organisms. Because of the inherent simplicity of this aim, ecology can be broken down into small sections that can be understood more easily.

Even though each of these individual studies is unique, and sometimes they seem to be contradictory, it is possible for ecologists to integrate these discrete investigations into a more comprehensive appreciation of how the Earth's organisms interact with their physical and biological environments.

It has never been more important for us to develop this understanding and appreciation, because human activity is changing the environment in uncertain ways, at a speed that has never before been experienced by members of the species *Homo sapiens*. These changes in the physical environment will affect human ecology dramatically. If human societies are to adapt to these effects in time to prevent serious problems, we will need to understand our own place in the ecology of our planet more fully than we ever have before.

To do this, we will need to understand the basic concepts of ecology, including the ways in which the physics and chemistry of the environment affect individual organisms; biological interactions such as competition and parasitism; the dynamics of populations; the structure and function of ecological communities; and changing patterns of land use.

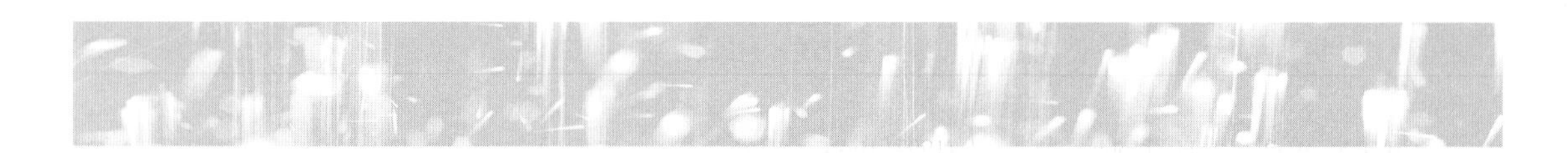

References

Aaenes, R., Saether, B.E. & Øritsland, N.A. (2000) Fluctuation of an introduced population of Svalbard reindeer: the effects of density dependence and climatic variability. *Ecography* **23**: 437–443.

Arseneault, D. & Payette, S. (1997) Landscape change following deforestation at the arctic tree line in Quebec, Canada. *Ecology* **78**: 693–706.

Ashmore, M. (1990) The greehouse gases. *Trends in Ecology and Evolution* **5**: 296–297.

Augspurger, C.K. (1984a) Pathogen mortality of tropical tree seedlings: experimental studies of the effects of dispersal distance, seedling density, and light conditions. *Oecologia* **61**: 211–217.

Augspurger, C.K. (1984b) Seedling survival of tropical tree species: interaction of dispersal distance, light-gaps, and pathogens. *Ecology* **65**: 1705–1712.

Barreto, G.R., Rushton, S.P., Strachan, R. & Macdonald, D.W. (1998) The role of habitat and mink predation in determining the status of water voles in England. *Animal Conservation* **1**: 129–130.

Bauwens, D., Hertz, P.E. & Castilla, A.M. (1996) Thermoregulation in a lacertid lizard: the relative contributions of distinct behavioral mechanisms. *Ecology* **77**: 1818–1830.

Berger, J. & Cunningham, C. (1998) Behavioural ecology in managed reserves: gender-based asymmetries in interspecific dominance in African elephants and rhinos. *Animal Conservation* **1**: 33–38.

Bertness, M.D., Leonard, G.H., Levine, J.M. & Bruno, J.F. (1999) Climate-driven interactions among rocky intertidal organisms caught between a rock and a hot place. *Oecologia* **120**: 446–450.

Bever, J.D., Westover, K.M. & Antonovics, J. (1997) Incorporating the soil community into plant population dynamics: the utility of the feedback approach. *Journal of Ecology* **85**: 561–573.

Bewley, J.D. & Black, M. (1994) *Seeds: Physiology of Development and Germination*. Plenum Press, New York.

Blaustein, A.R., Hoffman, P.D., Hokit, D.G. *et al.* (1994) UV repair and resistance to solar UV-B in amphibian eggs: a link to population declines? *Proceedings of the National Academy of Sciences of the United States of America* **91**: 1791–1795.

Blaustein, A.R., Kiesecker, J.M., Chivers, D.P. & Anthony, R.G. (1997) Ambient UV-B radiation causes deformities in amphibian embryos. *Proceedings of the National Academy of Sciences of the United States of America* **94**: 13735–13737.

Botsford, L.W., Castilla, J.C. & Peterson, C.H. (1997) The management of fisheries and marine ecosystems. *Science* **277**: 509–515.

Bourke, A.F.G & Franks, N.R. (1995) *Social Evolution in Ants*. Princeton University Press, Princeton, NJ.

Bowers, M.A. (1997) Influence of deer and other factors in an old-field plant community. In *The Science of Overabundance: Deer Ecology and Population Management* (McShea, W.J., Underwood, H.B. & Rappole, J.H., eds), pp. 310–326. Smithsonian Institution Press, Washington, DC.

Boyd, D.K. & Pletscher, D.H. (1999) Characteristics of dispersal in a colonizing wolf population in the central Rocky Mountains. *Journal of Wildlife Management* **63**: 1094–1108.

Bradshaw, A.D. (1971) Plant evolution in extreme environments. In *Ecological Genetics and Evolution* (Creed, R., ed.), pp. 20–50. Blackwell Scientific Publications, Oxford.

Breitburg, D.L., Loher, T., Pacey, C.A. & Gerstein, A. (1997) Varying effects of low dissolved oxygen on trophic interactions in an estuarine food web. *Ecological Monographs* **67**: 489–507.

Brosofske, K.D., Chen, J., Naiman, R.J. & Franklin, J.F. (1997) Harvesting effects on microclimatic gradients from small streams to uplands in western Washington. *Ecological Applications* **7**: 1188–1200.

Brown, J.H. & Munger, J.C. (1985) Experimental manipulation of a desert rodent community: food addition and species removal. *Ecology* **66**: 1545–1563.

Bucheli, E. & Leuchtmann, A. (1996) Evidence for genetic differentiation between choke-inducing and asymptomatic strains of the *Epichloë* grass endophyte from *Brachypodium sylvaticum*. *Evolution* **50**: 1879–1887.

Busato, P., Benvenuto, C. & Gherardi, F. (1998) Competitive dynamics of a Mediterranean hermit crab assemblage: the role of interference and exploitative competition for shells. *Journal of Natural History* **32**: 1447–1451.

Cambefort, Y. (1994) Body size, abundance, and geographical distribution of Afrotropical dung beetles (Coleoptera: Scarabaeidae). *Acta Oecologia* **15**: 165–179.

Campbell, N.A., Reece, J.B. & Mitchell, L.G. (1999) *Biology*, 5th edn. Benjamin Cummings (imprint of Addison Wesley Longman, Inc.), New York.

Carneiro, V.L., Fonseca, L.L., Andrade Filho, J.D. *et al.* (1993) Blood feeding activity of partially and totally engorged *Lutzomyia longipalpis* (Diptera: Psychodidae). *Memorias do Instituto Oswaldo Cruz* **88** (Suppl.): 269.

Carpenter, S.R. & Kitchell, J.F. (1988) Consumer control of lake productivity. *BioScience* **38**: 764–769.

Carpenter, S.R., Caraco, N.F., Correll, D.L. *et al.* (1998) Nonpoint pollution of surface waters with phosphorus and nitrogen. *Ecological Applications* **8**: 559–568.

Chapin III, F.S., Walker, B.H., Hobbs, R.J. *et al.* (1997) Biotic control over the functioning of ecosystems. *Science* **277**: 500–504.

Chown, S.L., Gremmen, N.J.M. & Gaston, K.J. (1998) Ecological biogeography of southern ocean islands: species–area relationships, human impacts and conservation. *American Naturalist* **152**: 562–575.

Clements, F.E. (1916) *Plant Succession: An Analysis of the Development of Vegetation*. Carnegie Institution of Washington Publication No. 520. Carnegie Institution, Washington, DC.

Cochran-Stafira, D.L. & von Ende, C.N. (1998) Integrating bacteria into food webs: studies with *Sarracenia purpurea* inquilines. *Ecology* **79**: 880–898.

Connell, J.H. (1961) The influence of interspecific competition and other factors on the distribution of the barnacle *Chthamalus stellatus*. *Ecology* **42**: 710–723.

Connell, J.H. (1978) Diversity in tropical rainforests and coral reefs. *Science* **199**: 1302–1310.

Corbet, G.B. & Harris, S. (1991) *The Handbook of British Mammals*. Blackwell Scientific Publications, Oxford.

Costantino, R.F., Desharnais, R.A., Cushing, J.M. & Dennis, B. (1997) Chaotic dynamics in an insect population. *Science* **275**: 389–391.

Cotgreave, P. & Stockley, P. (1994) Body size, insectivory and abundance in assemblages of small mammals. *Oikos* **71**: 89–96.

Cramp, S. & Simmons, K.E.L. (1977) *The Birds of the Western Palearctic*, Vol. 1. Oxford University Press, Oxford.

Cree, A., Daugherty, C.H. & Hay, J.M. (1995) Reproduction of a rare New Zealand reptile, the tuatara *Sphenodon punctatus*, on rat-free and rat-inhabited islands. *Conservation Biology* **9**: 373–383.

Crooks, K.R. & Soulé, M.E. (1999) Mesopredator release and avifaunal extinctions in a fragmented system. *Nature* **400**: 563–566.

Currie, D.J. (1991) Energy and large-scale patterns of animal- and plant-species richness. *American Naturalist* **137**: 27–49.

Dale, B.W., Adams, L.G. & Bowyer, R.T. (1994) Functional response of wolves preying on barren-ground caribou in a multiple-prey ecosystem. *Journal of Animal Ecology* **63**: 644–652.

Damuth, J. (1987) Interspecific allometry of population density in mammals and other animals: the independence of body mass and population energy use. *Biological Journal of the Linnean Society* **31**: 193–246.

Darwin, C. (1859) *The Origin of Species by Means of Natural Selection*. John Murray, London.

Davis, M.B. (1981) Quaternary history and the stability of forest communities. In *Forest Succession: Concepts and Application* (West, D.C., Shugart, H.H. & Botkin, D.B., eds), pp. 132–153. Springer-Verlag, New York.

Davis, M.B. (1985) History of the vegetation on the Mirror Lake watershed. In *An Ecosystem Approach to Aquatic Ecology: Mirror Lake and its Environment* (Likens, G.E., ed.), pp. 53–65. Springer-Verlag, New York.

Davis, M.B. (1989) Insights from paleoecology on global change. *Bulletin of the Ecology Society of America* **70**: 222–228.

de Grandé, L. & Bergeron, Y. (1997) Diversity and stability of understorey communities following disturbance in the southern boreal forest. *Journal of Ecology* **85**: 777–784.

Diamond, J.M. (1975) Assembly of species communities. In *Ecology and Evolution of Communities* (Cody, M.L. & Diamond, J.M., eds), pp. 342–444. Harvard University Press, Cambridge, MA.

Dunbar, R.I.M. (1998) Impact of global warming on the distribution and survival of the gelada baboon: a modelling approach. *Global Change Biology* **4**: 293–304.

Eltringham, K., Cooksey, I., Dixon, W. *et al.* (1998) Large mammals of Mkomazi. In *Mkomazi: The Ecology, Biodiversity and Conservation of a Tanzanian Savanna* (Coe, M., McWilliam, N., Stone, G. & Packer, M., eds), pp. 485–504. Royal Geographical Society, London.

Feldmeth, C.R. (1970) The respiratory energetics of two species of stream caddis fly larvae in relation to water flow. *Comparative Biochemistry and Physiology* **32**: 193–202.

ffrench, R. (1991) *A Guide to the Birds of Trinidad and Tobago*. A. & C. Black, London.

Findley, J.S. (1993) *Bats: A Community Perspective*. Cambridge University Press, Cambridge, UK.

Fitter, A.H., Fitter, R.S.R., Harris, I.T.B. & Williamson, M.H. (1995) Relationships between first flowering date and temperature in the flora of a locality in central England. *Functional Ecology* **9**: 55–60.

Fitzgibbon, G.D. & Lazarus, J. (1995) Antipredator behaviour of Serengeti ungulates: individual differences and population consequences. In *Serengeti II: Dynamics, Management and Conservation of an Ecosystem* (Sinclair, A.R.E. & Arcese, P., eds), pp. 274–296. Chicago University Press, Chicago.

Fox, B.J. & Brown, J.H. (1993) Assembly rules for functional groups in North American rodent communities. *Oikos* **67**: 358–370.

Fragaso, J.M.V. (1997) Tapir-generated seed shadows: scale-dependent patchiness in the Amazon rain forest. *Journal of Ecology* **85**: 519–529.

Franke, H.D. & Janke, M. (1998) Mechanisms and consequences of intra- and interspecific interference competition in *Idotea baltica* (Pallas) and *Idotea emarginata* (Fabricius) (Crustacea: Isopoda): a laboratory study of possible proximate causes of habitat segregation. *Journal of Experimental Marine Biology and Ecology* **227**: 1–21.

Gleason, H.A. (1926) The individualistic concept of plant association. *Bulletin of the Torrey Botanical Club* **53**: 7–26.

Gomez, J.M. (1994) Importance of direct and indirect effects in the interaction between a parasitic angiosperm (*Cuscuta epithymum*) and its host plant (*Hormathophylla spinosa*). *Oikos* **71**: 97–106.

Gordon, D.R. (1998) Effects of invasive, non-indigenous plant species on ecosystem processes: lessons from Florida. *Ecological Applications* **8**: 975–989.

Gorman, M.L., Mills, M.G., Raath, J.P. & Speakman, J.R. (1998) High hunting costs make African wild dogs vulnerable to kleptoparasitism by hyaenas. *Nature* **391**: 479–481.

Götmark, F. (1996) Simulating a colour mutation: conspicuous red wings in the European blackbird reduce the risk of attacks by sparrowhawks. *Functional Ecology* **10**: 355–359.

Goudswaard, K.P.C. & Witte, F. (1997) The catfish fauna of Lake Victoria after the Nile perch upsurge. *Environmental Biology of Fishes* **49**: 21–43.

Gould, J.L. & Keeton, W.T. (1996) *Biological Science*, 6th edn. W.W. Norton & Co., New York.

Grant, P.R. & Grant, B.R. (1992) Demography and the genetically effective sizes of two populations of Darwin's finches. *Ecology* **73**: 766–784.

Gray, A.N. & Spies, T.A. (1997) Microsite controls on tree seedling establishment in conifer forest canopy gaps. *Ecology* **78**: 2458–2473.

Greenwood, J.J.D., Gregory, R.D., Harris, S., Marris, P.A. & Yalden, D.W. (1996) Relations between abundance, body size and species number in British birds and mammals. *Philosophical Transactions of the Royal Society of London, Series B* **351**: 265–278.

Gullan, P.J. & Cranston, P.S. (2000) *The Insects: An Outline of Entomology*, 2nd edn. Blackwell Science, Oxford.

Haberle, S.G. (1999) Late Quaternary vegetation and climate change in the Amazon basin based on a 50 000 year pollen record from the Amazon fan, ODP site 932. *Quaternary Research* **51**: 27–38.

Hairston, N.G. (1980) The exponential test of an analysis of field distributions: competition in terrestrial salamanders. *Ecology* **61**: 817–826.

Harborne, J.B. (1997) *Introduction to Ecological Biochemistry*, 4th edn. Academic Press, London.

Harding, J.S. (1997) Strategies for coexistence in two species of New Zealand Hydropsychidae (Trichoptera). *Hydrobiologia* **350**: 25–33.

Hatcher, P.E., Paul, N.D., Ayres, P.G. & Whittaker, J.B. (1997) Nitrogen fertilization affects interactions between the components of an insect–fungus–plant tripartite system. *Functional Ecology* **11**: 537–544.

Heinrich, B. (1986) Thermoregulation and flight activity of a satyrine, *Coenonympha inornata* (Lepidoptera: Satyridae). *Ecology* **67**: 593–597.

Hill, J.K., Thomas, C.D. & Huntley, B. (1999) Climate and habitat availability determine 20th century changes in a butterfly's range margin. *Proceedings of the Royal Society of London, Series B* **266**: 1197–1206.

Hodge, S. & Arthur, W. (1997) Asymmetric interactions between species of seaweed fly. *Journal of Animal Ecology* **66**: 743–754.

Hogg, I.D. & Williams, D.D. (1996) Response of stream invertebrates to a global-warming thermal regime: an ecosystem-level manipulation. *Ecology* **77**: 395–407.

Houle, G. (1997) No evidence for interspecific interactions between plants in the first stage of succession on coastal dunes in subarctic Quebec, Canada. *Canadian Journal of Botany* **75**: 902–915.

Hudson, H.J. (1986) *Fungal Biology*. Edward Arnold, London.

Huey, R.B., Pianka, E.R. & Hoffman, J.A. (1977) Seasonal variation in thermoregulatory behavior and body temperature of diurnal Kalahari lizards. *Ecology* **58**: 1066–1075.

Huk, T. & Kuhne, B. (1999) Substrate selection by *Carabus clatratus* (Coleoptera, Carabidae) and its consequences for offspring development. *Oecologia* **121**: 348–354.

Hutchinson, G.E. (1959) Homage to Santa Rosalia, or why are there so many kinds of animals? *American Naturalist* **95**: 137–145.

Hutchison, B.A. & Matt, D.R. (1977) The distribution of solar radiation within a deciduous forest. *Ecological Monographs* **47**: 185–207.

Inouye, D.W. (2000) The ecological and evolutionary significance of frost in the context of climate change. *Ecology Letters* **3**: 457–463.

James, A.C. & Partridge, L. (1998) Geographic variation in competitive ability in *Drosophila melanogaster*. *American Naturalist* **151**: 530–537.

John, E. & Turkington, R. (1997) A 5-year study of the effects of nutrient availability and herbivory on two boreal forest herbs. *Journal of Ecology* **85**: 419–430.

Johnston, K.M. & Schmitz, O.J. (1997) Wildlife and climate change: assessing the sensitivity of selected species to simulated doubling of atmospheric CO_2. *Global Change Biology* **3**: 531–544.

Jones, C.G., Lawton, J.H. & Shachak, M. (1997) Positive and negative effects of organisms as physical ecosystem engineers. *Ecology*, **78**: 1946–1957.

Jumpponen, A., Mattson, K., Trappe, J.M. & Ohtonen, R. (1998) Effects of established willows on primary succession on Lyman Glacier forefront, North Cascade Range, Washington, USA: evidence for simultaneous canopy inhibition and soil facilitation. *Arctic and Alpine Research* **30**: 31–39.

Kaufman, D.M. & Willig, M.R. (1998) Latitudinal patterns of mammalian species richness in the New World: the effects of sampling method and faunal group. *Journal of Biogeography* **25**: 795–805.

Keesing, F. (1998) Impacts of ungulates on the demography and diversity of small mammals in central Kenya. *Oecologia* **116**: 381–389.

Kelly, D.W., Mustafa, Z. & Dye, C. (1996) Density-dependent feeding success in a field population of the sandfly, *Lutzomyia longipalpis*. *Journal of Animal Ecology* **65**: 517–527.

Kelly, P.E. & Larson, D.W. (1997) Dendroecological analysis of population dynamics of an old growth forest on cliff faces of the Niagara Escarpment, Canada. *Journal of Ecology* **85**: 467–478.

Kemp, A. (1995) *The Hornbills*. Oxford University Press, Oxford.

Kendeigh, S.C. (1982) *Bird populations in East Central Illinois: Fluctuations, Variations and Development over half a century*. Illinois Biological Monographs No. 52. University of Illinois Press, Champaign, IL.

Kingdon, J. (1997) *The Kingdon Field Guide to African Mammals*. Academic Press, San Diego.

Kirk, K.L. (1997) Life-history responses to variable environments: starvation and reproduction in planktonic rotifers. *Ecology* **78**: 434–441.

Knapp, R. & Casey, T.M. (1986) Thermal ecology, behavior, and growth of gypsy moth and eastern tent caterpillars. *Ecology* **67**: 598–608.

Kodric-Brown, A., Brown, J.H., Byers, G.S. & Gori, D.F. (1984) Organization of a tropical island community of hummingbirds and flowers. *Ecology* **65**: 1358–1368.

Kondratieff, B.C., Bishop, R.J. & Brasher, A.M. (1997) The life cycle of an introduced caddisfly, *Cheumatopsyche pettiti* (Banks) (Trichopter: Hydropsychidae) in Waikolu stream, Molokia, Hawaii. *Hydrobiologia* **350**: 81–85.

Lack, D. (1968) *Ecological Adaptations for Breeding in Birds*. Methuen, London.

Lawton, J.H. (1984) Non-competitive populations, non-convergent communities, and vacant niches: the herbivores of bracken. In *Ecological Communities; Conceptual Issues and the Evidence* (String, D.R., Simberloff, D., Abele, L.G. & Thistle, A.B., eds), pp. 67–100. Princeton University Press, Princeton, NJ.

Lawton, J.H. (1995) Book reviews. *Journal of Animal Ecology* **64**: 296–302.

Leadley, P.W. & Stocklin, J. (1996) Effects of elevated CO_2 on model calcareous grasslands: community, species, and genotype level responses. *Global Change Biology* **2**: 389–397.

Letcher, A.J. & Harvey, P.H. (1994) Variation in geographical range size among mammals of the palearctic. *American Naturalist* **144**: 30–42.

Letourneau, D.K. & Dyer, L.A. (1998) Experimental test in lowland tropical forest shows top-down effects through four trophic levels. *Ecology* **79**: 1678–1687.

Lieth, H. (1973) Primary production: terrestrial ecosystems. *Human Ecology* **1**: 303–332.

Lindemann, R.L. (1942) The trophic-dynamic aspect of ecology. *Ecology* **23**: 399–418.

Lindley, J.A. (1998) Diversity, biomass and production of decapod crustacean larvae in a changing environment. *Invertebrate Reproduction and Development* **33**: 209–219.

Little, L.R. & Maun, M.A. (1996) The 'Ammophila problem' revisited: a role for mycorrhizal fungi. *Journal of Ecology* **84**: 1–7.

Lloyd, A.H. & Graumlich, L.J. (1997) Holocene dynamics of treeline forests in the Sierra Nevada. *Ecology* **78**: 1199–1210.

Lockwood, J.L., Moulton, M.P. & Anderson, S.K. (1993) Morphological assortment and the assembly of communities of introduced passeriforms on oceanic islands: Tahiti versus Oahu. *American Naturalist* **141**: 398–408.

Lydon, J., Teasdale, J.R. & Chen, P.K. (1997) Allelopathic activity of annual wormwood (*Artemesia annua*) and the role of artemisinin. *Weed Science* **45**: 807–811.

Mackenzie, J.M. (1997) *The Empire of Nature*. Manchester University Press, Manchester.

Madsen, T. & Shine, R. (2000) Rain, fish and snakes: climatically driven population dynamics of Arafura filesnakes in tropical Australia. *Oecologia* **124**: 208–215.

Martin, T.E. (2001) Abiotic vs. biotic influences on habitat selection of coexisting species: climate change impacts? *Ecology* **82**: 175–188.

Massa, S., Sarra, P.G., Canganella, F. & Trovatelli, L.D. (1998) Protection of young chicks against *Salmonella kedougou* by administration of intestinal microflora. *International Journal of Food Microbiology* **40**: 123–126.

McGeoch, M.A. & Chown, S.L. (1997) Evidence of competition in a herbivorous, gall-inhabiting moth (Lepidoptera) community. *Oikos* **78**: 107–115.

Mehlman, D.W. (1997) Change in avian abundance across the geographic range in response to environmental change. *Ecological Applications* **7**: 614–624.

Miller, M.W. & Hay, M.E. (1996) Coral–seaweed–grazer–nutrient interactions on temperate reefs. *Ecological Monographs* **66**: 323–344.

Mitchell, R.J., Marrs, R.H., le Duc, M.G. & Auld, M.H.D. (1997) A study of succession on lowland heaths in Dorset, southern England: changes in vegetation and soil chemical properties. *Journal of Applied Ecology* **34**: 1426–1444.

Morley, R. (1998) Small mammals of Mkomazi. In *Mkomazi: The Ecology, Biodiversity and Conservation of a Tanzanian Savanna* (Coe, M., McWilliam, N., Stone, G. & Packer, M., eds), pp. 467–484. Royal Geographical Society, London.

Murray, D.L., Cary, J.R. & Keith, L.B. (1997) Interactive effects of sublethal nematodes and nutritional status on snowshoe hare vulnerability to predation. *Journal of Animal Ecology* **66**: 250–264.

Naeem, S., Thompson, L.J., Lawler, S.P., Lawton, J.H. & Woodfin, R.M. (1994) Declining biodiversity can alter the performance of ecosystems. *Nature* **368**: 734–737.

Nakajima, T. & Kurihara, Y. (1994) Evolutionary changes of ecological traits of bacterial populations through predator-mediated competition. I. Experimental analysis. *Oikos* **71**: 24–34.

Newsham, K.K., Fitter, A.H. & Watkinson, A.R. (1995) Arbuscular mycorrhizae protect an annual grass from root pathogenic fungi in the field. *Journal of Ecology* **83**: 991–1000.

Olff, H., de Leeuw, J., Bakker, J.P. *et al.* (1997) Vegetation succession and herbivory in a salt marsh: changes induced by sea level rise and silt deposition along an elevated gradient. *Journal of Ecology* **85**: 799–814.

Pagel, M.D., May, R.M. & Collie, A.R. (1991) Ecological aspects of the geographical distribution and diversity of mammalian species. *American Naturalist* **137**: 791–815.

Paige, K.N. & Whitham, T.G. (1987) Flexible life history traits: shifts by scarlet gilia in response to pollinator abundance. *Ecology* **68**: 1691–1695.

Pastor, J., Aber, J.D., McClaugherty, C.A. & Mellilo, J.M. (1984) Aboveground production and N and P cycling along a nitrogen mineralization gradient on Blackhawk Island, Wisconsin. *Ecology* **65**: 256–268.

Pearre, S. & Maass, R. (1998) Trends in the prey size-based trophic niches of feral and house cats (*Felis catus*). *Mammal Review* **28**: 125–139.

Pech, R.P., Sinclair, A.R.E., Newsome, A.E. & Catling, P.C. (1992) Limits to predator regulation of rabbits in Australia: evidence from predation removal experiments. *Oecologia* **89**: 102–112.

Petrie, M., Halliday, T. & Sanders, C. (1991) Peahens prefer peacocks with elaborate trains. *Animal Behaviour* **41**: 323–331.

Porembski, S., Brown, G. & Barthlott, W. (1995) An inverted latitudinal gradient of plant diversity in shallow depressions on Ivorian inselbergs. *Vegetatio* **117**: 151–163.

Poulin, R. (1998) *Evolutionary Ecology of Parasites*. Chapman & Hall, London.

Power, M.E., Matthews, W.J. & Stewart, A.J. (1985) Grazing minnows, piscivorous bass, and stream algae: dynamics of a strong interaction. *Ecology* **66**: 1448–1456.

Prasad, M.N.V. (1995) Cadmium toxicity and tolerance in vascular plants. *Environmental and Experimental Botany* **35**: 525–545.

Proulx, M. & Mazumder, A. (1998) Reversal of grazing impact on plant species richness in nutrient-poor vs. nutrient-rich ecosystems. *Ecology* **79**: 2581–2592.

Radtkey, R.R., Fallon, S.M. & Case, T.J. (1997) Character displacement in some Cnemidophorus lizards revisited: a phylogenetic analysis. *Proceedings of the National Academy of Sciences of the United States of America* **94**: 9740–9745.

Randolph, S.E. (1997) Abiotic and biotic determinants of the seasonal dynamics of the tick *Rhipicephalus appendiculatus* in South Africa. *Medical and Veterinary Entomology* **11**: 25–37.

Reznick, D.N. & Endler, J.A. (1982) The impact of predation on life history evolution in Trinidadian guppies (*Poecilia reticulata*). *Evolution* **36**: 160–177.

Rhodes, C.J. & Anderson, R.M. (1996) Power laws governing epidemics in isolated populations. *Nature* **381**: 600–602.

Roberts, C.M. (1987) Experimental analysis of resource sharing between herbivorous damselfish and blennies on the Great Barrier Reef. *Journal of Experimental Marine Biology and Ecology* **111**: 61–75.

Roberts, M.J. (1995) *Spiders of Britain and Northern Europe*. HarperCollins, London.

Robles, C.D. (1997) Changing recruitment in constant species assemblages: implications for predation theory in intertidal communities. *Ecology* **78**: 1400–1414.

Rygiewicz, P.T. & Andersen, C.P. (1994) Mycorrhizae alter quality and quantity of carbon allocated below ground. *Nature* **369**: 58–60.

Sait, S.M., Begon, M., Thompson, D.J. & Harvey, J.A. (1996) Parasitism of baculovirus-infected *Plodia interpunctella* by *Venturia canescens* and subsequent virus transmission. *Functional Ecology* **10**: 586–591.

Sanford, E. (1999) Regulation of keystone predation by small changes in ocean temperature. *Science* **283**: 2095–2097.

Schäppi, B. & Körner, C. (1997) *In situ* effects of elevated CO_2 on the carbon and nitrogen status of alpine plants. *Functional Ecology* **11**: 290–299.

Schimel, D.S. (1995) Terrestrial ecosystems and the carbon cycle. *Global Change Biology* **1**: 77–91.

Schmid, J. (2000) Daily torpor in the gray mouse lemur (*Microcebus murinus*) in Madagascar: energetic consequences and biological significance. *Oecologia* **123**: 175–183.

Schmitz, O.J. & Sinclair, A.R.E. (1997) Rethinking the role of deer in forest ecosystem dynamics. In *The Science of Overabundance: Deer Ecology and Population Management* (McShea, W.J., Underwood, H.B. & Rappole, J.H., eds), pp. 207–223. Smithsonian Institution Press, Washington, DC.

Schwilk, D.W., Keeley, J.E. & Bond, W.J. (1997) The intermediate disturbance hypothesis does not explain fire and diversity pattern in fynbos. *Plant Ecology* **132**: 77–84.

Seely, M.K. (1979) Irregular fog as a water source for desert dune beetles. *Oecologia* **42**: 213–227.

Sherman, P.W. & Morton, M.L. (1984) Demography of Belding's ground squirrels. *Ecology* **65**: 1617–1628.

Sibley, C.G. & Ahlquist, J.E. (1990) *Phylogeny and Classification of Birds: A Study in Molecular Evolution*. Yale University Press, New Haven, CT.

Silvertown, J.W. (1980) The dynamics of a grassland ecosystem: botanical equilibrium in the Park Grass experiment. *Journal of Applied Ecology* **17**: 491–504.

Sims, D.W. & Quayle, V.A. (1998) Selective foraging behaviour of basking sharks on zooplankton in a small scale front. *Nature* **393**: 460–464.

Sinclair, A.R.E. (1997) Carrying capacity and the overabundance of deer. In *The Science of Overabundance: Deer Ecology and Population Management* (McShea, W.J., Underwood, H.B. & Rappole, J.H., eds), pp. 380–394. Smithsonian Institution Press, Washington, DC.

Sinclair, A.R.E & Arcese, P. (1995) *Serengeti II: Dynamics, Management, and Conservation of an Ecosystem*. University of Chicago Press, Chicago.

Smith, J.L.D., Ahearn, S.C. & McDougal, C. (1998a) Landscape analysis of tiger distribution and habitat quality in Nepal. *Conservation Biology* **12**: 1338–1346.

Smith, S.D., Monson, R.K. & Andersen, J.E. (1997) *Physiological Ecology of North American Desert Plants*. Springer-Verlag, New York.

Smith, T.D. (1994) *Scaling Fisheries: The Science of Measuring the Effects of Fishing 1855–1955*. Cambridge University Press, Cambridge, UK.

Smith, W.K., Bell, D.T. & Shepherd, K.A. (1998b) Associations between leaf structure, orientation, and sunlight exposure in five western Australian communities. *American Journal of Botany* **85**: 56–63.

Spear, R.W. (1989) Late-Quaternary history of high-elevation vegetation in the White Mountains of New Hampshire. *Ecological Monographs* **59**: 125–151.

Stanford, C.B. (1998) *Chimpanzee and Red Colobus: The Ecology of Predator and Prey*. Harvard University Press, Cambridge, MA.

Stewart, D.A.B. & Samways, M.J. (1998) Conserving dragonfly (Odonata) assemblages relative to river dynamics in an African savanna game reserve. *Conservation Biology* **12**: 683–692.

Stone, G.N., Willmer, P.G. & Nee, S. (1996) Daily partitioning of pollinators in an African *Acacia* community. *Proceedings of the Royal Society of London, Series B* **263**: 1389–1393.

Straile, D. (2000) Meteorological forcing of plankton dynamics in a large and deep continental European lake. *Oecologia* **122**: 44–50.

Svenning, J.-C. (1999) Microhabitat specialization in a species-rich palm community in Amazonian Ecuador. *Journal of Ecology* **57**: 55–65.

Tallis, J.H. (1997) The pollen record of *Empetrum nigrum* in southern Pennine peats: implications for erosion and climate change. *Journal of Ecology* **85**: 455–465.

Thomas, J.A., Elmes, G.W., Clarke, R.T. *et al.* (1997) Field evidence and model predictions of butterfly-mediated apparent competition between gentian plants and red ants. *Acta Oecologia* **18**: 671–684.

Till, C. & Guiot, J. (1990) Reconstruction of precipitation in Morocco since 1100 AD based on *Cedrus atlantica* tree-ring widths. *Quaternary Research* **33**: 337–351.

Tilman, D. (1989) Competition, nutrient reduction and the competition neighbourhood of a bunchgrass. *Functional Ecology* **3**: 215–219.

Tilman, D. (1997) Community invasibility, recruitment limitation, and grassland biodiversity. *Ecology* **78**: 81–92.

Tilman, D., Wedin, D. & Knops, J. (1996) Productivity and sustainability influenced by biodiversity in grassland ecosystems. *Nature* **379**: 718–720.

Torgersen, C.E., Price, D.M., Li, H.W. & McIntosh, B.A. (1999) Multiscale thermal refugia and stream habitat associations of Chinook salmon in northeastern Oregon. *Ecological Applications* **9**: 301–319.

Torres-Contreras, H. & Bozinovic, F. (1997) Food selection in an herbivorous rodent: balancing nutrition with thermoregulation. *Ecology* **78**: 2230–2237.

Van Horne, B., Olson, G.S., Schooley, R.L., Corn, J.G. & Burnham, K.P. (1997) Effects of drought and prolonged winter on Townsend's ground squirrel demography in shrubsteppe habitats. *Ecological Monographs* **67**: 295–315.

van Marken Lichtenbelt, W.D., Vogel, J.T. & Wesselingh, R.A. (1997) Energetic consequences of field body temperatures in the green iguana. *Ecology* **78**: 297–307.

Veltman, C.J., Nee, S. & Crawley, M.J. (1996) Correlates of introduction success in exotic New Zealand birds. *American Naturalist* **147**: 542–557.

Vitousek, P.M., Aber, J., Howarth, R.W. *et al.* (1997a) Human alteration of the global nitrogen cycle: causes and consequences. *Issues in Ecology* **1**: 1–15.

Vitousek, P.M., Mooney, H.A., Lubchenco, J. & Melillo, J.M. (1997b) Human domination of earth's ecosystems. *Science* **277**: 494–499.

Webb III, T. (1987) The appearance and disappearance of major vegetational assemblages: long-term vegetational dynamics in eastern North America. *Vegetatio* **69**: 177–187.

Weiher, R. & Keddy, P.A. (1995) Assembly rules, null models and trait dispersion—new questions from old patterns. *Oikos* **74**: 159–164.

Whitney, G.G. (1994) *From Coastal Wilderness to Fruited Plain: A History of Environmental Change in Temperate North America from 1500 to the Present.* Cambridge University Press, Cambridge, UK.

Wiggins, G.B. (1977) *Larvae of the North American Caddisfly Genera (Trichoptera).* University of Toronto Press, Toronto.

Wiggins, G.B. (1978) Trichoptera. In *An Introduction to the Aquatic Insects of North America* (Merritt, R.W. & Cummins, K.W., eds). Kendall/Hunt, Dubuque, IA.

Willmer, P., Stone, G. & Johnston, I. (2000) *Environmental Physiology of Animals.* Blackwell Science, Oxford.

Willson, M.F. & Burley, N. (1983) *Mate Choice in Plants.* Princeton University Press, Princeton, NJ.

Wilson, J.B. & Roxburgh, S.H. (1994) A demonstration of guild-based assembly rules for a plant community and determination of intrinsic guilds. *Oikos* **69**: 267–276.

Wilson, J.B. & Whittaker, R.J. (1995) Assembly rules demonstrated in a saltmarsh community. *Journal of Ecology* **83**: 801–808.

Wolf, B.O. & Walsberg, G.E. (1996) Thermal effects of radiation and wind on a small bird and implications for microsite selection. *Ecology* **77**: 2228–2236.

Wood, B. & Gillman, M.P.N.A. (1998) The effects of disturbance on forest butterflies using two methods of sampling in Trinidad. *Biodiversity and Conservation* **7**: 597–616.

Wulff, J.L. (1997) Mutualisms among species of coral reef sponges. *Ecology* **78**: 146–159.

Yoshimura, Y., Kohshima, S. & Ohtani, S. (1997) A community of snow algae on Himalayan glacier: change of algal biomass and community structure with altitude. *Arctic and Alpine Research* **29**: 126–137.

Yun, K.W. & Maun, M.A. (1997) Allelopathic potential of *Artemesia campestris* ssp. *caudata* on Lake Heron Sand dunes. *Canadian Journal of Botany* **75**: 1903–1912.

Zamora, R. & Gomez, J.M. (1996) Carnivorous plant–slug interaction: a trip from herbivory to kleptoparasitism. *Journal of Animal Ecology* **65**: 154–160.

Zobel, D.B. & Antos, J.A. (1997) A decade of recovery of understory vegetation buried by volcanic tephra from Mount St Helens. *Ecological Monographs* **67**: 317–344.

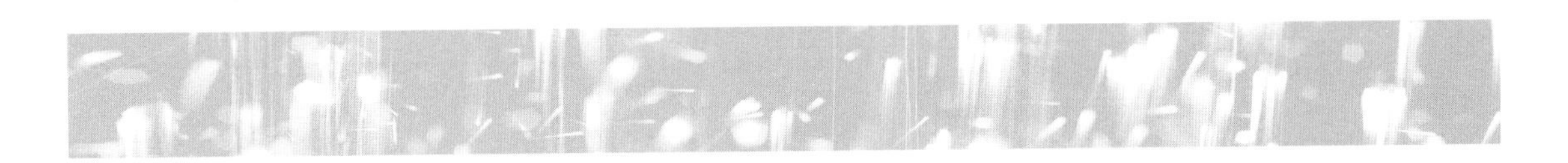

Index

Page numbers in *italics* refer to figures; those in **bold** refer to tables